Plants and Empire

Plants and Empire

*Colonial Bioprospecting
in the Atlantic World*

Londa Schiebinger

Harvard University Press
Cambridge, Massachusetts, and London, England | 2004

Copyright © 2004 by Londa Schiebinger
All rights reserved
Printed in the United States of America

Library of Congress Cataloging-in-Publication Data

Schiebinger, Londa L.
  Plants and empire : colonial bioprospecting in the Atlantic World / Londa Schiebinger.
    p. cm.
  Includes bibliographical references (p.   ).
  ISBN 0-674-01487-1 (cloth)
  1. Pride-of-Barbados (Plant).   2. Herbal abortifacients—History.
3. Slavery—Caribbean Area—History.   I. Title.

RG137.45.S35 2004
581.6′34—dc22   2004047364

*To the memory of the men and women
whose knowledge of fertility control
has been lost in the mists of time
and to the ravages of history*

Contents

Acknowledgments   *ix*

Introduction   *1*
"The Base for All Economics"   *5*
Plan of the Book   *12*

1 Voyaging Out   *23*
Botanistes Voyageurs   *25*
Maria Sibylla Merian   *30*
Biopirates   *35*
Who Owns Nature?   *44*
Voyaging Botanical Assistants   *46*
Creole Naturalists and Long-Term Residents   *51*
Armchair Botanists   *57*
The Search for the Amazons   *62*
Heroic Narratives   *65*

2 Bioprospecting   *73*
Drug Prospecting in the West Indies   *75*
Biocontact Zones   *82*
Secrets and Monopolies   *90*
Drug Prospecting at Home   *93*
Brokers of International Knowledge   *100*

3 Exotic Abortifacients   *105*
Merian's Peacock Flower   *107*
Abortion in Europe   *113*
Abortion in the West Indies   *128*
Abortion and the Slave Trade   *142*

4 The Fate of the Peacock Flower in Europe   150
   Animal Testing   153
   Self-Experimentation   156
   Human Subjects   159
   Testing for Sexual Difference   166
   The Complications of Race   171
   Abortifacients   177

5 Linguistic Imperialism   194
   Empire and Naming the Kingdoms of Nature   197
   Naming Conundrums   206
   Exceptions: Quassia *and* Cinchona   211
   Alternative Naming Practices   219

   Conclusion: Agnotology   226

   Notes   243

   Bibliography   286

   Credits   298

   Index   300

Acknowledgments

This book was a pleasure to write. It took me to exotic places and brought me into contact with wonderful scholars from many parts of the globe. I thank Roy Vickery of the Natural History Museum, London, for kindly opening for me Hans Sloane's herbarium (the first eight volumes contain Sloane's plants from Jamaica). The staff of the Laboratoire de Phanérogamie, Muséum National d'Histoire Naturelle, Paris, let me rummage through their many varieties of the *Poinciana* from specimen drawers stretching from ceiling to floor in room after room. I also thank John Symons at the Wellcome Library, London; the knowledgeable archivists and staff at the Bibliothèque Centrale Muséum National d'Histoire Naturelle, Paris; the Centre des Archives d'Outre-Mer, Aix-en-Provence; and the National Library of Jamaica, Kingston, who helped with images and arcane materials. Numerous colleagues directed me at crucial moments to valuable information. Jim McClellan generously shared his intriguing archival find concerning "the potato with two roots"; in addition he read and helpfully commented on the manuscript. Sue Broomhall called Mme. Claude Gauvard's work to my attention. Jerome Handler directed me to sources on midwifery in Barbados. Karen Reeds told me about Mark Jameson and his sixteenth-century garden of plants for gynecological complaints. Philip Boucher shared his expertise on various matters concerning Rochefort and du Tertre. Staffan Müller-Wille pointed me to Linnaeus' evaluation of Dahlberg and kindly read Chapter 5. Roberta Bivins called Sir William Jones's work to my attention. Claudia Swan shared information on the fascinating "root cutters." Lee Ann Newsom helped with technical questions concerning the biogeographical origins of the *Poinciana*. Brian Ogilvie read and commented very helpfully on Chapter 5. Linda Woodbridge provided the lovely citation from Ben Jonson. Clem Hawes came to my aid when I required an expert on Daniel Defoe's *Moll Flanders*. Sue Hanley read sections of Chapter 3 and answered questions about French law. Matthew Restall provided information on *Tabachin* and all things Spanish. Minnie Sinha provided careful comments on the theories and practices of imperialism. Alan Walker offered splendid insights into current scientific naming practices. Barbara Bush, Richard Drayton, Pamela Smith, and Emma Spary all provided much-needed advice and materials on the history of medicine, natural history, or the West Indies. Finally, a special thanks to Paula Findlen for her generous support for my many projects.

Much of this book was written during my year at the Max-Planck-Institut für Wissenschaftsgeschichte (1999–2000). Warm thanks to Lorraine Daston for her intellectual sparkle and kind hospitality. I also thank the Institute's library staff for tracking down and photocopying numerous sources. Support for that year was provided by the Alexander von Humboldt Foundation. I was the first woman historian to win the Alexander von Humboldt Research Prize; I know there will be many to follow. Travel to archives and time for writing were also supported by the National Science Foundation and the National Library of Medicine, National Institutes of Health. I am also grateful for the generous support for my research provided by Dean Susan Welch and A. Gregg Roeber at Pennsylvania State University. A warm thanks to Elizabeth Knoll, my patient editor, who made the journey through press a pleasant one. My thanks, too, to my research assistants: Mary Faulkner, Sarah Goodfellow, and Katherine Maas—the book is better because of you. Finally, I owe a special debt of gratitude to Ed Dumond who kept all my equipment running.

Sections of various chapters appeared in *Endeavour; Hypatia;* Caroline Jones and Peter Galison's edited volume *Picturing Science, Producing Art;* Lorraine Daston and Fernando Vidal's *Moral Authority of Nature;* Pamela Smith and Benjamin Schmidt's *Knowledge and Its Making;* my own and Claudia Swan's *Colonial Botany.* I thank these journals, editors, and presses for permission to reprint those materials.

Finally, it takes a village to sustain intellectual vigor. The students in my graduate seminar, "Colonial Exchange," pored over my manuscript and offered many wonderful corrections and suggestions—thanks to them all. My colleagues and friends—Rich Doyle, Laura Giannetti, Amy Greenberg, Gillian Hadfield, Ronnie Hsia, Mary Pickering, Guido Ruggiero, Sophie de Schaepdrijver, Susan Squier, and Nancy Tuana—all provided warmth, chatter, food, and sometimes even dance parties at crucial moments. To Robert Proctor, Geoffrey Schiebinger, and Jonathan Proctor, my love.

Plants and Empire

Introduction

> The Indians, who are not treated well by their Dutch masters, use the seeds [of this plant] to abort their children, so that their children will not become slaves like they are. The black slaves from Guinea and Angola have demanded to be well treated, threatening to refuse to have children. In fact, they sometimes take their own lives because they are treated so badly, and because they believe they will be born again, free and living in their own land. They told me this themselves.

In this moving passage from her magnificent 1705 *Metamorphosis insectorum Surinamensium,* Maria Sibylla Merian recorded how the African slave and Indian populations in Surinam, a Dutch colony, used the seeds of a plant she identified as the *flos pavonis,* literally "peacock flower," as an abortifacient.[1] This luxuriant plant still grows wild and in hedgerows and gardens throughout the Caribbean, and continues today to be known to many herb women and bush doctors as providing effective brews for inducing abortion.

Unlike the prolific peacock flower, Merian's work is rare and remarkable. A celebrated artist, the German-born Merian was one of very few European women to travel on her own in this period in pursuit of science. Women naturalists rarely figured in the rush to know exotic lands: Johanna Helena Herolt, Merian's eldest daughter, collected and painted insects and plants, much like her mother, in Surinam in 1711 when her husband traveled there to look after his business interests. Jeanne Baret became the first woman to circumnavigate the globe in 1776, but she sailed, disguised as a male, as an assistant to Philibert Commerson, the ship's botanist and the father of her illegitimate child. In the nineteenth century, women like Lady Charlotte Canning did sometimes collect botanical specimens, but almost always as colonial wives, traveling where their husbands happened to take them and not in pursuit of their own scientific programs.[2]

Merian's passage is also remarkable for what it reveals about the geopolitics of plants in the so-called "early modern world." Historians have

Figure I.1. Merian's *flos pavonis* (literally, peacock flower), which she described as a nine-foot-tall plant with brilliant yellow and red blossoms. Note the hearty seed (shown on the right through the opening in the pod) with which Merian claimed Arawak and African slave women aborted their children. Various parts of this plant are still used in many places in the Caribbean today as an abortifacient.

rightly focused on the explosion of knowledge associated with the scientific revolution and global expansion, and the frantic transfer of trade goods and plants between Europe and its colonies.[3] While much literature on colonial science has focused on how knowledge is made and moved between continents and heterodox traditions, I explore here instances of the *nontransfer* of important bodies of knowledge from the New World into Europe. In doing so, I develop a methodological tool that Robert N. Proctor has called "agnotology"—the study of culturally-induced ignorances—that serves as a counterweight to more traditional concerns for epistemology.[4] Agnotology refocuses questions about "how we know" to include questions about what we do *not* know, and why not. Ignorance is often not merely the absence of knowledge but an outcome of cultural and political struggle. Nature, after all, is infinitely rich and variable. What we know or do not know at any one time or place is shaped by particular histories, local and global priorities, funding patterns, institutional and disciplinary hierarchies, personal and professional myopia, and much else as well. I am interested in understanding how bodies of knowledge were constructed, but more interested in analyzing culturally produced ignorances of nature's body.[5]

This book presents the story not of a great man or a great woman, but of a great plant. Historians, post-colonialists, even historians of science rarely recognize the importance of plants to the processes that form and reform human societies and politics on a global scale. Plants seldom figure in the grand narratives of war, peace, or even everyday life in proportion to their importance to humans. Yet they are significant natural and cultural artifacts, often at the center of high intrigue. In the nineteenth century, the Bolivian government tortured and executed Manuel Incra, an Aymará Indian, for his part in smuggling seeds to the British of the *Cinchona officinalis* (the source of alkaloid quinine, also known as Peruvian bark), which cures malaria and other virulent fevers. Plants are also often entangled in high-stake politics. By the 1930s, the Dutch had undermined the South American monopoly on quinine by growing smuggled seeds on the other side of the globe in Java. When the Nazis occupied the Netherlands during World War II, one of their top priorities was to seize the world's stores of quinine, leaving the allies virtually none. As a consequence, more U.S. soldiers died from malaria during the war in the Pacific than from Japanese bullets and bayonets.[6]

Europeans have long moved plants around the world—in vast quantities and to great economic effect. Long before Christopher Columbus set foot in the Americas, plants in the form of spices, medicinal drugs, perfumes, and dyestuffs of all kinds sped along trading routes that stretched from the Far East into the Mediterranean. Columbus's discovery of America touched off a frenzy of plant movements: already on his second voyage, in 1494, Columbus brought sugarcane cuttings to Hispaniola, along with various citrus fruits, grapevines, olives, melons, onions, and radishes. Sugar—valued in Europe as a medicine as well as a sweetener—was acclimatized in the Americas around 1512 and was in full production in Hispaniola by 1525. The Spanish also introduced Chinese ginger into Mexico (in 1530), and by 1587 2 million pounds with a value approaching 250,000 ducats were being shipped annually into Seville. As time went on, plants also played a role in the political struggles surrounding slavery. Plantation owners transported breadfruit from Tahiti, for example, with the sole purpose of nourishing large slave populations cheaply and efficiently.[7]

Not all plants, of course, are or ever were equal. The plant whose history provides the leitmotif of this book, Merian's peacock flower *(Poinciana pulcherrima)*, is not a heroic plant of the historical stature of chocolate, the potato, quinine, coffee, tea, or even rhubarb, much used in the eighteenth century as a laxative.[8] Nonetheless it was a highly political plant, deployed in the struggle against slavery throughout the eighteenth century by slave women who used it to abort offspring who would otherwise be born into bondage. I lavish attention on this plant not because it is exquisitely beautiful and grows in stunningly inviting places, but because a number of naturalists independently identified it as an abortifacient widely used in the West Indies. Each observed Amerindians or slave women employ the plant effectively, and recorded that knowledge. Interestingly, the plant itself transferred easily: by 1700 it grew splendidly in botanical gardens across Europe. Philip Miller at the Chelsea Physic Garden outside London noted that "the seeds of this plant are annually brought over in plenty from the West-Indies." Proud of his gardening prowess, he announced that with proper management the *Poinciana* will grow much taller in England than in Barbados.[9] The flaming reds and yellows of this elegant flower, sometimes also called Barbados Pride or the Red Bird of Paradise, made it a favorite among Europeans.

Even though the peacock flower itself moved easily into Europe, knowledge of its use as an abortifacient did not. Why not? In a period when Europeans prized exotic botanicals—tulip bulbs, coffees, teas, chocolates, spices, and medicines of all sorts—what blocked the domestication of this potentially useful drug in Europe? As we shall see, trade winds of prevailing opinion impeded knowledge of New World abortifacients from reaching Europe. Here in this bit of history that did not come to pass we find a prime example of culturally cultivated ignorance—the unspoken but distinct configuration of events that converge to leave certain forms of knowledge unplucked from the tree of life.

"The Base for All Economics"

A resilient and long-standing narrative in the history of science has envisioned the flowering of modern botany as the rise of taxonomy, nomenclature, and "pure" systems of classification. The eighteenth century did, indeed, witness major developments in systematics (nomenclature and taxonomy) in many fields, including botany. These developments have often been portrayed as the coming of age of botany as a science when "knowledge about plants as plants [came to have] a value of its own apart from economic or medical considerations."[10] It is important, however, to see, as the great French comparative anatomist Louis-Jean-Marie Daubenton emphasized in the eighteenth century, and as William Stearn stressed more recently, how the "science" of botany continued to inform and be informed by what we today would call "applied" botany. Botany in this period was big science and big business, an essential part of the projection of military might into the resource-rich East and West Indies. It was precisely because finding and identifying valuable plants was so important to state purposes, Daubenton wrote, "that kings founded academic Chairs of Natural History." Curators of botanical gardens—both in Europe and its colonies—collected rare and beautiful plants for study and global exchange, but they specialized in cultivating those plants, such as *Cinchona*, crucial for European colonizing efforts in tropical climates.[11]

Historians of science, notably Marie-Noëlle Bourguet, Richard Drayton, John Gascoigne, Steven Harris, Roy MacLeod, David Philip Miller, François Regourd, and Emma Spary, have converged in emphasizing the importance of plants for the political and economic expansion of western European states. Across Europe, eighteenth-century political economists—from English and French mercantilists to German and Swedish cameralists—taught that the exact knowledge of nature was key to amassing national wealth, and hence power. "Plant mercantilists," as Bourguet and Christophe Bonneuil have called them, tightly controlled trade, attempting "as far as possible to multiply their own resources and diminish their tribute to foreigners." If resources were not to be had domestically, they were to be obtained through conquest and colonization. Colonies served as a fertile ground for the procurement and production of tropical plants that would not grow in harsh European climates. Colonies also served as captive markets for exports; states filled their coffers by monopolizing trade, imposing taxes on imports, and granting licenses for exports. Naturalists saw coffee, cacao, ipecacuanha (an emetic), jalap (a laxative), and Peruvian bark as moneymakers for king and country—and, often, for themselves. Peter Kalm, one of Carl Linnaeus' students wrote from London in 1748: "*Historia Naturalis* is the base for all economics, commerce, and manufactures."[12]

In order to understand how the study of a lowly periwinkle and its ilk can contribute to the wealth of nations, we must understand botany from an eighteenth-century point of view. Several botanical traditions coexisted in this period that only later became distinguished more sharply into applied botany (including economic and medicinal botany), horticulture and agriculture, and what we today call theoretical botany, especially nomenclature and taxonomy. Eighteenth-century Europeans defined *botany* as "that branch of natural history that distinguishes the uses, characters, classes, orders, genera, and species of plants" and the *botanist* as an "enquirer into the nature and properties of vegetables [who] ought to direct his [sic] view principally towards the investigation of useful qualities." Utility, according to Daubenton, author of the article "Botanique" in Diderot and d'Alembert's *Encyclopédie,* was and should remain prior to system. He lamented how botanists waste their time with nomenclature and taxonomies: "This defect in the study of botany is an obstacle to the advance of that science because it distracts us from its principal objective"—"use." Some botanists in this period even argued that utility, in this case the medicinal application of plants, provided "the foundation for natural classes." Economic botany was equally valued: Michel Adanson, well-known as a collector and taxonomist, judged that his success as a botanist resulted from the new method he developed for extracting dye from the indigo plant.[13]

Botanists today looking back will often distinguish economic, medical, and theoretical traditions, but it is important to realize that these were often talents embodied in a single botanist. The great Linnaeus, for example, like most botanists in this period, was also a practicing physician for whom the medical virtues of plants were of cardinal import. Now celebrated as the "father of modern taxonomy," the Swedish scholar often saw his taxonomic innovations as secondary to his many economic schemes. The eminent botanist William Stearn has pointed out that Linnaeus' binomial system of nomenclature first developed as a kind of shorthand to aid several of his economic botanical projects, most immediately for cataloguing Swedish fodders in order to enhance animal husbandry.[14]

Like others at the time, Linnaeus taught that the purpose of natural history was to render service to the state. Devoted to cameralist principles, he taught that national wealth could be aggrandized through the exact study of nature. Linnaeus' goal was to staunch the flow of bullion to Asia by growing valuable plants, including coffee, mulberry trees for silkworms, cotton, rhubarb, opium, and ginseng, within Sweden's borders. Hypothesizing that tropical plants were globally adaptable, Linnaeus hoped that he could "fool," "tempt," and "train" them to grow in Arctic lands and thereby create "Lapland cinnamon groves, Baltic tea plantations, and Finnish rice paddies."[15] He also hoped to overcome Sweden's devastating famines by finding new indigenous plants—fir bark, seaweed, burdock, bog myrtle, or Iceland moss (lichen), for example—that could feed hungry peasants.

This was not, however, a uniquely Swedish phenomenon. Sir Hans Sloane at one end of the eighteenth century and Sir Joseph Banks at the other both joined economic ventures to botanical exploration. As President of the Royal Society of London, President of the Royal College of Physicians, Royal Physician to George I, and a well-known enterprising spirit, Sloane introduced what he called the "stomachic"—hot milk chocolate—to England. The innovation was a big one: Cacao beans had traditionally been mixed with honey and hot peppers by the Mayan, Aztec, and Spanish; Sloane made this bitter drink palatable to the English by preparing cacao with milk and sugar—though the exact recipe was long kept secret.[16]

Historians of the eighteenth century have also begun to detail how botany—expertise in bioprospecting, plant identification, transport, and acclimatization—worked hand-in-hand with European colonial expansion. Early conquistadors entered the Americas looking for gold and silver. By the eighteenth century, naturalists sought "green gold." Rich vegetable organisms supplied lasting, seemingly ever renewable profits long after gold and silver ran out. Recognizing the value of naturalists to the colonial enterprise, the director of the Madrid Botanical Garden wrote, "a dozen naturalists and some chemists scattered in Spain's dominions . . . will offer an incomparably larger utility to the state than a hundred thousand men fighting for the enlargement of the Spanish empire." The very definition of "America" in Diderot and d'Alembert's *Encyclopédie* emphasized trade in sugar, tobacco, indigo, ginger, cassia, gums and aloes, sassafras, brazil wood (a dyestuff), guaiacum (a cure for syphilis), cinnamon, balsams of Tolu and Peru (against coughs), cochineal (a rich red dye), ipecacuanha, nutmeg, pineapple, and jalap along with the "water of Barbados." By the eighteenth century, sugar had become the most important cash crop imported into Europe from the Americas, but Peruvian bark was the most valuable commodity by weight.[17]

Mercantilism flourished through the fecund coupling of naval prowess to natural history. Eighteenth-century botanical exploration followed trade routes, as naturalists of all stripes found passage on trading-company, merchant-marine, and naval vessels headed for European territories abroad. Because most physicians and surgeons were trained in botany in this period, naval and royal physicians as well as East and West India Company surgeons stationed in Europe's far-flung colonies all contributed to

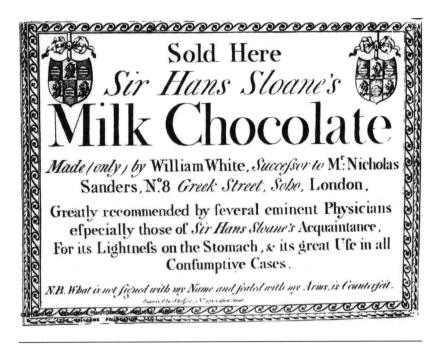

Figure I.2. Trade card featuring Sir Hans Sloane's milk chocolate as a medicine good for consumption and stomach ailments.

Figure I.3. Frontispiece to Charles de Rochefort's *Histoire naturelle et morale des Iles Antilles de l'Amérique*. Tainos and Caribs present a pineapple, parrot, and other exotic fruits, birds, and animals to a sexually ambiguous European monarch (perhaps Europa—note the bare feet). The scribe, perhaps Father Raymond, who prepared a Carib dictionary for Rochefort's book, stands behind the European figure, possibly translating the Americans' words.

worldwide plant collecting networks. Carl Thunberg, a Linnaean student and employee of the Dutch East India Company, collected plants in three categories: those fit for food, those useful as medicines, and those appropriate for "domestic and rural economy" (or agriculture, we would say). French botanist Adanson, employee of the Compagnie des Indes in Africa, made it his "first care" upon arrival in Senegal "to gather as many plants as possible for the King's Garden." In the first months of his 1749 voyage, he shipped chests carrying live saplings of "300 different trees" to botanists working at the distinguished Jardin du Roi of Paris. Botanists of the Dutch East and West India Companies sent regular cargoes of plants and seeds to botanical gardens and laboratories in Leiden and Amsterdam. British collectors furnished the Chelsea Physic Garden and later the Kew Gardens with dried and living specimens and exotic seeds from their extensive system of gardens stretching from Saint Vincent in the West Indies to Calcutta, Sydney, and Penang (Malaysia). Portugal, Spain, Holland, France, and England all secured vast colonial holdings by way of new markets in coffee, tea, sugarcane, pepper, nutmeg, cotton, and other profitable plants. Alice Stroup has emphasized that, after cartography, botany was the most highly funded science in late seventeenth- and eighteenth-century France. Between 1770 and 1820, Britain had 126 official collectors in the field and countless others involved in less formal ways (sending uncommissioned consignments of artifacts, plants, and remedies for gout and other common diseases).[18]

Science followed trade routes at the same time that naturalists—in the eighteenth century, mostly physicians—worked to improve commerce. Take, for example, the story of how coffee was introduced as a cash crop into Martinique. In 1714, a coffee plant was sent to Louis XIV as a gift from the mayor of Amsterdam and planted in the King's Garden in Paris. In 1716, young plants grown from the Dutch seeds were carried from the Jardin du Roi to Martinique by a certain M. Isemberg, a physician serving in the French Antilles. The doctor died shortly after his arrival and the plants failed. Another physician, M. Chirac, apparently Isemberg's replacement, brought a seedling raised from this same Dutch plant to the Antilles just four years later. According to contemporary accounts, he carried the tiny plant on a merchant ship, and when water became scarce, he was obliged to share his scant portion with the precious plant. Finally arriving in Martinique, he planted his charge with care in his garden. His success matched his hopes: the first year he gathered two pounds of seeds which he parceled out to other planters for cultivation. From Martinique, seeds were taken to Saint Domingue (currently Haiti, occupying the western part of the island of Hispaniola), Guadeloupe, and other neighboring islands in the Caribbean. Coffee cultivation—the second largest crop in the French Antilles in the eighteenth century—was thus originally nurtured by a colonial physician.[19]

In David Mackay's words, botanists at this time were "agents of empire": their inventories, classifications, and transplantations were the vanguard and in some cases the "instruments" of European order. Bruno Latour has argued further that the technologies of collection—both material and intellectual—extended the imperial power of European nations. Collecting served empires in at least three ways. First, botanists identified and catalogued precious plants in new territories, allowing their governments to secure a cheap supply of drugs, foods, and luxury items for domestic markets. Second, naturalists found domestic or colonial substitutes for luxury imports—for example, rhubarb, tea, coffee, or Peruvian bark—that had drained treasuries of precious metals. Thirdly, botanists employed their technical expertise to transport and acclimatize valuable plants to the soils of European territories around the world. The point was to bring coffee, indigo, cochineal, cassia, and sugarcane all into production within the boundaries of the empire itself. The Jamaican planter, Edward Long, put it succinctly, "commerce stands so largely indebted to physic [medicine], and its sister botany, not only for materials of import and export, but [for] the abilities of men employed in collecting those materials." Europe's naturalists not only collected the stuff of nature but lay their own peculiar grid of reason over nature so that nomenclatures and taxonomies, as we will see in what follows, often also served as "tools of empire."[20]

The botanical sciences served the colonial enterprise and were, in turn, structured by it. Global networks of botanical gardens, the laboratories of colonial botany, followed the contours of empire, and gardens often served its needs. The sixteen hundred botanical gardens that Europeans had founded worldwide by the end of the eighteenth century were not merely idyllic bits of green intended to delight city dwellers, but experimental stations for agriculture and way stations for plant acclimatization for domestic and global trade, rare medicaments, and cash crops. The director of the Jardin Royal des Plantes in Paris, the famous Georges-Louis Leclerc, comte de Buffon, made experimental forestry a central concern for his garden in order to restock forests at home and build vessels for merchant fleets and naval engagements abroad. Linnaeus, Buffon, and Joseph Banks all reigned over metropolitan botanical gardens in the eighteenth century, sitting like anointed monarchs at the centers of vast botanical empires.[21]

Plan of the Book

This book focuses on the movement, mixing, triumph, and extinction of different knowledges in the course of eighteenth-century encounters between Europeans and the peoples of the Caribbean. "Caribbean" here does not refer to the Caribbean Sea and its islands as we now understand them, but to a larger geographical area that Peter Hulme has called "the extended Caribbean," including the coastal regions and islands stretching from Jamestown, Virginia, to Bahia, Brazil, as defined by eighteenth-century Atlantic shipping routes, shared ecosystems, and common patterns of colonization and (eventually) enslavement. To set some limits, I will focus primarily on encounters between French, British, and Dutch/German naturalists and the indigenous and enslaved peoples of Jamaica, Surinam, and Saint Domingue—the jewel of the French Crown's colonies and the most profitable Caribbean colony in this period. I regret not including the Spanish Caribbean islands in this study, but I simply do not have the required background. The new literature on the self-confident Spanish creoles in South America, who challenged learning and natural historical practices imported from Europe, is highly interesting. Because Spanish soldiers and settlers took Amerindian wives more often than other Europeans, traditions may have survived that were extinguished in the areas I have examined, and it is intriguing to think that there may be stories of abortifacients still to discover in Spanish-speaking islands.[22]

This book considers the "long eighteenth century" beginning in the 1670s when the plantation system based on slave labor and sugar cultivation began to dominate the Caribbean, and when Europeans—whether French or not—implemented the *Code noir* (1685), a set of laws regulating the status and conduct of slaves and, to some extent, masters' power over slaves. This was also a time when Jean-Baptiste Colbert in France seized administration of the colonies and France embarked on a plan of state-run colonial science, centered in the Parisian Académie Royale des Sciences and the Jardin du Roi. It was also a time when terms such as *botaniste* and *botany* passed into widespread usage.[23] This study finishes toward the end of the eighteenth century with the revolutionary upheavals in North America, Europe, and Saint Domingue. This latter period sees the abolition of slavery in the French colonies in 1794 (reinstated in 1802) and the abandonment of the slave trade in the British colonies in

1807. There were many differences in voyaging and colonizing in the earlier and the later parts of the eighteenth century. By the 1760s, for example, Europeans were more secure in their holdings and more confident in their colonizing efforts—ships were faster, food supplies more reliable, and more soldiers and colonists survived the tropics.[24]

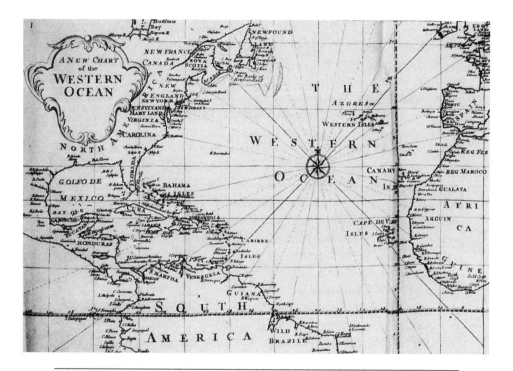

Figure I.4. The "Western Ocean," from Sir Hans Sloane's *Voyage to the Islands*... The map shows the Atlantic world as defined by eighteenth-century trade routes and port towns. These were territories accessible only by boat, and what we today call the Caribbean was part of a larger travel/trade network extending from Surinam and Guiana in the south to Florida in the north, Europe and Africa to the east. The "Caribbe Isles," as Europeans referred to them, included the easternmost islands of the West Indies, now known as the Lesser Antilles or Leeward and Windward Islands. Note the prominence of Hispaniola—today's Haiti and Dominican Republic, known in the eighteenth century as Saint Domingue in the west and Santo Domingo in the east, and the prominence of Spain, still a major power in Sloane's day. The western coast of Africa, whence came the slave labor that fueled the emerging sugar industry, is also featured.

Scribacious naturalists saw themselves writing not elegantly composed romances or histories, but "naked and simple truths in the same order in time and place" about events as they had occurred in the wild. Chapter 1 of this book investigates who studied plants in the eighteenth century—European *botanistes de cabinet,* voyaging botanical assistants, *voyageurs naturalistes* (a term coined by Daubenton; Jean-Baptiste-Christophe Fusée-Aublet used the term *botanistes voyageurs*)—in order to understand differing field practices, experiences, and points of view characteristic of diverse sorts of naturalists. Although European voyaging was largely a male affair, women figured in many of these categories.[25] It is also important to note that this book is built upon textual sources and for that reason provides most immediate access to European experiences and practices. Africans and Amerindians, however, were also naturalists and voyagers in this period, and I have tried where possible to draw out the knowledges and experiences of African slave, Arawak, Taino, and Carib naturalists albeit as reported through the eyes of Europeans.

In the early days of scientific voyaging, Europeans sought out and tended to value Amerindian naturalists' "simples" (or medicinal plants) and healing regimes that aided displaced Europeans' efforts to survive the unique diseases of hot and humid tropical areas. They also marveled at the many wonders of the New World: Thomas Trapham, a physician, reported from Jamaica on "shining trees" that bear such "starry lights" that a person could read by it, or the "jumping seeds" that, moved by an "internal spring of motion," jumped "up and down" and continued these "frolicks many days." By the eighteenth century there were few unadulterated indigenous plants, humans, or knowledges to be collected in the West Indies: peoples and plants, languages and knowledges had churned, mingled, and melded for over two hundred years. Europeans carried small arks of plants, animals, and peoples with them wherever they went. The slaver Richard Ligon, for example, shipped from Saint Jago off the coast of Africa for sale in Barbados "negroes, horses, and cattle." He also carried seeds of "Rosemary, Time, Winter Savory, Sweet Margerom, Pot Marjerom, Parsley, Penniroyall, Camomile, Sage, Tansie, lavender, Lavender Cotten, Garlicks, Onyons, Colworts, Cabbage, Turnips, Redishes, Marigolds, Lettice, Taragon, [and] Southernwood" which prospered well in their new home. Already in the eighteenth century, Nicolas-Louis Bourgeois noted the fusion of Amerindian and West African medical traditions that made it difficult, if not impossible, to distinguish between them. By this time period also certain peoples and their knowledges were already extinct. Alexander von Humboldt faithfully recorded what he could from the Amerindian peoples, the Caribbees, Guaypunabis, Marepizanoes, and Manitivitanoes, but these "vanquished nations" had gradually disappeared, leaving no other signs of their existence than a few words of their language mixed with that of their conquerors.[26]

Today "indigenous" often has the ring of something original and unsullied, and is used to refer to things non-Western. "Indigenous knowledges" are often set in opposition to "scientific knowledge" and sometimes even romanticized as panaceas for Western ills. As the eighteenth-century British physician Patrick Blair reminded his readers, however, "indigenous" meant simply "home-bred."[27] Thus, the term appropriately describes domestic European knowledge as much as knowledges collected from faraway and romantic places.

If confusion surrounds the term "indigenous," it engulfs the term "creole." Inhabitants of the United States today tend to think of "creole" as referring to pleasantly spicy food, people of mixed African and European origins, and languages that are "mixed" combinations of two or more separate traditions. To my guide in Jamaica, it meant a person of mixed race ("mulatto," she said quietly). To readers of the H-Caribbean listserv it had a multitude of meanings from "a culture with a Spanish flavor" to simply "black" (a person of African descent) to, more fashionably, a "hybridity" of cultures. Some people wanted to specify types of creoleness, as in "black French creole" or "white French creole." Others used it as a term to distinguish light-skinned from darker-skinned persons of African descent.

In this book, I will use "creole" in its eighteenth-century sense to mean a person of either European or African descent born in the New World. The term thus distinguished persons of such lineage from aborigines, on the one hand, and from persons born elsewhere and transported to the New World, on the other hand. The term could refer equally to humans, animals, or vegetables of such ancestry; it was not unusual to speak of "creole pigs" or "creole corn."[28] Eighteenth-century usage tended to be precise because creoles (creatures born in the islands though of foreign descent) had significantly higher survival rates than peoples, plants, or animals uprooted from their homelands and transported to a new and often harsh climate.

Chapter 2, "Bioprospecting," explores encounters in what I call the "biocontact zones" of the Caribbean and the means by which Europeans learned about New World flora. Much bioprospecting in this period centered on *materia alimentaria, materia luxuria,* and *materia medica* (materials for foods, luxuries, and medicines). The tropical regions of the world, occupying only 6 percent of Earth's surface, contain the greatest diversity of the world's flora. It is estimated that of the 250,000 species of higher plants living on Earth today 20 percent (or 35,000–50,000) are found in the tropical rainforests of the Amazon. Despite centuries of investigation, less than one-half of 1 percent of all flowering plant species has been studied for pharmacological use.[29] In the eighteenth century, the race to mine Caribbean and South American forests was run against the minimal chances of survival for individual prospectors; today it is run against the threat of large-scale destruction of bioresources. E. O. Wilson has estimated that by the early 1990s humans had destroyed more than half of the original forest cover in wet tropical regions, and that at current rates, tropical forests are being destroyed at a rate of 80,000 square miles per year.[30]

In addition to seeking "green gold," prospectors today often seek "genetic gold." Drug prospecting today generally takes one of two tacks. Exuding a certain technological machismo, many drug companies simply assay thousands of samples per week; their assumption is that chemical identification of active ingredients will eventually lead to the development of new drugs. Other companies prefer to deploy ethnobotanical techniques, seeking clues concerning drugs for specific ailments from local botanists or traditional healers. Ethnobotany is a term coined in 1896 by the American botanist John Harshberger to describe "studies of plants used by . . . aboriginal people."[31] Though the term may then have been new, ethnobotanical practices were already very old.

The genetic resources possessed by peoples and nations in the tropics (areas that tend to be rich genetically but poor economically) were not protected by international agreements until 1992. The Rio International Convention on Biological Diversity of that year has been applauded by many (and criticized by others) for making biological resources a matter of nation-states' property rights. Before 1992, plant resources were generally considered part of Earth's "common biological heritage"—they were there for the taking by whoever had the ingenuity, know-how, and capital to extract them and develop them into profitable commodities. (Prior to 1992, only a few plants had been lifted out of this "common heritage" and protected by the 1930 U.S. Plant Patent Act to safeguard Western plant breeders and their profits.) Today, the International Convention on Biological Diversity is intended to preserve not only plants but the lands where the plants grow. Thus states rich in biodiversity can negotiate a return of profits from multinationals allowed to "prospect" in their territory.

Some schemes have been developed to channel profits from drug development into conservation programs that aim to preserve the plants, people, and knowledges associated with the areas bioprospectors hope to mine in the future. One celebrated example is the Merck/INBio joint venture in Costa Rica. By agreement, the U.S. pharmaceutical giant Merck receives rights to Costa Rica's rich bioresources, while INBio (National Biodiversity Institute, a nongovernment, nonprofit scientific research institute) in Costa Rica receives access to advanced U.S. screening technologies and training for their researchers. Costa Rica also receives royalties earmarked for conservation. These royalties differ according to how much human knowledge comes with the plant: for plant samples alone Merck pays Costa Rica royalties of 1–6 percent of net sales; for biomaterials that come with some information about potential medical uses Merck pays 5–10 percent; and for plants with known medical uses Merck pays 10–15 percent.[32] Similar deals were struck in Surinam in 1993.

Few in the eighteenth century agonized over who owns nature. With breathtaking audacity, in 1493 the Borgian pope Alexander VI ceded the western hemisphere to Spain and the eastern hemisphere to Portugal with the line of demarcation passing 100 leagues west of the meridian passing through the Azores and Cape Verde. By the seventeenth century, the English East India Company (founded in 1599) and Dutch East India Company (founded in 1602), backed by state-supported navies, had overpowered Portuguese and Spanish trading and military posts. These sleek joint-stock companies along with savvy state ministers (especially in France) developed a mercantilist formula for commercial gain that tapped into the riches of empire as never before. While Europeans often recognized each others' monopolies and claims (while continuously conniving to overpower them), they tended to assume that non-European peoples had no proprietary claims to lands, resources, or knowledges. In the early modern world, the spoils of green monopolies fell to those who could police them.[33]

Monopolies and strident bioprospecting bred a global culture of secrets. In Chapter 2 I investigate the difficulties Europeans encountered entering foreign and uninviting terrains, and the ways they sought to wrest secrets from often unwilling and wily informants. This chapter deals with barriers—physical, conceptual, perceptual, and prejudicial—between knowledge traditions, and explores issues surrounding what we today call intellectual property rights, asking "who owns nature?" in various eighteenth-century contexts.

Chapter 3 takes us to the heart of the book where two of my central concerns come together. First, how did gender relations in Europe and its West Indian colonies guide European naturalists as they selected particular plants and technologies for transport back to Europe? Whereas gender has received some attention in eighteenth-century studies of exotic peoples, it has received relatively little attention in the histories of colonial science.[34] Throughout this period, European powers set up botanical networks—at home and abroad—that manipulated global botanical resources for their own economic interests. How did gender relations mold what plants and knowledge circulated through these networks?

I have identified abortifacients and anti-fertility agents as one class of plant-based drugs that did not circulate freely between the West Indies and Europe. I have chosen this class of plants, and specifically the peacock flower, to address my second concern: that of developing a detailed case study of agnotology. This chapter takes us into the murky world of abortion practices in the West Indies. For centuries, abortion has aroused strong feelings and controversy. Women seeking abortions have frequently been vilified for "crushing the conceived hopes of children within their entrails." It is difficult often to say exactly what was known about abortion and abortifacients in the past because such knowledge has often been suppressed in the present. The University of Michigan's copy of a 1925 German handbook on abortion, for example, bears a handwritten note that reads "put in locked case when catalogued." Attitudes like these complicate the study of abortifacients. The English physician Andrew Boord was typical in writing in 1598 that there are "many receipts of medicines [that cause 'abhorsion' as he called it], or . . . extreme purgacions, potions, and other laxatine drinkes, of which I dare not speak of at this time, least any light should have knowledge by which willful abortion may come of the multitudenesse of the flowers of a woman."[35]

Though I initially shied away from the subject (fearing such a highly politicized issue), it soon became clear that abortifacients represent an important aspect of gender politics in the voyages of scientific discovery. Chapter 3 discusses how widespread abortion practices were in the West Indies and how abortion was used among slaves as a form of political resistance.[36] As we shall see, there are some surprises here.

Agnotology—the study of cultural ignorances—drives the historian onto shaky ground. History typically charts events as they unfold; documenting nonoccurrences of probable developments requires new strategies and methodologies. My argument here is that well-developed knowledge of abortifacients did not transfer from the West Indies into Europe.

Chapter 4, "The Fate of the Peacock Flower in Europe," focuses on the usual mechanisms by which exotic medicines were introduced into Europe. Thanks to Andreas-Holger Maehle's *Drugs on Trial,* we understand the process by which a colonial drug, such as Peruvian bark, passed into the official *Pharmacopoeia* of London, Paris, and Amsterdam.[37] If we take Peruvian bark or smallpox inoculation as models of the kind of proof and persuasion required to introduce new and controversial medical practices into Europe, we can identify the normal pathways of eighteenth-century drug development. With these expectations in mind, we can chart how abortifacients did or did not move along those pathways.

While abortifacients were increasingly ignored in this period, women's distinctive health needs were not completely neglected. Eighteenth-century naturalists collected and physicians experimented extensively with emmenagogues (used to induce the menses), for example, which were important *materia medica* for women. Protracted testing with menstrual regulators was carried out by John Freind in London in 1720 and reported in his *Emmenologia.* In contrast to the testing of emmenagogues, I have found little testing—chemical or clinical—with abortifacients. This is curious because physicians performed what we would call therapeutic abortions to save mothers' lives throughout this period. (Sloane reported, for example, that he used "the hand" for this purpose—see Chapter 3.)[38] Although physicians often recorded their use of other medicines, carefully noting dosages and patient reactions, they rarely noted their use of abortifacients. Toward the end of the eighteenth century, physicians started testing savin—European women's abortifacient of choice—but only as an emmenagogue.

A final chapter, "Linguistic Imperialism," takes up a different aspect of colonial botany, while continuing to trace the cultural history of Merian's peacock flower. This chapter investigates the politics of botanical nomenclature. As Europeans explored new territory around the globe, they often named lands, rivers, plants, and so forth as if each were Adam himself, the first human to see the birds of the air and fish of the sea. In 1492, Christopher Columbus landed on the island the Tainos called Quisqueya and named it *La Isla Española* (or Hispaniola). He named the island known to the Caribs as Karukéra (Island of Beautiful Waters) Guadeloupe after the sanctuary Santa Maria de Guadalupe de Estremadura in Spain. The island of Alliouagana (Land of the Prickly Bush in Arawak) Columbus called Santa Maria de Montserrate for the Blessed Virgin of the monastery at Montserrate near Barcelona. Dominica was so called because Columbus landed there on a Sunday. The Portuguese named Barbados for *Barbata,* a kind of bearded fig that grows there. Farther north, a place formerly known as Wingandacoa was renamed Virginia, after England's supposedly virgin queen, Elizabeth I. Indigenous names did sometimes survive. Surinam derives from the Surinen, the country's first inhabitants, and Jamaica or Xaymaca (an Arawak name meaning the land of wood and water) persisted despite Columbus's efforts to christen it Santiago when he first sighted the island in 1494.[39]

This chapter explores the politics of eighteenth-century botanical nomenclature and sucks us into the quagmire of *nomina dubia, nomina ambigua, nomina confusa,* and *nomina nuda* (doubtful, ambiguous, confusing, and "naked" names).[40] A rich diversity of traditional names was funneled in this period through the intellectual straits of Linnaean nomenclature to produce standardized naming. Botanical Latin was made and remade in the eighteenth century to suit naturalists' purposes. If Latin, the language of European learning, was to become the standard language of botanical science, it might have incorporated customary names from other cultures as plants from those cultures entered Europe. It might also have preserved a sense of the biogeography of plants by marking plants with their places of origin. But plants more often were named for European botanists and their patrons. Naming practices celebrated a particular brand of historiography—namely, a history celebrating the deeds of great European men. It is remarkable that Linnaeus' system itself retold—to the exclusion of other histories—the story of elite European botany.

A narrative of imperial nomenclature describes well the fate of thousands of plant species that came under European perusal in the eighteenth century. Of the 149 new genera presented in the Spanish naturalist Gomez Ortega's *Prodromus,* for example, 116 celebrated the deeds of great men of Europe. *Carludovica palmate* (the palm from which Panama hats were later made), for instance, stands for King Carlos III and his wife Ludovica (the only female honored in this work).[41] This chapter explores also the exceptions to this account. How, for example, were the other thirty-three plants in Ortega's *Prodromus* named?

A final section of Chapter 5 investigates efforts made in the eighteenth century to order botanical nomenclature along indigenous lines. Alternatives to the Linnaean system did exist but were not developed. Investigating these eighteenth-century issues is particularly timely in our own age, given current heated debates among botanists concerning ending the mandatory use of Latin in the botanical *Code* (the only *Code* still requiring Latin) and switching instead to English. Tarcisco Filgueiras, a Brazilian botanist, has defended the continued use of Latin, arguing that it is not a language in the ordinary sense of the word but rather an artificial, simplified code for international communication among botanists. He has judged the proposed shift to English "arrogant" since it requires non-English speakers to learn a new language. "If any modern language is elected as the official language for international communication," he argues, "a certain culture, a certain political and economic system, and even certain races will be privileged."[42] J. McNeill, who has championed the switch to English, responds that English is already the common scientific language, and that retaining Latin nomenclature only requires non-English speakers to learn yet another language. Further, he argues, European languages are rooted in Latin and hence Europeans retain an advantage over persons from, say, Japan, Indonesia, Peru, or China. McNeill readily admits that privileging English represents merely a new form of imperialism. But, he adds, this reflects current practices within scientific communities.[43]

This book invites readers to board unstable, rickety ships and to hazard entering dark and steamy rainforests. The difficulties of voyaging in the eighteenth century should not be underestimated. Janet Schaw, a "Lady of Quality," bemoaned the "constant rains" and "ooze" that day and night penetrated her cramped cabin. The vessel that carried her from Scotland to the West Indies in the 1770s was so heavily loaded with goods and illegal cargo (including a company of emigrants smuggled aboard to increase the ship's overall profit) that it "broached to," listing completely onto one side during a heavy storm. All supplies on deck were lost. There was, she reported, "no such thing as being warm." The broad ocean between the Americas and Europe also spoiled much scientific cargo; many painstakingly collected specimens were lost completely. On his return voyage to England, Sloane lost a valuable large yellow snake and an iguana when the snake was shot dead by an apprehensive sailor and the lizard, surprised by a seaman while running along the gunnels of the ship, leapt overboard and was drowned. Nicolas-Joseph Thiery de Menonville, a French botanist who risked his life to steal the valued cochineal insect from the Spanish, had the misfortune to watch his nopals (the cactus on which the insects lived) rot and die in the short voyage between New Spain (Mexico) and Saint Domingue (Chapter 1). Disasters were not only of the natural kind: Richard de Tussac, who botanized for sixteen years in Saint Domingue saw 2,000 of his illustrations go up in flames when Cap Français was torched during the Haitian Revolution.[44]

Aublet, the French king's botanist, wrote in the 1770s from Guiana, amid the vast wild coast of South America. Conditions were so dreadful that for nearly a century (from 1852 to 1946) this area served France as a penal colony. He captured the desperation of the naturalist's lot when he wrote:

> You have to have been in a jungle to understand how dangerous it is to enter. The thorny trees, the tangle of razor sharp plants, the poisonous snakes, the deep pits of water all can maim, while the ever present danger of attack from fierce animals and angry fugitive maroons [runaway slaves] leaves the lone traveler fearing for his life . . . The slaves and Indians that one must take along as guides and porters are a source of constant anxiety. You have to win their respect, fear, and love, and to guess their designs and plots so that they do not abandon you in the jungle or kill you. You have to arm the guides and carriers, and often you are the only white amongst ten or twenty armed slaves—who do care not at all for Europeans. If these dangers are not enough, they are joined by a multitude of small torments in the form of ticks, mosquitoes, and other insects whose sting causes ulcers and make the journey miserable. On top of all of this is the constant suffocating heat and drenching rains. Under these conditions, it is necessary to prepare a description of the parts of the plant that cannot be transported or which will be altered in transport, to write of the different circumstances in which the plant is found, of the elevation, the soil, the dimensions of the plant, its qualities, and the nuances of its colors.

"All this," he exploded, "simply to gather the flowers and seeds of a few new plants."[45]

1

Voyaging Out

> Saint Domingue produced enough sugar, cotton, indigo, coffee, and cacao to satisfy amply the cupidity for gold and ambition for fortune.
> CHARLES ARTHAUD, 1787

Who were the colonial botanists of the eighteenth century? What sort of person risked rough seas, torrid lands, fevers, serpents, and "savages" merely to collect a few plants, even if for king and country? Frans Stafleu has identified a prominent divide among European botanists of the eighteenth century: the *botanistes de cabinet* like Linnaeus, who tended fine gardens and natural history cabinets often at university posts in Europe, and *botanistes voyageurs* like Jean-Baptiste-Christophe Fusée-Aublet, who ventured into tropical rainforests—confronting thorny trees, venomous snakes, drenching rains, ticks, mosquitoes, runaway slaves, and suffocating heat to search for useful and profitable plants. European naturalists were a varied lot: some were men of God, most were physicians; some paid their own passage, most were sent by trading companies, kings, or scientific academies. Some were mature and settled, most were young and unmarried. The vast majority were male. Linnaeus, who ordered many of his students into the field, insisted that they be "penniless bachelors," young and equally willing to sleep "on the hardest bench as in the softest bed." Some undertook voyages of a year or two in duration, hoping to amass fame or fortune and return quickly to Europe. Others went to the colonies, settled, and remained for a lifetime.[1]

Historians have tended to focus on the heroic figure of the European voyager, and I, too, will be saying a lot about European botanists voyaging as directors of scientific enterprises to far-off and unknown lands in search of nature's bounty. I will focus here on Sir Hans Sloane, Maria Sibylla Merian, and others who emerge as major protagonists in subsequent chapters of this study. As backdrop for our more focused questions concerning European knowledge and use of abortifacients, it is important to understand who these voyagers were in terms of social and educational background; it is significant, for example, that the majority of botanists in this period were trained as medical doctors. We also find that differences in how natural history was structured and funded in voyagers' home countries—whether centralized and funded by the government as in France, enabled by trading companies as in Holland, or dependent on private entrepreneurs as in England—often influenced how collecting was done and what sorts of goods were collected. European voyaging naturalists tended to publish novelesque accounts of their observations and adventures, and historians can reconstruct much about their experiences and field practices. To best represent the large number of naturalists involved in colonial botany, I have chosen one as a "type specimen"—a representative individual—for each category of naturalist that I have identified.

Stafleu emphasized a deep divide in perspectives and practices between what I like to call armchair naturalists, who coordinated and synthesized collecting from sinecures in Europe (university posts, heads of gardens, and so on), and voyaging botanists, who visited foreign lands. We find, however, a number of other important divides among European colonial botanists. Place of birth, for instance—whether born and bred in Europe or the Americas—molded naturalists' loyalties to knowledge traditions, peoples, and plants as they botanized in the Caribbean basin. Historians have recently drawn attention to the role of *European creoles,* born and educated in the colonies, who especially in Spanish domains formed self-confident elites and came to challenge Linnaean taxonomies and other botanical practices exported from the metropolis to the Americas.[2] Creoles of English and French origin elsewhere in the Caribbean were often schooled in Europe and thus tended to share the knowledge and values of their colleagues in the Old World. We also find that many European naturalists who took up *long-term residency* in the colonies developed a deeper knowledge of and affection for places and plants than voyagers who passed through in a matter of weeks or months. A final type of European naturalist highlighted in this chapter is the *voyaging botanical assistant*—the largely unsung illustrators, guides, finders, and support staff of the European naturalist who advanced the massive task of cataloguing nature. Because these assistants rarely published we know less about them. It is possible, though, to piece together aspects of their stories, and I have chosen to recount the adventures of Jeanne Baret, assistant to the French botanist Philibert Commerson.

Focus on these varieties of European naturalists should not obscure other voyagers and their contributions to knowledge of West Indian flora. *African slaves,* forced to abandon their homelands, were often experts in tropical flora and carried with them plants and knowledge of their uses to the Caribbean. Slaves in the West Indies, "salt water" or creole, also often sought out and experimented with flora indigenous to the area, as did the *Amerindians* who understood well the flora and fauna of their native soils. Amerindians, too, were often voyagers known to have moved plants from place to place in the Caribbean basin for many hundreds of years prior to European contact. They were also experts in the uses of plants as foods, medicines, and construction materials, for example, in hammocks, canoes, baskets, and other goods needed for daily life. These latter two groups— African slaves and Amerindians—we will meet in the next chapter, where I will discuss how Europeans enlisted their expertise in the colonial botanical enterprise.

Botanistes Voyageurs

Stereotypical eighteenth-century "voyagers" were heroic individuals who set off for unknown lands and returned with the fruits of their adventures in relatively short periods of time. As male and female "type specimens" for this category, I have chosen the English physician and future president of the Royal Society of London, Sir Hans Sloane, and the German-born naturalist and celebrated artist, Maria Sibylla Merian. Sloane and Merian were contemporaries, and allow for a nicely matched pair for comparison. Sloane traveled for fifteen months to Jamaica in 1687; Merian voyaged for twenty-one months to Surinam in 1699. Each produced a gloriously illustrated report: Sloane in 1707 and Merian in 1705.

Born in 1660 to a wealthy land-owning Scottish family in the north of Ireland, Sloane, like many youngest sons, found few prospects at home. He made his way to London at the age of nineteen to study medicine, chemistry, and botany and, thanks to his class standing, was introduced to some of the finest naturalists of his day—men such as John Ray and Robert Boyle. Like most botanical voyagers, Sloane was a medical man. The French (but not the Dutch or English) had fielded a large number of naturalist priests in the second half of the seventeenth century (including Jean-Baptiste du Tertre, Charles Plumier, Jean-Baptiste Labat, and Louis Feuillée), but by the eighteenth century this class of naturalists had largely been supplanted by physicians and surgeons.[3] Medicine was still strongly allied with botany: Antoine de Jussieu, director of the Royal Garden in Paris was a physician, for example, as were his brothers Bernard, who also worked at the King's Garden, and Joseph, who collected plants in South America from 1735–1771.

Medicine had been an itinerant profession in the Middle Ages, and young physicians still in the early modern period considered a grand European tour important to their education. Following this tradition, Sloane set off in 1683 for France and the Jardin du Roi, the great natural historical institution then run by Joseph Pitton de Tournefort. Like many in this period, Sloane picked up a quick degree while abroad, in his case from the University of Orange, one of the few universities in France where Protestants were allowed to study. Linnaeus likewise submitted a thirteen-page doctoral dissertation and received a medical degree during his eight-day residence at the University in Harderwijk in the Netherlands. University positions in this period were often "entailed," passing from father to son whether or not the latter was qualified.[4]

Sloane returned to London to practice medicine and was elected fellow of the Royal Society in 1685 (at age twenty-five) and shortly thereafter Fellow of the Royal College of Physicians, positions that depended as much on class status as on scientific distinction. Two years later the Duke of Albemarle, newly appointed governor of Jamaica, invited Sloane to serve as his personal physician on the island. Sloane jumped at this chance for a paid voyage to Jamaica. There were many foundational questions concerning nature to be answered. Ray, for instance, wanted to know "whether there be any species of plants common to America and Europe." Martin Lister, a naturalist, queried "whether there are any naked snails in Jamaica, I meane such as are naturallie without shells at Land as with us." Sloane himself hoped to find some specific remedies as sensationally useful as Peruvian bark *(Cinchona),* which had been introduced into the London *Pharmacopoeia* only in 1677. Full of youthful enthusiasm, Sloane later wrote, "this voyage seemed likewise to promise to be useful to me, as a physician: many of the Antient and best physicians having travell'd to the places whence their drugs were brought, to inform themselves concerning them."[5]

An astute businessman, Sloane arranged also to have himself appointed physician to the British West Indian Fleet, in charge of all surgeons, at an annual salary of £600 (a handsome sum; an ordinary seaman might make £1.5 per month in peacetime and £2.25 per month in wartime). An additional £300 was paid Sloane in advance for supplies. This arrangement, whereby a naturalist signed on for nonscientific work in order to travel to exotic flora, was typically how British and Dutch botanists traveled throughout the seventeenth and early eighteenth centuries. For the Dutch especially, science followed trade routes. Most naturalists shipped out as physicians or surgeons on East or West India Company vessels. Jacobus

Bontius traveled to Batavia (now Jakarta) as physician, apothecary, and surgical inspector to the Dutch East India Company territories in 1626; Engelbert Kaempfer, the German-born physician, studied the flora of Japan while in the service of the Dutch company in the 1690s; Paul Hermann, originally from Halle, traveled in 1672 as medical officer for the same company to the Cape of Good Hope and on to Ceylon (Sri Lanka); Carl Thunberg, Linnaeus' student, shipped out as surgeon aboard a Dutch East India ship to the Cape of Good Hope in 1770 and traveled eventually to Japan. Hendrick Adriaan van Reede tot Drakenstein, who produced the magisterial *Hortus Indicus Malabaricus* (1678–1693), was unusual in being a military man, but he worked for the same company. In addition to publishing this twelve-volume work describing 740 plants of Malabar (the region of southwest India where Vasco da Gama landed in 1498), van Reede requested that all governors in Holland's western quarters—Bengal, Surat, Persia, and the Cape of Good Hope—send "annually by the homeward-bound ships from Ceylon to the fatherland all kinds of seeds, bulbs, or roots of the trees, plants, herbs, flowers, etc., which each of you is able to collect in his district for a whole year" for the sake of "curiosity and also medicine." In addition to working for trading companies, British botanists financed their voyages by various other means: the Reverend Hugh Jones was dispatched to Maryland as both an Anglican minister and botanist for the London Temple Coffee House Botany Club; William Woodville privately financed his botanical research in Jamaica; Benjamin Moseley wrote treatises on the medical use of sugar, coffee, and other plants while surgeon-general of Jamaica. Not until the late eighteenth century did the British government begin financing voyages and supporting botanical research; in this they consciously followed the successful French model (see below).[6]

Sloane sailed from Portsmouth in September 1687 on a forty-four-gun frigate that accompanied two merchant ships and the duke's yacht to the Caribbean. Except for seasickness, his voyage was uneventful, and in December the party landed in Port Royal, Jamaica. Sloane recorded his purposes, field practices, and findings in his large two-volume *Voyage to the Islands of Madera, Barbadoes, Nieves, Saint Christophers, and Jamaica*, published eighteen years after his return to England. After perfunctory remarks concerning his diligent efforts to "advance natural knowledge," Sloane emphasized that the purpose of his undertaking was to make British colonists self-sufficient by instructing them in the uses of plants "growing sponte or in gardens" for medicines and foods. "It is very hard to carry thither," he noted, "such European simples as are proper for the cure of all sorts of diseases," especially since most drugs lose their virtues in transport. A second and equally important purpose was to enrich English pharmacopoeia back home. Sloane noted that even though useful West Indian herbs did not grow naturally in England, "many of them and their several parts" were brought over and used in medicines every day, and that if people were inquisitive, many more could be brought "to the great advantage of physicians and patients." He offered that her Grace, the Duchess of Beaufort, in her leisure hours from her more serious affairs of state had successfully raised many tropical plants in her gardens by means of "stoves and infirmaries" (see below).[7]

Above and beyond purposes aimed at promoting public good, Sloane had proposals of his own, namely, prospecting for new drugs that might turn him a profit. Before returning from Jamaica, he invested the greatest part of the fortune he had acquired there in "the bark" (a Jamaican equivalent of the Peruvian bark). Peruvian bark or quinine was a "specific," a remedy intended to cure a particular ailment, in this case ague and other malignant recurring fevers. Sloane carried a cargo of this remarkably reliable bark back to England and promoted it by prescription in his fashionable London practice. He continued prospecting for new profits from drugs throughout his lifetime. In 1709 he wrote to William Byrd in Virginia asking whether that colony could provide commercial quantities of ipecacuanha (an emetic fetching 30 shillings per pound) so that the two of them in a joint venture might seize the considerable profits to be made from this popular medicine from the Portuguese and Spanish. In 1732 Sloane contributed £20 annually as a subscriber to a project to establish a colony in Georgia; Mr. Millar, a surgeon, was hired by the group to search for new drugs and to coordinate the cultivation of useful plants in the prospective colony. Sloane also had interests in plant-collecting expeditions in Barbados and other Caribbean islands.[8]

During his "leisure-hours" in Jamaica, when not serving the duke or other British planters, Sloane botanized, searching "the several places I could think afforded natural productions." His first concern was to standardize information in a fashion useful to colleagues around the globe. Sloane noted how he measured plant parts "by my thumb, which, with a little allowance, I reckoned an Inch." He considered it useless to be more exact, because the leaves of vegetables even of the same sort, the wings of birds, the toes of frogs vary so much from individual to individual. Similarly, he found it difficult to provide exacting descriptions of the glorious new colors he saw—so wonderful that they required "new names to express them." He also cautioned that plants tended to blossom in different colors, depending on whether they were grown in Europe or in Jamaica, because of differences in the soils and temperatures of the two environments.[9]

Following common practices, Sloane did not botanize alone; he employed the Reverend Mr. Moore, "one of the best designers I could meet with there," to prepare illustrations of plants, animals, fish, birds, and insects *in situ*. The two of them eventually made the tedious journey across the central highlands to the north coast of the island on horseback, and like most Englishmen transplanted to the Torrid Zone, complained of the extreme heat, the snakes, the lack of convenient lodging, and the danger of ambush in the bush—all of which compromised accuracy in observations.

Sloane made a point of collecting information about plants previously unknown to him from "the Inhabitants, either Europeans, Indians or Blacks." In doing so, he was well aware that by 1687 neither the people nor many of the plants found in Jamaica were indigenous to that place. All were voyagers—some willing (such as Sloane himself), some forced (the "Blacks" of whom he spoke), some accidental (opportunistic plants and even rats carried along with cash cargoes). According to Sloane, nature itself—the people, plants, and landscape of the island—had been forged in the crucible of Spanish domination. The natives of the island, he recounted, were "all destroyed by the Spaniards," either by guns or by smallpox. The Amerindians Sloane knew in Jamaica had been brought there as slaves, taken by the Spanish "by surprize from the Musquitos or Florida." The Spanish had also transported to the island Africans as slaves "from several places in Guinea" and many fruit trees, useful foods, and medicinal plants from South America. These plants "throve wonderfully, and now grow [in Jamaica] as it were sponte." When the English conquered Jamaica in 1655, the Spanish were forced to abandon plantations flourishing with exotic and valuable plants. They were also forced to forsake their "Blacks and Indians" who had learned "the skill of using" these plants for foods and medicines.[10]

The Dutch and English, too, Sloane noted, had made their contribution to this mélange of people, plants, and knowledge, bringing from Surinam to Jamaica their skill in the use of tropical flora. Sloane found in Europeans' dogged movement of plants and peoples an explanation for the uniformity in nature and its uses across the West Indies: "many of the virtues of plants [he discovered] agree with the observations of authors writing from other parts of the West Indies."[11]

Collecting what he could during his fifteen-month stay in the tropics, Sloane prepared live plants and dried specimens for transport back to England. Altogether Sloane managed to amass 800 new plants, among them what became the type specimen for *Theobroma cacao* (Theobroma meaning "food of the gods"). While in Jamaica, Sloane also collected a future wife. Six years after his return to England, he married Elizabeth, daughter of John Langley, an alderman of London, who had inherited her father's fortune. Elizabeth was, perhaps more attractively, the widow of Fulk Rose, Esq., who had owned thriving sugar plantations in Jamaica (where Sloane and Elizabeth Langley Rose probably first met). Elizabeth inherited the widow's third of the income of the Rose estates. This massive fortune, combined with his ongoing economic interests, supported Sloane's science and moved him to the center of London's scientific establishment, a position he retained for more than fifty years.

Maria Sibylla Merian

Sloane and Merian are both central to the story of abortifacients; each identified the peacock flower, later known as the *Poinciana,* while in the Caribbean. To my knowledge, Merian was the only European woman who voyaged exclusively in pursuit of her science in the seventeenth or eighteenth centuries. Amazingly, it is possible to list other women botanic travelers in a single page. Merian was accompanied by her daughter, Dorothea Maria, who served as her assistant. Merian's other daughter, Johanna Helena, also collected *naturalia* in Surinam, traveling in 1711 with her husband, who administered an orphanage there. Jeanne Baret sailed with Louis-Antoine de Bougainville (for whom the showy flowering vine was named) as Philibert Commerson's botanical assistant (see below). In the 1770s Lady Anne Monson (Chapter 5) accompanied her husband, a colonel, from England to the East Indies and there indulged her passion for natural history, but she did not set her own itinerary. And in the 1790s, Maria Riddell, an ardent naturalist, accompanied her father William Woodley, governor of Saint Kitts and the Leeward Islands, to the West Indies. Riddell stole precious minutes to publish her *Voyages to the Madeira, and Leeward Caribbean Isles with Sketches of the Natural History of these Islands,* recording the uses of various plants: a cactus used by "West India ladies" to dye their ribbons, gauzes, and cheeks; a wild licorice whose berries could be strung like beads for necklaces and bracelets; and an Indian arrowroot that served as a remedy against dysentery and the venom of poisoned arrows. Nonetheless her duties as a wife and daughter took priority and, by her own account, left her little time to botanize.[12]

It became more common for women to travel in the nineteenth century. Sarah Bowdich, for one, accompanied her husband, Edward, to Africa in 1823 as the female collaborator in this "creative couple." Mrs. Bowdich, as she was called, had intended to serve as illustrator to her scientific husband, but Edward died of malaria in Gambia. With three small children to care for and no money for a speedy return to England, she took up where her husband had left off, collecting plants and arranging his papers for publication. Although Bowdich, like many voyaging naturalists, lost numerous engravings, books, and dried plants to the raging seas on her return passage to Europe, she continued her work, taking her remaining specimens to the Jardin du Roi in Paris for study. She is today hailed as the first woman to collect plants systematically in tropical Africa; she eventually published an account of the flora of the Cape Verde Islands and the area in and around Banjul.[13]

Merian, born in 1647, was indeed bold to travel to Surinam in search of exotic insects, accompanied only by her twenty-one-year-old daughter, whom she had trained from childhood as a painter and assistant. Bodily and moral imperatives kept the vast majority of Europe's women close to home. Medical men differed in their views about the physical effects of the tropical climate on female physiology. Some, like Thomas Trapham, a Jamaican physician, emphasized that the air in the West Indian islands—made hot by the sun and moist by the moon—was exceedingly agreeable to women, "beneficial to their living, including their conceptions and facilitating their births." Although as the "moistest sex" women suffered greatly from worms, the "vivifying hot" and "multiplyingly moist" air made women fecund like "all other produces in the Indies." Writing from Saint Domingue, Nicolas-Louis Bourgeois celebrated the tropics as a paradise for women: the heat of the climate eased childbirth and allowed women to live to an old age. Thunberg, traveling in the tropics of the East Indies, agreed that women held out better than men against the dysenteries and putrid fevers that put Europeans in early graves, but, he added, "those who come from Europe with rosy cheeks lose this species of beauty in a short time, and are afterwards as pale as corpses."[14]

More often, however, physicians emphasized the grave dangers of travel to the Torrid Zone. In the seventeenth century, many taught that crossing the equator led to infertility. For this reason Dutch women were reluctant to migrate to Brazil. Summing up medical opinion at the end of the eighteenth century, Johann Blumenbach, a German physiologist, emphasized that white women taken to very warm climates succumbed to "copious menstruation, which almost always ended, in a short space of time, in fatal hemorrhages of the uterus." Many women feared that if they gave birth in the tropics, they would deliver children resembling the native peoples of those areas. The intense African sun, it was thought, produced black babies regardless of the parents' complexions. The French physician Jean-Baptiste-René Pouppé-Desportes warned that women aged more quickly under the intense tropical sun; their menstrual cycles also ended at an earlier age, leaving them open to a host of dangerous maladies. Even tropical products were sometimes cause for alarm: women in France, for example, feared that if they consumed too much chocolate, they ran the risk of birthing black babies.[15]

Apart from the real and imagined dangers involved, one important reason women did not travel is that they were never hired by trading companies, scientific academies, or governments as voyaging naturalists. Like Alexander von Humboldt and a few others, Merian financed her own voyage. Unlike Humboldt, however, Merian had no inheritance to rely upon and instead raised the money for her trip by selling a large collection of her paintings and specimens in addition to subscriptions to her prospective Surinam book. Although Merian did not have official sponsorship, her travels to this strange and sometimes unwelcoming land were facilitated, to some extent, by her long-standing connections with the Labadists, a religious group with whom Merian had close contact and which had members living also in Surinam.[16]

Merian sailed from her home in Amsterdam, although she had been born in the free imperial city of Frankfurt am Main. The German princely states and free imperial cities did not hold colonies like many of their European neighbors, despite the strong shipping traditions of Hansa cities such as Lubeck and Hamburg. Nonetheless naturalists from German-speaking areas were well represented among scientific voyagers. Paul Hermann, who finished his life as professor of botany at Leiden, was perhaps the most prominent, but many, like Merian, found their ways to port cities and joined the ecumenical investigation of the world's flora and fauna. Early on, Spanish monarchs had requested German mathematicians aboard their ships to chart the movement of the stars, the position of the oceans, and the flow of the tides because of their precision as astronomers and astrologers.[17]

Merian did not fit the profile of the usual male naturalist—young and unmarried—in this period. Although unmarried, having divorced her artist husband, Johann Andreas Graff, at a relatively young age in order to follow her own interests, at age 52 when she set sail, Merian was considerably older than most voyagers. She was also not trained in medicine as were many voyaging botanists in this period. Maria Sibylla numbered among the few artists who voyaged with scientific expeditions. These included Aimé Bonpland, who traveled with Humboldt, drying herbarium specimens in small smoky huts closed off against the humidity of the tropics; Pierre Jossigny, who voyaged with Commerson; and Alexander Buchan and Sydney Parkinson, who voyaged with Joseph Banks and died probably of malaria or dysentery on the voyage.

Merian had been born into a prominent family of painters, engravers, and publishers. From an early age she served an informal apprenticeship with her stepfather, guild painter Jacob Marrel, after her own father, the famous Matthäus Merian the Elder, died. A contemporary, Joachim von Sandrart, confirmed that "in her home, Merian received good training in sketching and in painting (both oil and water-color) all manner of flowers, fruit, and birds, and in particular . . . worms, flies, mosquitoes, spiders." She also learned to engrave copper plates and mix paints. It was not unusual for women to serve as botanical illustrators (despite restrictions such as the Nürnberg Painters' Code of 1596, which prohibited women from using oils). Barbara Dietzch and Maria Moninckx worked in Merian's style in Amsterdam. Moninckx painted with her father, Jan, for Agnes Block, who owned extensive private gardens, and for Caspar Commelin, director of the Amsterdam Botanical Garden. In France, Madeleine Basseporte recorded on vellum the rare plants at the Jardin du Roi from 1735 until her death in 1780.[18]

Art, not medicine, gave Merian entrée to natural history exploration. She did not travel in the service of a botanist, as did many other artists and illustrators, but instead set her own scientific agenda. Having studied insects since the age of thirteen in Frankfurt and Nürnberg, she moved in 1691 to Amsterdam, the hub of Dutch global commerce, to study the extensive collections of rarities from the East and West Indies, especially the natural history collections of the mayor, also director of the East India Company, and his nephew. Merian was disappointed, however, that these collections contained only dead specimens; what interested her was the metamorphosis and life cycles of caterpillars. She thus set out to do her own research: "This all resolved me to undertake a great and expensive trip to Surinam (a hot and humid land) where these gentlemen had obtained these insects, so that I could continue my observations."[19]

Although Merian's scientific biography differs from that of many naturalists, it also resembles theirs in a number of revealing ways. Throughout her life, Merian, like Sloane, combined science and art with commerce. In Nürnberg, Frankfurt, and Amsterdam she established thriving businesses—selling fine silks, satins, and linens painted with flowers of her own design from which she earned a comfortable living. Her colors (for which she had developed new techniques to enhance their durability) became so renowned that she was commissioned by an army general to paint his field tent with all manner of birds and flowers—even on the battlefield, the general wanted to feel that he was enjoying the quiet of a garden house.[20]

Like many male naturalists, Merian also joined commercial interests to her scientific voyage. In the same way that Sloane sought a substitute for the valuable Peruvian bark in Jamaica, Merian sought other varieties of caterpillars in Surinam that, like silkworms, might produce fine thread. Silk was, in this period, big business. In 1700, the Academy of Sciences in Berlin tried (unsuccessfully) to fund their scientific endeavors through a silk monopoly; Merian's own stepuncle was in the silk trade in Frankfurt. Silk indeed became important in colonial manufactory: in the late eighteenth century, the "Lady Governess" of the English East India Company in India, for instance, directed a plantation of mulberry trees at the female orphanage in Madras where at least one hundred girls were profitably engaged producing silk.[21]

In Surinam Merian found one potential silkworm (today identified as the *Rothschildia aurota*) that feeds on what she called a "China tree." The caterpillar produced an "ochre colored thread" that she believed could "produce good silk and bring a good profit." She sent samples of the caterpillars—so "fat that they roll"—back to Holland, though to my knowledge she never brought this anticipated moneymaker into production. Her commercial interests in Surinam focused on supplying her specimen trade back in Amsterdam. Before leaving Surinam, she arranged with a local man to continue to supply her with all manner of butterflies, insects, fireflies, iguanas, snakes, and turtles, many preserved in brandy, for sale in Amsterdam, but by 1705 he was dead.[22]

Like Sloane and other male naturalists, Merian relied on Amerindians and African slaves—whom she referred to as "my slaves" *(myne Slaven)*—for aid in finding choice specimens and for safety in travel. Her slaves hacked openings for her in the dense rain forest, dug up roots, helped her tend her botanical garden, paddled her and her assistants upriver, and supplied choice maggots, fireflies, and shells. For two years she and her daughter collected, studied, and drew insects and plants of the region, gathering specimens early in the cool of the day and preparing them in the evening. When writing commentaries to her plates featuring her finds, she added—as was common in this period—"information from the Indians." These included uses of plants in medicine (cotton and senna leaves cool and cure wounds), foods (a recipe for cassava bread), buildings, clothing, and jewelry. According to the historian Natalie Zemon Davis, Merian brought her "Indian woman" with her to Amsterdam. Nothing more is known about this woman.[23]

Overcome with malaria, Merian was forced to leave the Torrid Zone in 1701, sooner than she had intended. Upon her return to Amsterdam, she, like other heroic voyagers, began compiling her travel account, in this case her *Metamorphosis insectorum Surinamensium,* portraying the reproduction and development of numerous insects and the many plants they feed upon "never before described or drawn." Invoking familiar tropes of voyaging literature, she advertised her book as "the first and strangest work painted in America": "This work is rare and will remain rare . . . since the trip is costly and the heat makes living [in Surinam] extremely difficult."[24]

Biopirates

In contrast to English and Dutch voyaging naturalists, who followed trade routes catching rides and taking whatever positions they could to pursue botany, French voyaging botanists were sent out by the government. Jean-Baptiste Colbert, finance minister to Louis XIV, attempted to follow the successful Dutch model by founding the Compagnie des Indes Occidentales in 1664 (following upon the failure of the Compagnie des Isles de l'Amérique in 1648). Unlike Dutch companies, however, the French company was a commercial enterprise created and supported by the government; as Stewart Mims has shown, 3 of the 5 million livres financing the company were furnished by the king.[25]

Within a decade this company, too, had failed. Colbert nevertheless remained steadfast in his resolve to advance the French state through commerce and to harness the expertise of naturalists to this effort. In step with the centralization of French state power in the seventeenth century, Colbert established the Académie Royale des Sciences and expanded the Jardin Royal des Plantes Médicinales in Paris, both of which sent out naturalists. The King's Garden, directed in turn by Tournefort, Antoine de Jussieu, and Georges-Louis Leclerc, comte de Buffon, was the primary training ground for botanists and the headquarters for rationalizing worldwide French botanical and agricultural production. Colbert's goal was a familiar one: to decrease France's reliance on imports and to increase its production of luxury goods for export. James McClellan and François Regourd have labeled this highly bureaucratized system the "scientifico-colonial machine"—a system that worked to centralize the material and intellectual resources of France's colonies by drawing natural resources from the peripheries to the center. Richard Drayton has argued that similar alliances between Crown patronage and natural historical bioprospecting did not form in Britain until late in the eighteenth century.[26]

The French Royal Garden of Medicinal Plants was founded in 1626 by Guy de La Brosse, physician to Louis XIII. The purpose of the garden was to conserve all types of useful medical herbs for the instruction of medical students as well as to cultivate ornamental exotics for the luxury trade. Taking the simpler name Jardin du Roi in 1718 (and later known as the Jardin Royal des Plantes), this royal establishment hosted courses on botany, chemistry, eventually anatomy, zoology, forestry, agronomy, metallurgy, and all branches of natural history. Almost every French botanical voyager—to the man—passed through the portals of this institution. Some studied also at the rival garden in Montpellier, created in 1593, where one section was reserved for medicinal herbs and the other for acclimatization experiments. The Parisian garden also housed the Cabinet du Roi, an entrepôt for specimen exchange throughout Europe. In France, the king's men collected for the king.[27]

One early voyager sponsored by the King's Garden was Tournefort, professor of botany there, sent in 1699 by Louis XIV to the Levant to discover plants (France did not have direct commercial access to the medicinal drugs of the Middle East), metals, and new minerals, to verify the coastal maps of the area for the French navy, and to bring back information for French diplomats. The Greek and Arabic drug trade was still important in the seventeenth century, and Tournefort made a pilgrimage to Mount Ararat in eastern Turkey, where Noah's Ark had allegedly come to rest after the Flood and a place widely believed to be the birthplace of terrestrial flora and fauna. Deeply disappointed, Tournefort wrote, "everything published by travelers about the mountain is false. There are neither monks nor hermits, neither solitaries nor recluses. It is a horrible mountain . . . half of it is hidden by snow all year round; the rest is bare and sandy."[28] Nonetheless, Tournefort brought back 1,356 plants from the region.

Even the priest voyagers, always in search of useful plants in addition to souls, were sent by the king. Charles Plumier, of the order of Minim, traveled to the Antilles in 1689 with the title *Botaniste du Roi* (the king's botanist). He died during his fourth voyage to the Americas, searching for *Cinchona* trees in Peru.[29]

Second to the Jardin du Roi in plant reconnaissance was the Parisian Royal Academy of Sciences, created by Colbert in 1666 to enhance the

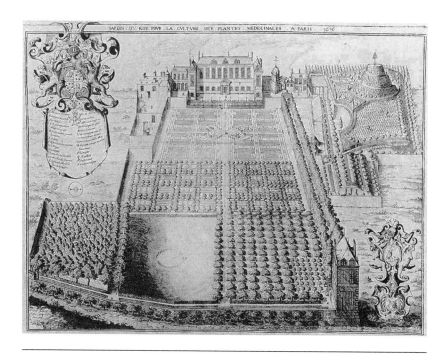

Figure 1.1. The King's Garden, founded in Paris in 1626. Louis XIII created this garden for the cultivation of medicinal plants at the urging of his apothecary, Guy de La Brosse. It later became known as the Jardin du Roi. Under Louis XIV it blossomed as the leading European center for botanical and natural historical research.

power and privilege of the state. In 1735, this academy commissioned the celebrated expedition led by Pierre Bouguer, Charles-Marie de La Condamine, and Louis Godin to the equatorial regions of South America to measure the length of a degree of the meridian near the equator in order to determine Earth's size and shape (by comparing it to similar measurements taken at the North Pole). The Frenchmen were among the first foreign scientists to penetrate into the interior of Spanish Peru: for centuries Spain had closely guarded the secrets of its American natural resources. Although accompanied by emissaries of the Spanish governor, La Condamine found opportunities to spirit away seedlings of the precious Peruvian *Cinchona* and rubber trees *(caoutchouc)*, to test Amazonian curare (on chickens in Cayenne—see Chapter 4), and to settle the question of the real existence of the Amazons (the warlike women said to inhabit the wilds of the river named after them). La Condamine took Joseph de Jussieu to serve as his "botanical eyes." La Condamine's hope was a familiar one: to procure (legally or not) *Cinchona* and rubber trees for production in some part of the French empire. La Condamine wrote of his prospecting for the valuable Peruvian bark:

> On the 3rd June I spent the whole day on one of these mountains [near Loja in present-day Ecuador]. With the aid of two Indians of the region whom I took with me to serve as guides, I was able to collect no more than eight or nine young plants of Quinquina [*Cinchona*] in a proper state for transportation. These I had planted in earth taken from the spot in a case of suitable size and had them carried with care on the shoulders of a man whom I kept constantly in my sight, and then by canoe. I hoped to leave some of the plants at Cayenne for cultivation and to transport the others to the King's Garden in France.[30]

Despite his care, the plants did not prosper (La Condamine did not know that *Cinchona* grows only at high altitudes). Nonetheless, while gathering geographical information, he also collected seeds of a number of potentially valuable plants at "every step" on his journey—he mentioned ipecacuanha, simaroba, sarsaparilla, guaiacum, cacao, and vanilla—and kept his eyes open for others yet unknown to Europeans.

European powers—the Spanish, Portuguese, Dutch, English, and French—jealously guarded their natural resources, their "green gold." Such monopolies and secrets bred a counterforce of bioespionage and piracy. Spain's approach was to close the borders of its vast South American holdings and scrutinize interlopers. If La Condamine had been found carrying *Cinchona* from Spanish dominions into French territories, he would have been imprisoned, mutilated, or executed, even though the French expedition had entered Spanish domains in South America legally (if not honorably) with passports from Philip V, King of Spain.

Others, however, did not. Nicolas-Joseph Thiery de Menonville, a royal French botanist traveling some forty years after La Condamine, entered New Spain (Mexico) to steal the prized cochineal, a small beetle that produces a rich scarlet dye. I call him a biopirate rather than a bioprospector, because Thiery de Menonville—whose clandestine voyage was sponsored by the French ministry of the navy—had no passport or papers for legitimate passage into Mexico and he intended to extract the cochineal by any means necessary. Had he been caught, France would have done nothing to save him. The French government promised him 6,000 livres for his services (an extraordinarily large sum), but could not officially recognize his illegal activities. Thiery's mission was to naturalize the cochineal insect to Saint Domingue, thus breaking for the first time the lucrative 250-year Spanish monopoly on cochineal production. In the eighteenth century, the Spanish produced 1.5 million pounds of the insect per year and by 1784 were receiving £500,000 annually for the dye in Europe. The French, in particular, paid dearly for the cochineal used at their famous Royal Gobelins Tapestry works. Should Thiery have succeeded in naturalizing cochineal to French soil in Saint Domingue, this new colonial crop would have showered France with riches equal to those produced by sugar, coffee, indigo, and cacao.[31] This colony was, after all, the most prosperous in the New World, providing by 1789 two-thirds of France's foreign revenues. Trade in cochineal would only enhance those revenues.

Cochineal had been cultivated from precolonial times by the Mixtec and Zapotec peoples in Oaxaca, Mexico to provide color for their houses and cottons. The valuable dye continued to be produced throughout the colonial period—albeit now for the Spanish—by Oaxacan Indians working on small peasant plots. The highly prized red and orange dyes were made from the dried, pulverized bodies of a beetle (the females only) with the Latin name *Dactylopius coccus* that is indigenous to Mexico and feeds parasitically on two genera of cacti, the nopal *(Nopalea)* and the prickly pear *(Opuntia)*. By the 1870s, cochineal was replaced almost entirely by synthetic dyes, but it again became popular in the twentieth century as a colorant for foods, cosmetics, and beverages after synthetic red dyes (based on petrochemicals) were found to be carcinogenic. Peru now produces 85 percent of the world's supply of cochineal; it takes 70,000 insects to make one pound of dye.³²

Thiery's travel account, *Voyage à Guaxaca* (Oaxaca), bristles with the high intrigue of a "mission impossible." Engaged in the psychological strategies of advanced war games, Thiery played on what he perceived to be the pride and laziness of the Spanish and, when possible, pitted Spaniard against Spaniard, African against Amerindian, officer against officer. After receiving instructions from the French ministry, the "new Argonaut," as he called himself, embarked on his sixty-six-day journey from France to Saint Domingue in 1776. From there he planned to take passage to Havana (Cuba) on a slaving ship, but found instead a French merchant ship headed to that port. Before leaving he procured a passport from the Intendant of Saint Domingue stating that he was both a botanist

Figure 1.2. The manner of gathering and cultivating cochineal in Oaxaca, Mexico. Amerindians—both men and women—gather the cochineal insects from the cacti upon which they breed. Piles of insects are fired just enough to kill them and then laid in the sun to dry.

and physician. Though one of the few botanists in this period not trained as a physician, he envisioned that traveling under this guise would render him "more luxuries" and, more important, "less suspicion." He set off from Saint Domingue in January 1777 with some food, clothes, and a number of vials, flasks, cases, and boxes of all sizes. Driven by his dream, "all future obstacles vanished from sight, and already I felt it in my power to possess the precious treasure which I sought."[33]

Thiery de Menonville herbalized in Havana while waiting for a vessel to carry him into New Spain. Traveling as an upper-class European (often flashing his gold-capped cane, diamond ring, good manners, and fine connections), he gained entrée to the governor's home and circle of friends. The bonds of class brought him a passport into Veracruz, his next port of call. The ship's captain, however, not sharing the confraternity of wellborn men, charged Thiery 100 hard dollars—an exorbitant sum—for passage. Thiery's first priority upon touching Mexican soil was to win over the people. His Cuban passport, however, was ruled invalid, and Thiery desperately required papers if he was to travel to Oaxaca, where he had learned the finest cochineal was to be had. Botanizing while waiting, the Frenchman identified the "true" jalap of Mexico (a valuable drug), which immediately endeared him to the inhabitants of Veracruz. Thus he honed his core deception: outwardly he played the herbalizing physician in search of drugs, while inwardly he mocked the "idleness and ignorance" of people who paid a high price to import jalap when it grew "under their very noses."[34]

Thiery's herbalizing provided the pretext for his repeated forays into the countryside, attended only by his "Negro," who carried his portfolio, hatchet, mattock, and breakfast. As time went on, he won the admiration of persons of every rank and became known as the "French doctor." This charade, he wrote, "allowed me to conceal my ultimate project."[35]

Despite his elaborate ruse, the viceroy of New Spain refused his French rival papers and ordered him to leave the country, stating that he did not wish "to open to strangers the secrets of the country" (though Thiery's exact plans were probably unknown). Undeterred, the French nobleman resolved to travel, despite the possibly fatal consequences, without papers and on foot into the heart of foreign territory. He elaborately planned his journey: he would leave Veracruz in such a way as to appear to be taking a walk rather than a journey, he would travel on back roads, and lodge in the poorest Indian huts along the way. If discovered, he would pretend to have lost his way. Knowing that his French dress, ignorance of the road, and bad Spanish would arouse suspicion, he decided to declare himself a Catalan from the frontiers of France, which would explain why he spoke French well but Spanish poorly. To further his success, he vowed to go "always neatly dressed, bearing some trinkets and small gifts, affecting good humor, and, importantly, to pay liberally for all I take." Thus resolved, he left at nine o'clock in the evening, scaled the town ramparts, and bade the city adieu.[36]

Lodging with Indians, paying grandly for the few eggs and tortillas they could spare, and asking aid of African and Indian guides, Thiery reached his "golden fleece" *(toison d'or)*—the cochineal fields of Oaxaca—with his heart pounding in expectation and fear. Offering that he was a physician in need of the cochineal to make an ointment for the gout, he purchased the live insects and some of the cacti they feed on from various African and Indian proprietors. These he stored in chests of appropriate size with partitions and locks that he had commissioned for the transport of his treasure. He found also ripe pods of the prized vanilla, which he marked and then hid by mixing them and some green shoots from the root into a jumble of ordinary plants in his box. Always ready with a story and a strategy, Thiery returned to Veracruz congratulating himself on having, in the course of his journey, eluded two viceroys, six governors, thirty *alcaldes* (local officials), and 1,200 customs guards. In only twenty days, he had journeyed "240 leagues [720 miles] . . . over roads so bad as often to be almost impassable, under a burning sun, in a wretched country, without resources, and among a people whose language I did not speak; in a country, in short, where I was destitute of a protector or any connections, and where every public officer was a threat to me."[37]

At Veracruz, he packed his many treasures for the relatively short sea voyage east across the Gulf of Mexico to Saint Domingue. Despite his precautions, several sailors discovered him tending his cacti and recognized the contraband insects living on them. Thiery again claimed that the cochineal, along with the cacti, vanilla, and jalap, were ingredients for his ointment for the gout. When queried further, he had to admit that these were not the "whole secret" of his potent mixture, but that he compounded them with incense, gold dust, silver leaf, and—he added in a half whisper—"some blessed lint that had touched the relics of Santo Torribio." To secure the crew's trust he threw some additional "Latin into the mix." His deception spared him once again the severe consequences otherwise meted out to smugglers.[38]

Upon his return to Saint Domingue, Thiery was celebrated for the service he had rendered the state and rewarded with the title *Botaniste du Roi* and a permanent pension of 6,000 livres per annum—a sum double that granted members of the Académie Royale des Sciences in Paris. Thiery established a nopal cactus plantation in the botanical garden at Port-au-Prince and also sent cacti and insects to the garden of the Cercle des Philadelphes, the scientific academy located in Cap Français, the "Paris of the Antilles." Thiery planned to distribute "free, without distinction of rich or poor," to each inhabitant of the island the treasures of his voyage: the nopal and cochineal, vanilla, jalap, indigo, and cotton. While Indians and former slaves cultivated cochineal in New Spain, Thiery imagined that in the French colonies, this work would best suit a particular class, the *gens de couleur* (people of color), whom Thiery remarked were growing in numbers everyday. The *gens de couleur* were, in Thiery's view, "almost

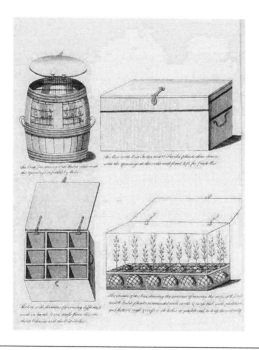

Figure 1.3. Eighteenth-century European boxes fashioned for transporting plants. Nicolas-Joseph Thiery de Menonville's cases were similar to the one shown in the upper right corner, designed by John Ellis. The box is closed to protect the plants from sea salt and opened at both ends and the front to allow for the circulation of fresh air.

Frenchmen," meaning that they had the industry and savvy to succeed in producing this potentially rich merchandise on their small holdings that could not support cane, tobacco, indigo, roucou, or other cash crops.[39]

The dye produced from Thiery's insects was tested by chemists at the Jardin du Roi in Paris and proved of very high quality. The Frenchman, however, died of "malignant fever" in 1780, just two years after his return to Saint Domingue, at the age of forty-one. His successors at the Port-au-Prince garden, M. Joubert de la Motte, a royal physician from France, and his assistant, the creole M. Chotard, were unable to coax the cochineal into production, bringing Thiery's plan to naught. The *grana fina* cochineal for which Thiery de Menonville had risked his life died out in Saint Domingue. His work, however, led to the discovery of the *grana silvestre,* a wild variety of the insect native to the French Caribbean island that was soon put into production. By 1787, the Academy of Sciences in Paris had confirmed that the dye produced in their American colony was nearly of the same quality as the Mexican cochineal.

Thiery's biopiracy opened a new branch of French commerce in Saint Domingue for a few years. He succeeded where Linnaeus in Sweden, the English, and the North Americans had failed. In 1759 the English Society of Arts had offered a prize of £100 to anyone who could produce a quantity of not less than twenty-five pounds of Jamaican-grown cochineal, but no one even came close. The British later attempted to naturalize cochineal in "Hindostan" (India), but again without success. The North Americans also failed in their attempt to produce the insect in South Carolina. The Dutch succeeded in establishing cochineal production in Java, but New Spain (Mexico) long remained the primary producer. Production in Saint Domingue halted with the upheavals and flames of the Haitian revolution.[40]

Who Owns Nature?

In a remarkable passage, Thiery de Menonville discussed his act of piracy, his perilous theft *(un larcin périlleux),* as some called it. "To have stolen the cochineal," he wrote, "would, in my opinion, have been an act of social injustice, as far as regards the cultivator whose garden I might have despoiled." This injustice he sought to avoid by purchasing the cacti and insects—at whatever price the owner might ask—thus, in his view, royally compensating the poor Indian or former slave for their goods. His chief concern was not to injure individual cultivators. As for Spain—from whom he pirated the plants and insects—he felt no obligation whatsoever. He reasoned that he, "as a prototype of a different nation," shared equal rights with the Spanish to "nature's favors." If the Indians had collectively refused his request to purchase the cochineal, "I should then have considered myself, *as in the case of war*, absolved from the restrictions of social laws" and would have carried away by stratagem what was denied by entreaty.[41]

Who owned nature in the eighteenth century? This was a question Europeans were concerned with (and in the West Indies we have access only to their point of view). Let's skip to the present for some comparative temporal insights.

The Rio International Convention on Biological Diversity of 1992 placed biological and genetic resources in the custody of nation-states and subject to their property-right laws (though this sometimes conflicts with community rights of people whose territories span modern national boundaries). John Merson has stated that prior to the 1990s, plants and other biological resources were considered part of the "global commons."[42] Nature, in other words, was there for the taking. In the eighteenth century, unspoken notions concerning a global commons applied only to nature and its resources outside Europe: European trading companies and states claimed exclusive rights to the natural resources of the territories they could hold militarily. As we will see, secrecy was often the only weapon Amerindians and Africans in the West Indies had to wield against European bioprospectors (Chapter 2).

Thiery felt a responsibility to individual cultivators—whether Amerindian or African-Mexican—and sought to compensate them for their accumulated knowledge and resources (or so he says), even though his piracy could potentially have undermined their livelihood. With Spain, however, he considered himself in a state of war, and, if he was clever enough to steal the prize, he felt justified in doing so.

Thiery de Menonville was not the first (nor would he be the last) of the French biopirates. The sixteenth-century druggist, Pierre Belon, scoured the Levant for the secrets of Arabic drugs, ointments, and salves. Medical men in Constantinople, whose narrow streets were peopled with spies and counterspies, closely guarded their secrets and often eliminated indiscreet inquirers. Belon eventually managed to "tame" or acclimatize the poppy and other medicinal plants (the cork oak and strawberry tree) to France. But perhaps he knew too much; he was murdered in 1564 in the Bois de Boulogne, where he had been installed by the Valois king, Henri II. His death was attributed to a passing prowler, but Belon's friends were convinced he was assassinated.[43]

Thiery may have been more directly inspired by Pierre Poivre, a French botanist who successfully conducted covert raids in the 1750s and again in the 1760s and 1770s into the Dutch East Indies to steal pepper, nutmeg, cinnamon, and cloves for acclimatization and cultivation in his gardens on the Isle de France (the island of Mauritius in the Indian Ocean). Historian Patrice Bret has perceptively remarked that botanists were "foot soldiers" in the "botanical wars" for tropical resources; finding and authenticating valuable plants was naturalists' work. Following Poivre's successes acclimatizing Dutch colonial plants in the Indian Ocean, the French government commissioned incursions into the Spanish empire in order to enrich their West Indian holdings in Cayenne and Saint Domingue. Jean-Baptiste Leblond, another royal naturalist, was paid a pension of 6,000 livres per year from 1786–1802 by the general controller of finance to find *Cinchona* in Spanish territories in Guiana and thus break the monopoly on that indispensable anti-malarial. While underway, Leblond also prepared reports on cotton, indigo, pepper, cinnamon, cloves, and breadfruit, and sent many plants and animals to the King's Garden in Paris.[44]

Voyaging Botanical Assistants

Heroic voyaging botanists, largely upper-class, were supported by numerous lower-class assistants whose work secured their success. John Woodward in his instructions to voyagers specified in detail how to collect, label, and record the uses of plants as described by the people of a particular country. This, he noted, "may be done by the hands of servants; and that too at their spare and leisure times." Assistants thus carried and fetched; they netted fine specimens, drew, recorded, and generally contributed to the massive task of cataloguing nature's bounty. Even displaced Africans served Europeans in this capacity: one Thomas Richmond, a field assistant to Joseph Banks on his voyage with James Cook in 1768, was well on his way to becoming the first African to circumnavigate the globe, but he died in the snows of Tierra del Fuego off the southern tip of South America.[45]

I have selected Jeanne Baret as the type specimen for voyaging botanical assistants, even though she was extraordinary in becoming the first woman to circumnavigate the globe. Not the independent researcher Merian was, Baret—or as she signed her name, Barret—traveled in the capacity of a valet (for which purpose she disguised herself as a man) to Philibert Commerson, physician and royal naturalist on Bougainville's voyage (1766–1769). According to the account Baret gave after she was discovered, she had "tricked her master [Commerson] by presenting herself dressed as a man . . . at the dock [in Rochefort] at the moment the vessel was being boarded." Baret's tale, as it is often told, recounts how the 116 young French sailors grew suspicious over the long months of the voyage, remarking that the young "valet" neither grew a beard nor was seen to relieve himself. Despite their suspicions Baret was not found out until the French ships were halfway around the world. It was only when the *Cythéréens* (Tahitians) circled her, chanting *"ayenene, ayenene"* (girl, girl), that the Europeans discovered that she was, indeed, a woman.[46]

As fanciful as this story may be, Commerson certainly knew her and knew she was a woman. In 1764, two years before the voyage, he had engaged the then twenty-four-year-old Baret, who according to Bougainville was "neither plain, nor pretty" as his housekeeper (Commerson was thirty-seven years old at the time). As the king's naturalist, Commerson was allotted 2,000 livres to engage a personal servant and an illustrator for the duration of his voyage round the world. It seems unlikely that, as was later claimed, he would have chosen just anyone standing at the pier on the day of departure for this technically demanding and intimate work. He nonetheless proclaimed his innocence, writing sometime later that this "courageous young woman who, taking the clothing and temperament of a man, had the curiosity and audacity to circumnavigate the globe, accompanied us without us knowing it." He could not have done otherwise; a French royal naval ordinance of 1689, reconfirmed in 1765, forbade officers and sailors to invite women to pass the night on board ship or to come on board for anything more than a short visit. Officers who failed to abide by this order were suspended for a month and sailors were sentenced to fifteen days in chains.[47]

It is difficult to evaluate the relationship between Commerson and Baret. Commerson hired Baret as his housekeeper two years after his wife's death in childbirth. Curiously, he hired her when she was some five months pregnant, just as she was becoming less able to do heavy domestic labor. Baret registered her pregnancy August 22, 1764, in Digoin outside Lyons, as required by law, but refused to name the father. Shortly thereafter both Commerson and Baret moved to Paris, where she began her work for him on September 6th, for which she was paid 100 livres per year. The child, Jean-Pierre Baret, was born in January of the next year and put out to a wet nurse (a typical form of infanticide in this period). The child died some months later.⁴⁸ Commerson had left his son from his marriage in the care of his brother-in-law, a priest living in Toulon-sur-Arroux, when he moved to Paris and took up residence with Baret near the Jardin du Roi, where he hoped to advance his career as a naturalist.

Before Commerson set sail with Bougainville in 1767, he, like Merian, prepared his will, a document that again suggests an intimate relationship between the botanist and his assistant. Interestingly, in his will, Commerson made provisions for his body to be given to science, taken after his death to the nearest anatomical theater for dissection, a fate usually reserved in Catholic France for executed criminals. His skeleton and body parts were all to be used for study, save for his heart, which was to be placed in a marble sepulcher next to the body of his deceased wife. He bequeathed his manuscripts and property to his young son, and he left his household linens and furnishings along with 600 livres to Jeanne Baret.⁴⁹ He also left instructions that she was to be allowed to remain in his house for a year after his death in order to put his natural history collections in order before sending them to the Cabinet des Estampes du Roi. Despite the differences in their ages and class backgrounds, it is clear that Commerson and Baret knew each other well. On board the *Étoile*, Commerson and his valet both suffered (or feigned to suffer) terribly from seasickness, so that it seemed reasonable for them to lodge together at night in his cabin.

The story of Baret's voyage and discovery was recounted at some length in journals kept by various crew members. Curiously, Commerson's own account in the ship log read only "18 July 1768 the physician M. Commerson's domestic was discovered to be a girl who until now passed as a boy." Bougainville's journal entry was short and matter-of-fact, perhaps because he sailed with the *Boudeuse*, the other vessel in the two-ship caravan. According to the combined accounts, no one suspected a thing until after about a month when the ship company's "gentle repose" was interrupted by murmuring that a girl was on board. The sailors eyed Commerson's valet with suspicion: he had a small and heavy build, large buttocks, elevated chest, small round head, freckled face, tender and clear voice, and a dexterity and delicacy of hand that threw his sex into question. Charles-Félix-Pierre Fesche, a volunteer with Bougainville aboard the *Boudeuse*, noted after this episode that Baret took the precaution of binding her breasts "to hide her sex."⁵⁰

The captains of the two vessels ignored the situation, but day by day the rumors grew louder. Commerson was told that his valet could no longer pass the night in his cabin but must take an ordinary hammock in the stern with the other five servants on board. There the sailors, "pushed and pulled by curiosity," pressed themselves on the valet, but he was "cruelly insensible to their offers and entreaties." At the same time, the "false man" continued to assure the crew that he was not at all of the "feminine sex" but was in fact a eunuch, or, as Baret reportedly put it, "made by accident like those [men] whom the great sultans make the guardians of their seraglios." She claimed, in other words, that the ambiguities of her sex had resulted from castration.[51]

After this incident, "our man," the reports continued, worked as hard as possible "to appear that which he claimed to be." He worked "like a Negro." The voyagers witnessed him [Baret] accompanying his master on all his expeditions amid the snows and icy hills of the Straits of Magellan, carrying provisions, weapons, and portfolios of plants. Baret's arduous labor had the desired effect, and the crew's suspicions subsided "due to lack of evidence." However, when the ships arrived at Nouvelle Cythère, a "savage" named Boutavèry immediately singled out the valet from the crowd of sailors as somehow different. His suspicions were confirmed when she went ashore the next day to herbalize with her master; the "savages" apparently immediately saw what the Europeans could not—that she was clearly female. When the Tahitians began carrying her off to "present to her the honors customary in the island," Baret had to be rescued by the ship's officer of the guard.[52]

Back on board, Bougainville reported that "I was then obliged, according to the ordinances of the king, to assure myself that the suspicions were founded." After sailing undiscovered for a year and a half, Baret finally confessed that she was a woman; she said that she had deceived her master at Rochefort and that she had also some years earlier served a Genevan gentleman also disguised as a valet. She continued that she was born in Burgundy and soon orphaned. Finding herself in "a distressed situation," she resolved to disguise her sex. She confessed that she knew when she boarded that the ship was voyaging around the world, and that such a voyage had raised her curiosity. Bougainville judged her to be the first woman to circumnavigate the globe and added, "I must do her justice to affirm that she has always behaved on board with the most scrupulous modesty."[53]

After her discovery, Baret abandoned her disguise. Bougainville noted that he "took measures so that nothing disagreeable would happen to her." It is unclear if he meant during the remainder of the voyage or in court upon her return to France. She herself left nothing to chance and carried two loaded pistols with her at all times on board ship.[54]

Commerson was not disciplined. One observer of those days wrote that he could have said much about Commerson's "embarrassment," but he did not. Another observer wrote that "because of his [Commerson's] age and knowing of the scandal such an incident would cause on a long voyage and because his actions contravened a royal ordnance," Commerson probably did not know Baret's sex when she boarded the ship. But, this observer continued, Commerson knew of it as early as Montevideo (in present-day Uruguay) because he did not let her mix with the inhabitants there.[55]

Nothing is reported of Commerson and Baret's relationship after her discovery except that she continued to be dedicated to him and accompanied him herbalizing. Commerson died in 1773 on the return voyage on the Isle de France, and in 1774 Baret married Jean Dubernat, a former petty officer of the royal trading post on the island. She later returned to France, where in consequence of Bougainville's intervention, the courts pardoned her. Bougainville judged the case innocuous; "her example," he wrote, "is not likely to be contagious." Rather than disciplining her, the royal navy granted her an annual pension of 200 livres in 1785, referring to her as a *"femme extraordinaire."*[56]

Why the male disguise? Women often donned masculine garb when they entered domains where women dared not tread. Dressed as men, Anne Bonny and Mary Read, for example, lived as pirates in the West Indies in the 1710s. Other women disguised themselves as males in order to join the military, to marry women of their choice, or to enter various areas of science and medicine. At the close of the eighteenth century the future prize-winning mathematician Sophie Germain took courses at the newly opened Ecolé Polytechnique in Paris (which, like most European universities, was closed to women at the time) under the pseudonym Antoine-August LeBlanc. At the turn of the nineteenth century, a young woman's guardians sent her to the University of Edinburgh dressed as a boy. After being awarded a medical degree in 1812, "James" Barry joined the British Army, becoming the second-highest ranking medical officer in the colonial military establishment. Her true sex was not discovered until after her death. Similarly, the first woman to attend medical school in the United States in the 1850s, Elizabeth Blackwell, was counseled by a sympathetic professor to attend classes disguised as a man. A second French woman sailed around the world in the early nineteenth century accompanying her husband, a naval officer. Rose Marie Pinon de Freycinet embarked on this long voyage dressed as a sailor. By the time she was discovered, it was too late for the ship to turn back.[57]

Why did Commerson take Baret around the world? How did she contribute to colonial botanizing? Commerson chose to contravene French law and risked taking her as his valet because she was a useful and faithful servant. Captain Bougainville called her "an expert botanist." Commerson wrote that she traversed with agility the highest mountains and the deepest forests without complaint. "Armed with a bow like Diana," he continued, "armed with intelligence and seriousness like Minerva . . . she eluded the snare of animals and men, not without many times risking her life and her honor." He praised her for the many plants she collected, the herbaria she constructed, the many collections of insects and shells she curated. As a tribute to her, Commerson lent her name to one their finds: the *Baretia bonnafidia,* a plant known for its "doubtful sexual characteristics." Her name, however, fell into obscurity when botanists renamed this plant the *Turraea (Meliaceae).*[58]

Baret's story shows that in the eighteenth century women could be engaged in science, even in the close quarters of a scientific voyage. Like Edward and Sarah Bowdich, Baret and Commerson might be seen as a creative couple, a legacy of the guild system where husband and wife worked side by side, often over a lifetime, in pursuit of a common goal. She also represents the many invisible assistants serving men who held official posts—either as royal pensioners or professors—in science. Her dedication to Commerson was such that he affectionately called her his "beast of burden."[59] Nineteenth-century prudery suppressed the story of Commerson's *valet fille en homme* and it is just beginning to be retold today.

Baret was not the only woman to serve as a voyaging botanical assistant in this era. Charlotte Dugée, a free "mulatress" and royal illustrator born in Saint Domingue, accompanied Jean-Baptiste Patris, a royal physician in Cayenne, as an artist/illustrator as he explored the source of the Maroni in Guiana in 1766. After months of exploration, she went mad and ventured alone into the forest, where she disappeared without a trace.[60]

Creole Naturalists and Long-Term Residents

The historian of science Antonio Lafuente has discussed the rise of a distinctive "creole science" in the Spanish domains of New Spain (Mexico) and New Granada (Colombia, Ecuador, Panama, and Venezuela) after

1780, by which he means science done by Europeans born and educated in the New World who often fused local American with metropolitan European knowledges. The Spanish, unlike the Dutch, French, or English, conceived of their American holdings not as colonies but as integral parts of an extended monarchy. Accordingly, Spain laid the foundations for self-sufficient Spanish-speaking scientific communities in its American territories; this included founding universities, botanical gardens, printing establishments, and hospitals where the medical effectiveness of certain plants was tested. The first university in the Western hemisphere, today the Autonomous University of Santo Domingo, was founded in 1538, others were founded in New Spain (the Royal Pontifical University, today the Universidad Nacional Autónoma de México) in 1551 and in the Viceroyalty of Peru (the Universidad Nacional Mayor de San Marcos de Lima) also in 1551.[61]

One does not find organized "creole science" in the French and British Caribbean islands. The top echelons of the scientific communities in these islands were filled with men born and educated in Europe. Several Caribbean-born naturalists did emerge late in the eighteenth century, but they were educated in Europe (for example, James Thomson, born in Jamaica and educated in Edinburgh, and Jean-Baptiste Mathieu Thibault de Chanvalon, born in Martinique and educated in Paris). In the French holdings Médéric-Louis-Elie Moreau de Saint-Méry, a well-known creole who codified French jurisprudence in the Antilles, was also educated in Europe (in Paris). Some colleges were founded in New England and Virginia (Harvard in 1638, William and Mary in 1693, and Yale in 1701), but not in French, English, or Dutch holdings in the Caribbean, apart from Codrington College in Barbados, established in 1710 by the English to teach "divinity, physick, and chirurgery."

Although one cannot speak of creole science in the Dutch, English, or French West Indies, the character of botanists traveling to the Caribbean changed in the second half of the eighteenth century as settlements became more secure. In contrast to Sloane, Merian, and even Humboldt, who sojourned in these strange and marvelous lands for a year or two, many naturalists at the end of the century settled into life in the Torrid Zone for ten, twenty, or more years. Though these long-term residents often still married in Europe and returned there to die, many spent the best part of their scientific years in the colonies. Their knowledge of the area was deep; their first allegiances were not always to Paris, London, Amsterdam, or Edinburgh. Some founded local scientific societies, such as the botanical garden at Saint Vincent in 1765 (that doubled as a scientific society with a sixteen-foot telescope, thermometers, barometers, microscopes, and air pump); the Society for the Encouragement of Natural History and Useful Arts in Barbados in 1784; the *Surinaamse Lettervrienden* in 1786; and the Physico-Medical Society in Grenada in 1791. The Cercle des Philadelphes, founded in Saint Domingue in 1785 (a Masonic body of mostly medical men that enjoyed close ties to learned academies both in France and in Philadelphia) supported all aspects of science and began publishing *Memoires* in 1788, but the Cercle did not survive the Haitian revolution and the death of its founder Charles Arthaud (who was presumably killed in Cap Français in 1791).[62]

Saint Domingue and Jamaica hosted several established long-term resident naturalists in the eighteenth century—including Bourgeois and Pouppé-Desportes, and Thomson and John Quier, of whom I will say more in the following chapters. In this section I focus on the lesser-known attempts to chart the flora of Guiana and botanists who might be described as "company men." The French had successfully colonized Martinique, Guadeloupe, and Saint Domingue in the late seventeenth century, but did not establish major settlements in Guiana until after they lost Canada in the Seven Years' War. Christened "El Dorado" for its reputed riches, Spanish and English explorers sought a mythical cache of gold and precious jewels said to be hidden on an island in Lake Parimá. By the eighteenth century, however, explorers in the region turned their attention to the profits to be made from *Cinchona*, sugar, roucou, indigo, cacao, coffee, and cotton. Between 1764 and 1766, 12,000 colonists (mostly German) were sent to this area, three quarters of whom died soon after arrival.[63]

Jean-Baptiste-Christophe Fusée-Aublet, a career *botaniste voyageur* (as he called himself), was a company man who worked first for the Compagnie des Indes on the Isle de France and finally for the king in Guiana. His education was typical of eighteenth-century voyaging botanists— fiercely scientific and peripatetic. From his earliest youth he evinced an active interest in natural history, and especially in botany. In order to pursue natural history, he absconded from his Provence home, joined a squadron headed to Spain, and from there traveled to Grenada, Spain, where he studied pharmacy with Don Antonio Sanchez Lopez. Aublet eventually returned to France (Montpellier) to study medical botany and chemistry before moving on to Lyon, where he came to know Christophe de Jussieu, father of Antoine de Jussieu and his naturalist brothers who domi-

Figure 1.4. Frontispiece to Jean-Baptiste-Christophe Fusée-Aublet's *Histoire des plantes de la Guiane Françoise* (1775), vol. 1. The native guide, portrayed as an effeminate male, holds a ceremonial war club in his right hand, signaling his status as a headman. Interestingly, this Carib or Galibi chief holds a pen in his left hand, perhaps indicating the importance of his role in directing Aublet (crowned with laurels in the cameo portrait) to the "different interesting objects in the culture and commerce of French Guiana." In this image, Aublet highlights America's wealth of cacti and palms. He foregrounds the fruit of the maripa (commonly known as the American oil palm) and other palms.

nated French natural history for two generations. Seeking employment, he joined the staff of a Spanish army hospital and suffered through two military campaigns. Finding the work "extremely tumultuous and little instructive," he left for Paris and was soon employed at the Hôpital de la Charité, where Sloane, Pouppé-Desportes, and others had also trained. He eventually made his way to the Jardin du Roi, where for seven years he studied chemistry, pharmacy, mineralogy, botany, and zoology. He planned then to travel to Prussia to study under the celebrated chemist Johann Heinrich Pott (who contributed to the discovery of manganese), but was approached in 1751 at the age of thirty-one by the minister of the French navy to sign on as "botanist and first apothecary-compositor" with the Compagnie des Indes on the Isle de France.[64]

Trading companies had physicians, surgeons, medical botanists, and pharmacists, like Aublet, stationed along their routes. His job was typical: like Bontius in the Dutch East Indies, he was commissioned to set up a laboratory and garden to produce foods and medicines for his employers. Aublet sought, first and foremost, local replacements on the Isle de France for standard drugs that otherwise had to be shipped from Europe to the local hospital. These plants he also sent to his counterparts in the French territories of Bourbon and Pondicherry. Second, he devised a method of monitoring drugs supplied to company ship captains, who were known to resell surplus medicaments for their own profit.[65] Third, Aublet raised animals (such as cows) and grew plants (watercress, which he pronounced particularly salubrious) for distribution to the inhabitants of the island on condition that they return a part of their yield to the company. The company liked his work and increased his salary; the directors also sent a boatload of onions and seeds of other European staples for Aublet to introduce to the island.

Settling in, Aublet fell in love and had three children with a Sénégalese slave, Armelle, whose freedom he purchased from the company (Chapter 3). He wrote passionately against slavery and made a point of retaining the indigenous names of new plants that he found in the region (Chapter 5).

But his was not an easy life. Aublet incurred the wrath of Pierre Poivre, the influential administrator of the Isle de France, when he shot off memos to company directors denouncing as bogus the nutmeg Poivre had smuggled out of the Banda Islands (in the Dutch East Indies) for the company. According to Aublet, Poivre's nutmeg might resemble the "true nutmeg," but had no commercial value. (Nutmeg was so precious in 1191 that the Romans fumigated the streets with the fragrant spice for the coronation of Emperor Henry VI.) Poivre finally accused Aublet of killing the precious nutmeg seedlings he had hoped to acclimatize to French territories. Aublet also clashed with local apothecaries, who accused him of not supplying them with the medicines they required. He finally left the Isle de France, installing himself and his plants on the neighboring island of Bourbon (today Réunion). Despite the intervention of powerful patrons, Poivre got the upper hand and Aublet returned to France to quiet the controversies surrounding him.[66]

Five months after his return, Aublet was appointed royal apothecary-botanist to prospect for plants in Guiana, on the "wild coast" of South America. His mission was to inventory the plants and animals of the region, looking for anything that could turn a profit for France. He was also to chart rivers and waterways for the military. He sailed aboard the *Patriot* with M. de Behague, who was in charge of a mission to secure cotton for French industry (thus ending imports from India) in addition to cinnamon, nutmeg, and clove. All this was to be kept secret from the Dutch, who still monopolized trade in these spices.[67]

For two years (from 1762 to 1764) Aublet threw himself into his work, criss-crossing every "canton" in Guiana. His superiors praised his initiative and energy but not his character, which they found argumentative and difficult. His first run-in was with the Jesuits, who tried to block his efforts to penetrate to the interior of the country by refusing to supply him with Roucouyenne Indian guides (today known as the Wayana). Aublet, for his part, denounced the Jesuits for enslaving and exploiting the Indians on their missions. Despite these difficulties, within a year Aublet sent off his first scientific cargo to France. His botanical bounty was distributed to the ministers of the navy and finance, and to company officials. A. M. de Bombarde received a handwritten journal along with samples of Guianese flora (probably live specimens); the duc de Choiseul in Versailles and Buffon at the Jardin du Roi in Paris received one case of herbs; boxes of roots, diverse seeds, and fruits; two barrels of plants; and some animals.[68]

Like Merian, Aublet complained that his compatriots in the colonies were in it only for the money. This, he professed, was not what motivated true naturalists. It should be enough for them to have "the satisfaction of being useful, and if successful, a little celebrated." Like Merian also, Aublet left the Torrid Zone after a short time because of his health and what he perceived to be a total absence of the "conveniences of life."[69]

Upon returning to France, Aublet established a perfumery in Paris and manufactured an "essence of rose" much admired by Louis XV. He tried to recoup the specimens he had sent, especially to Bombarde, but was told that after Bombarde's death his extensive natural history cabinet had been dispersed. The historian Jean Chaïa has suggested that Bombarde probably gave Aublet's rich collection to Adanson, Bombarde's protégé. The herbarium that Aublet managed to reassemble by the time of his death was subsequently sold to Joseph Banks.[70]

Armchair Botanists

Many of Europe's leading naturalists—Linnaeus, Antoine and Bernard de Jussieu, and Buffon—never left Europe, but conducted large-scale trade in plants through far-flung colonial networks. Linnaeus, for example, following an extended land-voyage as a young man to Lapland, refused a position as physician to the Dutch East India Company in Surinam in 1737. Nonetheless, his extensive taxonomies depended on the colonial enterprise. By the end of his career, in fact, Linnaeus sat at the center of a vast scientific empire where, in the comfort of his home and gardens in Uppsala, he received specimens and news of new discoveries from some 570 Swedish and foreign correspondents. Jan and Caspar Commelin similarly enjoyed a worldwide network of correspondents through the gardens in Amsterdam, as did Paul Hermann and Herman Boerhaave in Leiden, the Jussieu brothers at the Jardin du Roi in France, Mary Capel Somerset at Beaufort, and Joseph Banks, eventually, at Kew. Bruno Latour has labeled these various European institutions "centres of calculation" where the vast empiry of empire bore fruit as science.[71]

Both Latour and Marie-Nöelle Bourguet have detailed how European naturalists armed travelers with rules and instructions so that ship captains, missionaries, and other amateurs could collect, record, and preserve specimens in a uniform manner. Linnaeus, for example, prepared a generation of students to travel as botanists aboard the ships sent out by Dutch, English, and Swedish East and West India companies and scientific academies. Fanning out like ambassadors around the globe, these students were often sent with specific instructions to supply particular observations, seeds, or plants. Disciplining and codifying practices in the field allowed the naturalists at home in Europe to "see" at a distance.[72]

European botanical gardens, home to the *botanistes de cabinet*, were founded in the sixteenth century primarily as "physic" gardens (gardens of medicinal plants) attached to universities and hospitals to cultivate "simples" for medicines and to teach medical botany to physicians and apothecaries. By the eighteenth century, these gardens were networked with experimental colonial gardens, stretching from Saint Vincent in the West Indies to Malabar, Sri Lanka, the Cape of Good Hope, and on to Java. Plants traveled across these networks in all directions, with many destined for European gardens. Plumier, for example, put into circulation in Europe 900 new species of plants from the West Indies; Joseph de Jussieu collected countless specimens during his thirty-six years in South America; and Humboldt returned from South America with almost 60,000 plants (collected and prepared by Bonpland) belonging to some 6,200 species, thus enriching Europe's botanical treasury by 5 percent.[73]

Naturalists' specimens not only enriched botanical gardens and European science but also served as a medium of exchange among naturalists. In the words of the historian Sverker Sörlin, they became "hard cash" in a new market, the "stock-exchange" of scientific goods.[74] Merian sold her specimens to the public, while French collectors, including La Condamine, Thiery de Menonville, and Leblond, exchanged choice specimens for patronage. Terms of employment at the King's Garden required that scientific cargo reach Paris at regular intervals.

Sloane's and Linnaeus' networks demonstrate well how this economy of specimen-exchange worked. Sloane's short foray into the wilds of Jamaica established him as a "broker" of natural historical knowledge. His cache of 800 plants amassed in the field ("most thereof new") served as his biocapital. Upon returning to London, Sloane settled both the plants and information concerning them into extensive European networks of collectors, gardens, and knowledge. He purported to show his treasures "freely" to all lovers of such curiosities, but his primary trading partners were drawn from the landed and scientific elites of Europe. In England, he traded with William Courten (also known as Charlton), with Sir Arthur Rawdon, who conveniently had collectors in Virginia (from where his gardener James Harlow had brought ships laden with cases of trees and herbs), with William Sherard, gardener to the Duchess of Beaufort, and many others. In France, he entered into a lifelong exchange with Tournefort, which opened to him the riches of the French botanical garden. In Danzig, he established connections with Jakob Breyne. For the remainder of his life he sought to expand his initial collection of plants. At Courten's death, Sloane acquired his entire collection (including an extensive herbarium), valued at £50,000. He also bought the 8,000 herbarium specimens of the Queen's botanist Leonard Plukenet, the international collection of the apothecary James Petiver, the collections of Hermann, who had botanized in Ceylon, James Cunningham's plants from China, and George Joseph Kamel's plants from the Philippines. By his death in 1753, Sloane held the largest collection of dried seeds, roots, wood, resins, fruits, and plant specimens in the pre-Linnaean era—a collection that came to form the nucleus of the British Museum. Sloane's private wealth and massive stockpile of specimens catapulted him into the top echelons of British science.[75]

Linnaeus, without money or noteworthy social standing, flourished in botanical networks, receiving much sought-after botanical specimens from correspondents all over the world in exchange for access to his international contacts. More important, he offered adherents the sense of contributing to a higher cause—his universal system of classification—and, at times, his correspondents garnered the greatest gift of all: immortality by having his or her name ensconced in Linnaeus' system of scientific botanical nomenclature (see Chapter 5).

Exotic flora could serve as capital for exchange, even at the lower end of the social ladder. In 1719, a French fugitive wrote colonial officials in Guiana from Surinam asking to be pardoned and repatriated in the French colony in exchange for "germinating coffee seeds," which he proposed to smuggle out of Dutch Surinam. The French had recently identified coffee as a possible colonial cultivar and had seeds sent from the Jardin du Roi to Martinique in 1716, as noted above. Authorities jumped at the chance to win this potentially lucrative crop also for Guiana and approved the man's request. The man was pardoned, and the seeds, received by M. d'Albon, commissioner of the navy, flourished. Plants were distributed to French plantations throughout Guiana, and coffee soon became a valuable export.[76]

Women, too, stockpiled and traded in specimens. Mary Capel Somerset, Duchess of Beaufort, for example, directed vast and costly gardens in the late seventeenth and early eighteenth centuries. Much like other early-modern women scientists who worked in what I have elsewhere called "noble networks," she gained access to cosmopolitan botanical exchange by virtue of her elevated social status. By the latter part of the eighteenth century, few individuals commanded the resources to compete with state-financed gardens like the Jardin du Roi and Kew. As in many areas of science, when botanical gardens became public, women no longer served as directors.[77]

The Duchess of Beaufort began collecting plants in the 1690s for her gardens at Badminton, Gloucestershire and at Beaufort House in Chelsea (bordering the Chelsea Physic Garden). Her collection included botanical curiosities from Barbados, Jamaica, Virginia, Guinea, the Canaries, the Cape of Good Hope, Malabar, Sri Lanka, Japan, and China, as well as specimens from continental Europe, England, Scotland, and Wales. The duchess was a hands-on patron; she corresponded and exchanged specimens and seeds of "foreign" plants with the principal botanists of her time, including Sloane, whose renowned herbarium contained twelve volumes of dried exotic plants from her garden. William Sherard, who served as tutor to her son between 1700 and 1702, wrote that these gardens would soon "outdo any in Europe, being furnished with all conveniences imaginable and a good stock of plants to which I added 1,500 and shall daily procure more from my correspondents."[78]

The duchess was no amateur gardener. Her lists and catalogues of plants cross-referenced her collection with the other leading gardens and major botanical works. Like many botanists of the time, the Duchess of Beaufort was interested in classification only insofar as it was useful for her to organize her gardens and catalogues. Hers was primarily an acclimatization garden, where she coaxed rare exotics, often from seed, into flower in England's inhospitable climate. Not only were seeds and plants sent to her from around the world, but she also sent out requests listing plants she desired. The first edition of William Aiton's *Hortus Kewensis* (1789) credited the duchess with the introduction of some sixty-four exotic species into England. As befit a woman of her era, she kept her flowers in "Health, Order and Decency."[79]

According to the historian Douglas Chambers, the duchess's head gardener was John Adams. Sloane, however, noted "an old woman" who worked under the duchess's supervision. Praising her gardens, Sloane wrote in 1703, "Her Grace [has] what she called an Infirmary or small green house, to which she moved sickly or unthriving plants, and with proper culture by the care of an old woman under her Grace's direction brought them to greater perfection then [sic] at Hampton court or anywhere."[80]

The duchess's brothers were also master gardeners. Henry Capel, Lord of Tewkesbury, whose specialty was fruit, had glass houses of oranges and "myrtiles" built near Kew House, which he had inherited through marriage. His holdings eventually became Kew Gardens, home of the Royal Botanic Gardens.[81]

Similar to Beaufort, Agnes Block, an Amsterdam patrician who owned a country manor in Nieuwersluis on the Vecht some thirty kilometers from Amsterdam, used her exotic flora and seeds to enter into correspondence with botanical gardens throughout Europe. She commissioned many well-known Dutch painters, including Merian and her daughter, Johanna Helena, to record her impressive inventory.[82]

Some women, however, declined to enter genteel networks of scientific exchange. In the first decade of the eighteenth century, we encounter an interesting exchange between Merian and James Petiver, a London apothecary and fellow of the Royal Society, whose natural history collection Sloane valued at £4,000. Petiver seems to have initiated the exchange sometime in 1703 by sending Merian "a present of insects." In a letter written the next spring, Merian solicited Petiver's aid in procuring an English translation of her *Metamorphosis*. Merian suggested that the work be accompanied by a dedication to the Queen of England, this being "rational, as coming from a woman to a personage of the same sex." Petiver prepared at least a partial translation in which he chose to order Merian's entries by types of creature (one chapter devoted to lizards, frogs, and snakes; a second to butterflies; and a third to moths—something that distressed her), but no English translation of Merian's book appeared in the eighteenth century.[83]

Merian's letters to Petiver were businesslike. She arranged to send him the copies of her beautifully illustrated *Metamorphosis* that he ordered. The folio cost eighteen Dutch guilders, she wrote to him, "but because you have ordered some seven of it, you shall have it for fifteen." She emphasized that she must be paid in guilders in advance, and that the books had to be picked up in Amsterdam. Once he had paid his bill, she continued, "I will deliver . . . the books just as I have done for other people." In this same letter of 1705, Merian broke off specimen exchanges with Petiver. She thanked him for his most recent gifts of "little animals," but sent them back, saying that she sought only samples that could reveal "the generation, production, and mutation of animals, how they produce one from another, and the manner of their food." Nonetheless Merian sent him some earthen jars of Surinam beetles for which she said she had "no use."[84]

Petiver and Merian's exchange was anything but direct. Letters took from three weeks to five months to arrive; some did not arrive at all. He wrote in English through his secretary and she answered in Dutch, German, and, through translators, in French and English. The translators did not always serve her well. Christopher Adolph, who rendered Merian's letter of Friday, October 19, 1705 into English, added: "According to your command, I have put into English the inclosed German letter as well as I could, and if it be not as well as you could wish, I hope you will pardon it because the original is none of the best stile, nor clearest sense but just as women's epistles generally are."[85]

The Search for the Amazons

Before leaving this chapter, I want to consider a curious example of knowledge transfer—from the European imagination into South America—that fueled European research projects. Anthony Grafton and others have emphasized how New World encounters shook the foundations of ancient texts and contributed to Europeans' new respect for empirical knowledge. We see, however, in the century-long search for the Amazons, those warlike daughters of Ares, the power ancient learning still held over European minds. There are many examples of ancient texts encumbering Europeans' scientific observations: Pliny the Elder's account of *Anthropophagi* ("man-eaters") encouraged many to spot such cannibals in the New World, especially among the Caribs. Similarly Pliny's description of *Homo troglodytes* launched, at Linnaeus' urging, a Swedish expedition to track down a specimen of this purported second species of human wherever it might exist. Even Humboldt, who was well aware that Europeans regularly "adorned" descriptions sent from the New World "with some features drawn from classic antiquity," searched for many years for the mythical Amazons.[86]

European reports concerning the reality of the New World Amazons stretch back nearly to Columbus himself. The river we today call the Amazon was named by the Spanish soldier Francisco de Orellana in 1541 after he witnessed pitched battles between tribes of female warriors along its banks. He took these warrior-women to be the true Amazons, describing them as each fighting with the strength of ten men. Naturalists, in particular La Condamine and Humboldt, were still searching to confirm or forever banish the notion of the existence of these improbable women in the 1730s and 1790s. This foray into the scientific search for the Amazons demonstrates the extent to which Europeans approached New World naturalia through cultural and gendered frameworks that not only set research priorities but often made travelers from Europe unable to absorb radically new information, a topic I look at more closely in Chapter 2.

La Condamine purported to approach the New World critically. He considered "the majority of the natives of South America liars, credulous, and prone to the marvelous" and often described them as intellectually "apathetic" and "stupid." Nonetheless, when investigating the validity of the myth of New World Amazons, which he described as a "republic of females who lived apart from men," La Condamine collected evidence from the "peoples of various nations" off and on for the eight years he explored South America. Key for him was that what he took to be independent reports of Amazon sightings all converged "in one common center, that is, the mountains in the heart of Guiana, a realm not yet known to either the Portuguese of Para or the French of Cayenne."[87] La Condamine judged these reports consistent and, importantly, unsullied by prior European contact, and hence authentic.

The French mathematician found corroborating evidence for the existence of the Amazons in a green stone called "Amazon stone" (a gemstone variety of microcline or feldspar). Amerindians living along the Topayos river had inherited them from their forefathers who, in turn, had obtained them from the *Cougnantainsecouima,* in their language signifying "women without husbands," among whom these stones were found in abundance. La Condamine also gave what he called a "moral" argument for the likely existence of these women and their aversion to men. He reasoned that if the Amazons existed anywhere, it was mostly likely in America, where women often accompanied their husbands to war and where the hardships of their domestic life might have provoked them to "shake off the yoke of their tyrants, to form an independent society, and to avoid that vilifying condition of slavery . . . in which they had previously lived."[88] He characterized their flight from oppressive husbands as analogous to that of slaves who, disgusted by European ill-treatment, frequently escaped from their masters into the jungle and lived together in bands.

La Condamine energetically sought out an old man whose father had actually seen the Amazons—a potential eyewitness, once removed. Upon arriving among the inhabitants of Coari in the wilds of what is today Amazonas (in central Brazil), he found the old man dead, but his son, himself already seventy years of age, assured La Condamine that "his grandfather had in reality seen them [the warrior women] pass by at the entrance of the Cuchiura river, that they came from the Cayamé which falls into the Amazon on the southern side, between the Tesé and the Coasi, and that the grandfather had spoken with four of them." This witness, who had passed down to his son the names of the four women, did not mention the supposed Amazon custom of cutting off a breast so that their bows lay flat on their chests, in La Condamine's opinion "too remarkable an occurrence to have escaped observation." The Frenchman supposed that the story of these women sacrificing a breast for martial accuracy resulted from European travelers' "fondness of the wonderful."

These tales told by Europeans, La Condamine reasoned, may even have found their way into Amerindians' own recounting of "sightings" of the Amazons.[89]

La Condamine finally located a native inhabitant of Mortigura, a missionary settlement in the vicinity of Para, who offered to show him a river, called Irijo, where he claimed, if one traveled upstream for several days, one would find the Amazons. Daunted by the great effort required to travel through dense jungle and high mountains, La Condamine let slip away his chance to become the first European to lay eyes on the true Amazons. He rationalized his decision by saying that because of their migratory habits, the women had most probably already moved on. Even more likely, he mused, this nation of husbandless women may have forsaken their ancient habits because they either had been conquered by another nation or at long last "lost the aversion of their mothers to the company of men."[90]

Traveling some forty years later, Humboldt, while still obsessed with the possible existence of such singular women, did not conclude, as had La Condamine, that these women had once actually roamed the wilds of Guiana. La Condamine, he averred, defended the reality of the *Cougnantainsecouima* merely to sate the lust of Parisians for marvelous tales. Yet Humboldt himself added some intriguing details to the Amazonian myth: they fashioned *sarbacans* (blowpipes through which poison arrows are shot) and gave them as presents to the men who slept with them (in the month of April only); and they discovered and applied with success the medicine hidden in their green stones (said to heal epilepsy, as well as liver and kidney disease).

But Humboldt was more concerned to recount how Europeans had created this myth than with the myth itself. Humboldt delineated several steps in the process by which Europeans "found" in America what pleased them and what existed only in their own imaginations. The myth began, Humboldt argued, in the works of sixteenth-century travelers—Amerigo Vespucci, Christopher Columbus, Gonzalo Fernández de Oviedo y Valdés, among others—who adorned their descriptions of the New World with features drawn from classical antiquity, "merely finding among the newly discovered nations all that the Greeks had earlier taught concerning the first ages of the world." Voyagers, he continued, also sought to flatter their patrons: thus Oviedo sought to compliment Cardinal Bembo's great knowledge of antiquity with his tales of Amazons, and Sir Walter Raleigh aspired to pique the "virgin" Queen Elizabeth's interest by describing a "warlike republic of women without husbands."[91]

Humboldt was dismayed that what began as a mere "ornament of style" and a "pleasure of the mind" became in the eighteenth century a "subject of grave discussion" and research. He explained the sightings of probable Amazons by the Amerindians not as purposeful deceits but as misinterpretations by Europeans. When Europeans asked about the existence of warlike women, Amerindians (attempting to oblige) described "women in different parts of America, [who] wearied of a state of slavery in which they were held by the men, united themselves together, like the fugitive Negroes." To preserve their independence these maroons became warriors. It was these women—bands of runaway wives—whom Europeans, Humboldt surmised, mistakenly identified as "Amazons." Humboldt also criticized Europeans who imagined Amerindian women defending their huts and even nuns defending their convents as the much-sought-after Amazons. He ended his discussion: "such was the disposition of men's minds, that in the long succession of travelers, who crowded on each other in their discoveries, and in narrations of the marvels of the New World, everyone chose to have seen what his predecessors had announced." Everard Ferdinand Im Thurn, stationed in Guiana in the nineteenth century, suggested that the Amerindians had learned about the Amazons from the Europeans and, after centuries of European enquiry, had integrated this appealing legend into their own stories about the uncharted interior of their country.[92]

Heroic Narratives

Amazon myths were but one gender dynamic running through voyagers' travels. Mary Terrall has suggested that eighteenth-century scientific voyagers fashioned their exploits into narratives that served to heighten a new and much-prized variety of heroic masculinity. These travel narratives featured "adventurer-scientists" who encountered life-threatening nature, wild natives, and hostile colonials all in the service of science. Terrall has argued that the figure of the heroic traveler was juxtaposed against the "feminized" domestic spectator, which, in combination with a number of other factors, increased gender bipolarization in eighteenth-century science and functioned, at least on one level, to strengthen the exclusion of women from science. Emma Spary, in her history of the Parisian botanical garden, has also emphasized the "heroic" nature of the romantic voyager: "the hard-working man willing to sacrifice his own safety for science." In an era when women were defined as too delicate to carry out the rigorous work of science, the celebration of its heroic qualities only strengthened

Figure 1.5. A European gentleman carried "by man's back" over the Andes. He reads to overcome the boredom of the many hours required to sit motionless lest his slightest movement send the porter, chair, and himself plummeting down the mountainside. Alexander von Humboldt complained that "human carriers" were too confident in their abilities and chose imprudently steep and narrow paths over raging rivers and cavernous valleys. Humboldt, who sketched this scene, rarely rode by man's back, preferring to walk the distance barefoot after thorns and stones had destroyed his shoes.

the notion that it suited men best. The figure of the heroic traveler did, in fact, loom large in eighteenth-century travel literature. Linnaeus in his *Critica botanica* emphasized that genera of plants should be named after botanists to commemorate their heroism in pursuit of science: Tournefort "traversed the heights of the Alps, the ravines of the Pyrenees" in search of plants; Clutius wandered "through the desert of Barbary"; Micheli botanized in stormy weather and died of pleurisy and pneumonia; Sherard spent "days and nights" completing his *Phytopinax,* "whence he contracted a wasting sickness and died"; and so on. Even modern-day travelers to the tropics can sympathize to a certain extent with the hardships encountered while traversing those areas. Travel was, and can still be, difficult for the unprepared or unfortunate.[93]

But other gender dynamics become apparent in this literature as well. In many instances, eighteenth-century travelogues placed European male travelers in what Europeans might interpret as traditionally feminine roles that also became closely associated with the French upper classes in the years immediately preceding the French revolution. In these cases, it was "native guides" in the countries they visited who took on masculine roles (at least as so interpreted in European society), while Europeans were afforded numerous and often exorbitant courtesies. Images of naturalists being carried on chairs over the Andes are well known. In October of 1801, for example, as winter approached, Humboldt, Bonpland, and their troupe of assistants and carriers crossed the Quindiu, considered the most difficult mountain to traverse in the Andes. Humboldt dramatized the difficulty of the passage: "the weather was severe, the rain continual, the mud thick and stagnant. Food was scarce along the long road and places to lodge nonexistent." The abundant thorns quickly shredded boots so that, unless naturalists were carried by "men's backs," they had to walk the great distance barefoot. "Few people of means go by foot in these conditions," Humboldt concluded, "they arrange instead to be carried in a chair tied to a man's back." The German naturalist-explorer instructed his readers that the locals in this part of the world spoke of going "by man's back" in the same way that others speak of going by horseback.

These carriers or *cargueros,* according to Humboldt, took voyagers across terrain not accessible even by mule. These men—both "mestizo" and white—were said to carry six to seven *arrobas* (seventy-five to eighty kilos) up steep mountain paths; the most robust of them carried nine *arrobas*. Humboldt reported that the Spanish government had once offered to build a better road into the region, but that the *cargueros,* for whom carrying had become a livelihood, protested, preferring wild and free lives in the forest to sedentary and monotonous existences in a village.[94]

La Condamine also painted a pretty picture of "helpless" European explorers in his description of the high rope bridges used by the Incas, which he hesitated to cross. "The natives," he wrote, "who are far from being naturally intrepid, pass such bridges on the trot, with their loads on their shoulders . . . and laugh at the timidity of the traveler who hesitates to venture."[95]

True, European gentlemen were refined and not accustomed to dirtying their hands in manual labor. Yet the developing conventions of colonial master/servant relationships rendered Europeans unduly dependent on their guides in the tropics. Michel Adanson, traveling in Senegal in the 1750s, was carried across a river on his man's shoulders. Coming to the Niger, Adanson hesitated to cross by foot because he feared falling into a hole. While his "Negro" sounded the bottom, he rested in a tree, being "tired of snakes and the water." Finally Adanson was coaxed to cross: "I mounted upon his shoulders," Adanson wrote, "with my gun in my hand, a few birds, and a bundle of plants. With his guide soon up to his neck in water, Adanson wrote, "I resigned myself to his skilful guidance, or rather to my own good fortune, and I let him do as he pleased. He waded through . . . with amazing resolution, without being in the least daunted, though he was obliged to swallow three large gulps of water [by then well above his mouth and nose] which for some time took away his breath."[96]

Female travelers also penned narratives of hardship. Merian described suffocating heat, thorny and impassible forests, and life-threatening illness. Overcome with fever, she wrote that for her insects, "I nearly paid with my life." The Scottish woman Janet Schaw, too, dramatized the hardships of her passage to the West Indies. Rolling seas and constant rains capsized their overladen boat which, when righted, oozed with sodden beds and water-logged trunks. All her provisions washed overboard, save for one trusty tea kettle and a large ham.[97]

If, as historians have suggested, academy scientists composed heroic narratives for themselves, they also styled them for traveling ladies. Jean Godin des Odonais, the official measurer for the 1735 Parisian academy's South American expedition, recounted the story of his wife's journey down the Amazon from her native Peru to his home in France. This tale shared all the elements of eighteenth-century heroic narratives: Madame Isabelle de Grandmaison Godin des Odonais, a Spanish noblewoman born in Peru, had the unhappy fate of watching her seven immediate traveling companions succumb to ferocious beasts, venomous snakes, thirst, hunger, and fatigue while she alone survived.

When the French equatorial commission left Quito, Jean Godin des Odonais stayed on, taking up a post as professor of astronomy and natural science at the College of Quito in 1739. Here he met and married the fifteen-year-old Isabelle. Godin left for Cayenne in 1749 to prepare passage for his family to France. The Portuguese refused him a passport to return to Quito to fetch his wife and extended household for the arduous journey to the Atlantic port city. Finally granted the required papers in 1758, Godin fell ill and sent word that his wife should meet him in Cayenne. She sold their household furniture and left Quito in 1769 accompanied by her two brothers, a physician, her "Negro," three female "mulatto or American" servants, and thirty-one American natives to carry her and her baggage. Slowly, the difficulties began. The village where the travelers were to pick up a boat was caught in the grip of a smallpox epidemic and the carriers, who had been paid in advance, all ran away. The party found two Indians in the village free from the contagion and waited while they constructed a boat. They piloted her downstream for two days before abandoning her. With no one to steer the craft, Madame Godin and her party went ashore and built themselves a hut, waiting some twenty days for a canoe that never arrived. They eventually fashioned a raft, but it immediately capsized and they lost all their supplies. Tired of the river, the party entered the forest. Wearied from walking, their feet torn by thorns, their provisions exhausted, they fell to the ground and awaited death. After three or four days, Isabelle, stretched on the ground alongside the corpses of her brothers and other companions, resolved to go on. Delirious, she at last found some water, wild fruit, and fresh partridge eggs. She was finally discovered by some Indians who gave her passage in their canoes to his in Cayenne. Godin and his wife later sailed for France, reaching his estate in 1773. As a female, Madame Godin des Odonais was a privileged and rare traveler. She was motivated, however, by the love for her husband, not for science.[98]

Travel in the Torrid Zone was, indeed, perilous. Nature was remarkably sex-neutral in dealing with Europeans who ventured into the tropics; men were no more spared than the "fair sex." Africa was known as the "white man's grave"; India, as the "the European graveyard." Thunberg, traveling in the East Indies in the 1770s, discussed the frailty of the European body in unfamiliar environs ("our Houses of Clay") while recording that 158 men out of a crew of 400 perished before his ship rounded the Cape of Good Hope (many from scurvy). According to Philip Curtin's estimate, European merchants and troops in West Africa died at a rate of 483 per thousand per year throughout the eighteenth century. Conditions in the West Indies, Guiana, and Surinam were not as severe, but bad enough. In Jamaica, death rates among Europeans ranged between forty to ninety per thousand per year, spiking to 200 per thousand in Kingston in the early part of the eighteenth century. Disease, especially yellow fever, was so endemic that European populations failed to increase in Jamaica. Humboldt, traveling in the Caribbean a century after Sloane and Merian, still bemoaned his being "condemned" to mosquitoes, the want of food, the impossibility of drying botanical specimens in a moist and humid rain forest, but most of all to the notorious "ploughman insect" that burrowed into his fingers and toes. After two days of agony, Humboldt found a "mulatto woman" who removed the insects and their egg sacks one by one by digging into his skin with a sharpened stick. Finding this cure more painful than the insects themselves, Humboldt located a "Javita Indian" who made an infusion of the bark of the *uzao* (a shrub with small leaves like those of the cassia) that promptly killed the pests. Humboldt was so fearful of this particular pestilence that he and his men never traveled without *uzao* bark in their boats.[99]

Travel books from this time are filled with stories of troops in the New World suffering from unknown diseases. Even in the temperate climate of northern New France (Canada), unidentified diseases decimated newcomers. One captain in the area tried to hide the number of his men who were sick. "We did greatly fear that the people of the country would perceive our weakness" and take advantage of it. His men were afflicted with "swollen knees, sinuses shrunk together, teeth spoiled, gums rotten and stinking." The captain consulted his guide, who recommended the juice and sap of the leaves of a certain tree, in his language called *ameda* (sassafras?). Two women were sent to fetch some of it and the guide showed the captain how to boil the bark and leaves of the tree together, drinking some of the decoction every other day, and applying the dregs of it to the afflicted legs. The captain's men refused to drink the unknown medicine until two hardy souls took it and were quickly delivered from their sickness. They marveled that the drink cured in five or six days what "all the physicians of Montpellier and Lovaine [Louvain] . . . with all the drugs of Alexandria" could not have done in a year.[100] This story had a happy ending, but many did not.

Even elite scientists—who had every possible physical advantage—faced hardships. La Condamine recounted the "martyrs" to science: of the five from the Académie Royale des Sciences expedition to the Arctic sent out in the mid eighteenth century, only one survived; Nicolas-Louis de La Caille died during his travels to the Cape of Good Hope, a "martyr to astronomy"; Abbé Chappe d'Auteroche, an academician, died in California in 1769. La Condamine registered the deaths of his own companions during the 1735 expedition to the equator: Couplet, the most robust and one of the youngest, was carried off three days after his arrival at Quito by a putrid fever; Seniergues, the surgeon, was killed in a scuffle at Cuenca; Pierre Bouguer died of an abscess of the liver; Louis Godin, younger than Bouguer, survived the latter but two years; de Morainville, who remained in Quito, died when he fell from the scaffolding of a church he had designed; Hugo married at Quito and was never heard from again. Joseph de Jussieu, like Mabillion, went insane. La Condamine himself suffered from deafness and paralysis in his lower limbs, and a number of the expedition's servants, "white as well as of colour," perished—two of them violently.[101]

In addition to the rigors of the climate and dangers of the road, human-made obstacles also impeded travel. Travelers waited months for passage on ships and for passports through the territory of unfriendly countries. La Condamine waited at Cayenne six months for passage back to Europe, and an additional two months at the Hague for the passports required to cross the Low Countries into France. Thiery de Menonville similarly waited six months in Havana for a ship to take him to Veracruz, Mexico.

Natural and human-made hardships also prevented many of these naturalists' treasures from arriving in Europe. In 1753, Adanson in Africa lost his plants, which he plaintively reminded the reader "had occasioned the voyage" in the first place. His precious cuttings and seeds were all destroyed by severe cold on their way from Brest (in Bretagne) to Paris. In 1777, the beleaguered Thiery de Menonville lost a cargo of precious plants he had dispatched to the King's Garden in Paris when the ship sank. The most devastating experience was that of Joseph de Jussieu, who botanized for the French King's Garden for three decades in and around Quito and Lima. He returned to France a broken man, sick and having lost his memory at the end of his life. His many crates of specimens that he had collected, dried, and packed for shipping were all left in foreign ports. Nothing but a few memoirs remained of his work.[102]

Then, too, samples that naturalists managed to get back to Europe were not always in the best of shape. Collected at great distances from ports, often during the wrong season, and insufficiently preserved to withstand a prolonged sea voyage or adulterated for greater commercial gain, many of the specimens were useless when they arrived in Europe.[103] John Ray complained, for example, that the 800 new species Sloane proudly carried to England had no roots—a plant part Ray considered vital for classification.

European governments, trading companies, and scientific establishments fielded a number of naturalists with diverse goals and outlooks in the Caribbean. Surveying colonial resources, they sought to discover and domesticate plants that might add to the nation's (or their own) wealth and power either on European soil itself or, when the rigors of that climate spoke against it, in the rich and fecund soils of one of Europe's tropical colonies. In so doing Europeans, who had little experience with flora in the Torrid Zone, often relied on the expertise of peoples indigenous to those areas. How did Europeans cull secrets of nature from tropical lands? How did they wrest secrets from local informants? What sorts of knowledges did Europeans choose to collect, and what did they leave behind?

2

Bioprospecting

> It is quite by accident and only from savage nations that we owe our knowledge of specifics [medicines]; we owe not one to the science of the physicians.
>
> P.-L. MOREAU DE MAUPERTUIS, 1752

Pierre-Louis Moreau de Maupertuis, president of the Berlin-Brandenburg Akademie der Wissenschaften, surely overstated his case when he claimed that European drug discoveries depended on either "accidents" or the knowledge of non-European peoples, whom the Europeans often characterized as "savages." How were new drugs identified in the eighteenth century? How did useful new medicines such as ipecacuanha, jalap, and Peruvian bark arrive at London shops and Parisian hospitals?

The search for new and profitable drugs in the eighteenth century was not unlike bioprospecting today. Then as now, the European search for effective pharmacopoeia was fueled by the vast fortunes to be made. We should remember that chocolate was first popularized in England as a medicine—good for stomach ailments and consumption—by Sir Hans Sloane. Edward Long, Sloane's compatriot in Jamaica seventy years later, reminded his readers that those men and women who immigrated to the West Indies did so not for the purpose of compiling natural histories, but "avowedly for the purpose of accumulating money." Most European émigrés, Long continued, considered these islands only a temporary abode and had little interest in them except for the riches they could afford.[1]

The European search for new medicines was also fed by mercantilist efforts to make European countries pharmaceutically self-sufficient, thereby checking the flow of bullion to foreign countries and eventually creating positive trade surpluses. Pierre-Henri-Hippolyte Bodard, the French physician and botanist, decried the annual loss of revenue to foreign countries, a loss aggravated by the French upper classes' preference for "objects difficult to obtain." He estimated the loss in revenue from saffron alone at 20,000 livres annually and from the new Peruvian bark at 7,380,000 livres per year. Other exotics, such as rhubarb, tea, cacao, and jalap, similarly drained European coffers. In England, too, physicians called for patients not to send "to the East Indies for Drugs that they may fetch better out of their own Gardens."[2]

In many instances, however, useful drugs—Peruvian bark, for example—would not grow in Europe and efforts were made to cultivate these plants in the colonies. Nicolas-Louis Bourgeois, who served as secretary of the Chamber of Agriculture during his twenty-eight-year residence in Saint Domingue, advocated in the 1780s the use of local drugs for local complaints. "Why have recourse to foreign drugs," he asked, "when it is believed that there are so many here [in the Antilles] of use?" There is "a mass" of simples, he continued, in Saint Domingue that need only be examined: "one will find among them more of virtue than of those that come at great expense from far away." Many of these drugs grown in the colonies were imported into Europe with the government making a tidy sum in duties. Success in learning how to substitute indigenous plants for exotics had other benefits as well. French botanist Michel-Étienne Descourtilz, who arrived in Saint Domingue at the height of the (Haitian) revolution in 1799, was spared when other Europeans were massacred because he was one of the few who knew how to replace imported medicines with indigenous cures—especially valuable knowledge after the hospital's pharmacy burned to the ground.[3]

The most pressing motivation for investigating tropical medicines, however, was to keep European troops and planters alive in the colonies. Colonial botany was crucial to Europe's successful control of tropical areas, where voyagers from temperate zones became sick and died in alarming numbers. Bourgeois noted that finding new drugs was not merely a matter of curiosity but necessary "to cure our maladies and provide new assistance." Europeans moving into the tropics encountered illnesses completely unknown to them; their standard pharmacopoeia, sometimes old and useless from transport on damp and unsound vessels, frequently proved ineffective against new torments. Bourgeois complained that "apothecaries and surgeons buy their remedies from ship captains coming from France, but the vegetable remedies last hardly one year, and when they are old, they do more harm than good." Bourgeois complained further that drugs that are "scrupulously" monitored in France were not regulated at all in the colonies; anyone could open a pharmacy [*boutique des drogues*].[4]

This chapter attempts to bring to light the knowledge and practices of Amerindian naturalists, the peripatetic Arawaks, Tainos, and Caribs, who moved much knowledge and many plants from place to place in the Caribbean basin, and of African naturalists—both males and females—who transported African flora and knowledge of its uses with them to the West Indies. Most of these naturalists remain faceless and nameless, and we have access to their practices primarily through the filter of European texts. With all eighteenth-century natural history we face the problem of imperfect sources removed from us by several centuries. But Caribbean slaves and Amerindians present a special case. Because they left no written documents, knowledge of their cultures and medicines, including their extensive use of plants, depends here almost entirely on European naturalists' accounts.

This chapter launches an examination of Amerindian and African slave medical practices and continues our exploration of European naturalists' field practices as they collected plants either for profit or for medicines in the West Indies. Was it the trained scientific European eye that allowed for new drug discoveries? To what extent did Europeans in the West Indies depend on African slave and Amerindian naturalists for information concerning what later came to be known as tropical medicine, as Maupertuis suggested above? How did naturalists from Europe wrest this information—upon which their very survival often depended—from populations they had conquered and enslaved?

Drug Prospecting in the West Indies

Before the onset of rampant racism in the nineteenth century, many Europeans valued the knowledges of indigenous Americans, Africans, and peoples of India and the East Indies. Richard Drayton has argued with respect to eighteenth-century England that racist tendencies were tempered by a recognition that inhabitants in the colonies—in the West Indies, Amerindian, transported African, or African or European creoles—often held knowledges worth recruiting. This was also true in earlier centuries: Harold Cook has discussed how seventeenth-century Dutch physicians in Java valued local medical information, and Richard Grove has written that the *Hortus Indicus Malabaricus* (1678–1704) prepared by Hendrik Adriaan van Reede tot Drakenstein, the East Indies Company governor in Malabar, was "a profoundly indigenous text," a compilation of South Asian botany "without equal." One might argue that in respect to natural history, an epistemological shift took place over the course of the sixteenth and seventeenth centuries away from Europeans relying on the "summa of ancient wisdom" (Dioscorides, Pliny, Galen) toward their valuing (or at least appreciating) the authority of native peoples encountered through global expansion. European physicians no longer defined their task as simply verifying the effectiveness of ancient medicines (or merely identifying local substitutes); instead, they took as their starting point for empirical investigations the drugs, dyes, and foodstuffs suggested to them by native "informants."[5]

How did Europeans prepare for bioprospecting expeditions? Before leaving Europe, learned naturalists pored over documents from all across Europe seeking information concerning the flora they were likely to encounter in the course of their voyage. Sloane was typical in this regard. Before setting out for Jamaica, he collated all data available in Europe concerning plants of tropical areas (in both the East and West Indies) so that he would recognize new plants when he encountered them. Upon arriving in the West Indies, Sloane turned to locals for information, collecting "the best information" concerning the natural products of the country from "books and the local inhabitants, either European, Indian or Black." Maria Sibylla Merian also set great store by local informants in Surinam. In recording the medical uses of particular plants, she often ended her entries with the words: "They told me this themselves."[6]

Who were the local informants to whom Europeans turned for information concerning useful foods and medicines? When Christopher Columbus arrived in Hispaniola in 1492, the island was inhabited by approximately 1 million Tainos and Caribs—both of whom had migrated into this area from South America sometime before 400 B.C. These peoples established gardens (called by Tainos *conucos*) for the cultivation of their most prized foodstuffs and medicinal herbs. Spanish physicians sometimes enjoyed especially close ties with the descendants of these voyaging Amerindian naturalists. Antonio de Villasante, for example, learned the virtues of plants in Hispaniola from his Christianized wife, Catalina de Ayahibx, a Taino chief (or *cacica*).[7]

By the sixteenth century, Taino and Carib populations in the Caribbean had been decimated by conquest and disease. The Caribs had run the Arawaks out of the Lesser Antilles; the Spanish had crushed both peoples. A 1660 peace agreement between the English, the French, and the Spanish exiled the remaining warring Caribs to the islands of Saint Vincent and Dominica. A report issued in 1687 found only 111 Caribs living in Martinique. The larger islands, such as Jamaica and Hispaniola, heavily inhabited by Europeans and Africans, were left with perhaps even smaller populations of Amerindians.[8] European physicians nonetheless gleaned what information they could from the few survivors.

Figure 2.1. "The natural inhabitants of the Antilles of America, called savages" under a papaya tree. Jean-Baptiste du Tertre described these "savages" (whom he did not identify further) as the most content, happy, peaceful, honest, and healthy people on earth. Their flat foreheads and noses were not a defect of nature but, among them, considered a mark of beauty and often fashioned by their mothers. Du Tertre added that the Americans were more ignorant than the Europeans, but less vicious. Note the water-tight basket, used to carry provisions in canoes, in the woman's left hand (the mélange of Caribs and Arawaks on Dominica still make such baskets); note the war club in the man's left hand, the bow and arrows in his right hand.

In the eighteenth century as earlier, medicine and botany were closely allied, and botany remained a required element in European medical education. Jean-Baptiste-René Pouppé-Desportes, royal physician in Cap Français, Saint Domingue from 1732 until his death in 1748, typifies voyaging naturalists who settled into life in the Caribbean. In efforts to increase the efficacy of his cures, he supplemented his mainstay of remedies sent from the Hôpital de la Charité in Paris with local "Carib simples." Because the first Europeans who came to the Americas, Pouppé-Desportes wrote, were afflicted by illnesses completely unknown to them it was nec-

Figure 2.2. A Carib woman points to the *roucou* tree, whose flowers yield a highly prized red dye. Charles de Rochefort reported that the Caribs, who cultivated these trees in their gardens, used the roots to give their meats and sauces a rich saffron color and fragrance. Europeans in this period eagerly gathered information about plants and their uses from males and females native to the Caribbean.

essary to employ remedies used by "the naturals of the country whom one calls savages." In the third volume of his *Histoire des maladies de Saint Domingue,* this long-term resident presented what he called "an American pharmacopoeia," offering an extended list of Carib remedies used to cure diseases. Europeans had begun producing *Pharmacopoeia,* official compendia of medicinal drugs for major cities, in the sixteenth century in an effort to secure uniformity in remedies. Pouppé-Desportes' *Pharmacopoeia* is one of the first to record Amerindian remedies. As was typical of these works, he cross-referenced plant names in Latin, French, and the vernacular Carib. Offering synonyms (across languages and cultures) was a common European practice, especially in the pre-Linnaean era, but only rarely were Amerindian names systematically included (see Chapter 5).[9]

Even more important than adopting Amerindian cures, Pouppé-Desportes urged that Europeans emulate the native Caribbean way of life. Europeans living in Saint Domingue would not need medicines, he wrote, if they lived as "frugally and tranquilly as the savages." But, and his is a common refrain among medical men serving in the colonies, since Europeans (and he means males, who made up the majority of the European population in the West Indies) are "given to excesses in both their eating and liquor, they often require strong remedies." During his sixteen years in the island, Pouppé-Desportes continued to collect and test local remedies that he then classified according to their medicinal properties.[10]

Amerindian populations, however, continued to decline, and by the 1780s Bourgeois complained that "of the prodigious multitude of natives of which [the Spanish chroniclers] speak, not a single of their descendants can be discovered whose origin has been conserved pure and without mixture." By the 1790s, a chronicler of the revolution in Saint Domingue bemoaned the fact that no trace of a "single native" remained.[11]

With the decline of indigenous populations, slave medicines took on an unexpected importance in the West Indies, even though in the first half of the eighteenth century Africans on the big sugar islands were no more native to the area than were Europeans (at least 80 percent were born in Africa). Unlike Europeans, however, Africans knew tropical diseases, their preventions, and their cures. Lieutenant John Stedman, a Scottish mercenary living in Surinam, for example, worked alongside a number of African slaves. One old "Negro," named Caramaca, had given him the threefold secret of survival: 1) never wear boots but instead harden your bare feet (which Stedman did by incessantly pacing the deck of his boat); 2) discard the heavy European military jacket and dress as lightly as possible; and 3) bathe twice a day by plunging into the river. Some of these were distressing to Europeans, especially the latter, given their distaste for bathing. In contrast to his intimate relations with African populations in Surinam, Stedman had only vague trading relations with the Arawaks and Caribs—both of whom had been driven inland into the mountainous areas far from Dutch settlements.[12]

Bourgeois, a longtime resident of Saint Domingue, was one of those Europeans who appreciated slave medicines. Considering health a matter of state importance, he eulogized the "marvelous cures" abounding in the islands and remarked that *les nègres* were "almost the only ones who know how to use them"; they had, he wrote, more knowledge of these cures than the whites *(les blancs)*.[13]

It is impossible to know with any precision how much African herbal knowledge was transferred into the New World by voyaging slave naturalists. Displaced Africans must have found familiar medicinal plants growing in the American tropics, and they must have discovered—through commerce with the Amerindians or their own trial and error—plants with virtues similar to those used back home (see Conclusion). Bourgeois confirms that there were many "doctors" [*médecins*] among the Africans, who "brought their treatments from their own countries," but he does not discuss this point in detail.[14]

Bourgeois also praised the skill of slave doctors. "I could see immediately," he wrote, "that the Negroes were more ingenious than we in the art of procuring health . . . Our colony possesses an infinity of Negroes and Negresses [*nègres & meme des négresses*] who practice medicine, and in whom many whites have much confidence. The most dangerous [plant] poisons can be transformed into the most salubrious remedies when prepared by a skilled hand; I have seen cures that very much surprised me." What most surprised Bourgeois was that the Africans rebuffed two mainstays of European medicine: bleeding and purging. "If left to themselves, the Negroes do not bleed patients or administer enemas."[15]

Confidence in African naturalists' cures was so high among whites that when Sir Henry Morgan, lieutenant governor of Jamaica, became dissatisfied with Sloane's treatment of his disease, he sent for a "black doctor." Late in the eighteenth century, Jamaicans still sought the cures of "Negro Doctors." James Knight wrote that "many secrets in the art of physick, may be obtained from the Negro Doctors, were proper methods taken, which I think is not below our physicians to enquire into, as it may be of great service to themselves, as well as mankind."[16]

Even as late as 1799, amid the chaos of the Haitian revolution, Michel Descourtilz learned many of his cures from a "mulatress," a creole naturalist of mixed race. Well into the nineteenth century, French physicians praised women of African origins living in Saint Domingue for their extensive knowledge of medicines. Initially suspicious, the mulatress told Descourtilz nothing, but eventually wooed by his drawings of the plants from the Artibonite valley, "which she coveted," Descourtilz was able to obtain many recipes from her. These he "corrected" as he experimented with them. His efforts, he wrote, "were crowned with very satisfying results." Descourtilz is one of the few naturalists who collected information on abortifacients (and aphrodisiacs)—one might suspect much of it from this woman (see Chapters 3 and 4).[17]

Attitudes among Europeans across the Caribbean, of course, were not uniform. Pierre Barrère, a French royal botanist working in Cayenne along the coast of Guiana from 1722 to 1725, did not think much of Amerindian cures. The good health the Indians (he named twenty-four peoples living there) enjoy, he claimed, resulted from their careful diet, frequent bathing and moderate indulgence in pleasure. In a word, he wrote, "our Indians are completely ignorant of how to compound medicines. The few remedies they know they have learned from the Portuguese and other Europeans." He nonetheless recorded several Amerindian plant names and their medical uses. David de Isaac Cohen Nassy, a Jewish physician working in nearby Surinam in the latter part of the eighteenth century, remarked that the "Negroes" played a large role in the health of the colony with their "herbs and claimed cures." But he also noted that their cures were "more valued among the Christians than among the Jews." Sloane had a similarly poor opinion of slave medicine. Although he took care to collect what the Africans in Jamaica told him, he did not find their cures in any way "reasonable, or successful." What they knew, he wrote, they had learned from the Indians.[18]

Even in an era in which many Europeans valued the knowledge of Amerindian and African naturalists, mythologies of drug discoveries suggested that knowledge traveled up a rather anthropo- and Eurocentric Chain of Being, from animals (with their instinctive cures), to Amerindians, to the Spanish, and, according to Charles-Marie de La Condamine, ultimately to the French. La Condamine, who traveled extensively in what is today Ecuador and Peru, recounted the ancient legend that South American lions suffering from fevers found relief by chewing the bark of the *Cinchona* tree. Observing its curative powers, the Indians, too, began treating malaria and other "quartan" (recurring) fevers with the bark. The

Spanish then learned of the cure from the Indians, and the French, the self-appointed keepers of universal knowledge in the age of enlightenment, learned of it from the Spanish.[19]

A number of eighteenth-century histories of medicine subscribed to the notion that brute animals were the first discovers of many beneficial cures. La Condamine's countryman Pouppé-Desportes offered two further examples in this genre—one from Martinique and another from Saint Domingue. In the first, a potent antidote to snakebite had been discovered by the lowly grass snake. "Unhappy enough to live on the serpent-infested island of Martinique," wrote Pouppé-Desportes, the snake learned to employ a certain herb when attacked by a venomous serpent. The effect was so wonderful that the natives called the plant *herbe à serpent* or serpent herb. In similar fashion, Amerindian naturalists had discovered the excellent qualities of the "sugar tree" by observing wild pigs shredding the bark of this tree with their tusks when hurt and rubbing their injuries with its sap. For this reason this sap was called "wild pig balm." As Edward Long in Jamaica put it, "brutes are botanists by instinct." He reasoned that humans, too, in their "rude state" possessed a similar instinct to recognize herbs, balms, and salves necessary for their preservation. Robert James in London concurred and, like Maupertuis, disparaged European physicians for their ineffectualness in developing lifesaving cures. Drugs, he wrote, are discovered by "savages" by a "natural instinct perceivable both in man and beast," or by "madmen," by whom he meant the alchemists who, from time to time, "blundered" upon some cure by accident.[20]

Long was very much a racist in how he looked at local cures. Although he agreed that Negro cures often worked "wonderfully" where even European art had failed, he did not admit to any creativity among Africans. "Negroes," he charged, apply their herbs "randomly" and, like monkeys, receive their skill only from "their Creator, who has impartially provided all animals with means conducive to their preservation."[21]

Biocontact Zones

Scholars have offered various explanations for Europeans' massive success in conquering the New World. These explanations range from Europeans' superior guns, powerful horses and vicious attack dogs, to the accidental and strategic spread of diseases such as smallpox, and even Europeans' highly developed record keeping. In her study of French natural history, Emma Spary, following a Latourian model, has discussed how botanists and gardeners at the Jardin Royal des Plantes in Paris, such as André Thouin, instructed and manipulated voyagers in order to speed new specimens into European urban centers to enhance both the power and wealth of nations. Successful acclimatization of plants for agriculture, medicine, or the luxury trade required that plants be sent into France with precise information about their cultivation, virtues, and uses. Spary has postulated that botanical men at the center needed a "rigidly structured, unvarying, universally agreed method of describing and inscribing" so that the useful plants could be known and successfully positioned within European frameworks of agri- and horticulture.[22]

What the naturalists at the Jardin du Roi in Paris, the Hortus Medicus in Amsterdam, or, later, at the Kew Gardens in London could not control and standardize was the "contact zone"—where Europeans negotiated with informants of all kinds and those with medicinal and natural historical expertise in particular. Mary Louise Pratt has defined this "space of colonial encounters" as "the space in which peoples geographically and historically separated come into contact with each other and establish ongoing relations, usually involving conditions of coercion, radical inequality, and intractable conflict."[23] Here I wish to specify *biocontact zones* in order to set the contact between European, Amerindian, and African naturalists in a context that highlights the exchange of plants and their cultural uses.

The notion of a contact zone, as it is often understood, is not unproblematic. To isolate the contact between Europeans and non-Europeans as the primary site of analysis improperly constructs an overly rigid notion of non-Europeans as "others." Furthermore, contact is not restricted to a bounded zone; Europeans had contact at many junctures over the course of a voyage; there were encounters among persons of different classes and professions within Europe. There were certainly the encounters of young men who, driven by hunger to seek refuge in urban port cities, were kidnapped and shipped out as sailors on trading company ships. (Edward Long called the kidnappers "mantraders"; Carl Thunberg called them "man-stealers.") Abroad, encounters between Europeans and other Europeans were often as troubled as those with non-Europeans. Merian, for example, had unhappy relations with Dutch planters in Surinam. "They mock me," she wrote, "because I am interested in something other than sugar." In her turn, Merian criticized the planters for failing to cultivate plants other than sugarcane and for their harsh treatment of the Amerindians.[24]

Here I will focus (despite my caveats) on biocontact zones in the West

Indies and look closely at strategies Europeans deployed to entice information about useful plants from their informants, and how Europeans were often, in turn, manipulated by the very people they sought to control. Inside Europe, naturalists are known to have carefully choreographed complex systems of patronage—coaxing favors while also disciplining one another with varying degrees of success. Europeans in the field often found it more difficult to manipulate the informants they wished to exploit and control. They often simply refused to cooperate.[25] As we shall see, botanical exchange in these transcultural zones was fraught with difficulties.

"Noise"—or intellectual interference—in biocontact zones was often deafening. Loudest perhaps was the cacophony of languages. Europeans usually only scratched the surface of local peoples' knowledges of plants and remedies because they were often unable or unwilling to speak local languages. Edward Bancroft (a Massachusetts privateer and double agent who sent missives in invisible ink during the American Revolutionary War) worked in Guiana as a young man in the 1760s. He bemoaned the fact that he was "but little acquainted with the Indian languages," which were necessary for "acquiring that knowledge of the properties, and effects of the several classes of Animals, and Vegetables, which experience, during a long succession of ages, must have suggested to these natives." Though he endeavored to overcome these difficulties through the use of interpreters, he remarked that these efforts were largely "in vain."[26]

La Condamine, Pouppé-Desportes, and Alexander von Humboldt were all keenly interested in local languages. La Condamine spoke what he called "the Peruvian language" and even owned a 1614 "Quichoa" (Quechua) dictionary, which he used to study etymologies (of "quinquina," or quinine, for example). Dictionaries of various South American languages (Taracso, Quechua, Nahuatl, and Zapoteco) had been available since the late sixteenth century, prepared mainly by Spanish Jesuits. These often substantial volumes were to be "useful to persons traveling in the area for commerce, to cultivate the land, or to win souls." Concerning the language of the ancient Peruvians, La Condamine noted that by his day (1730s and 1740s) it was "strongly mixed" with Spanish. The French mathematician judged the languages of South America harshly, writing that "many possess energy and bear traces of elegance, but they are universally barren of terms for the expression of abstract and universal ideas, such as time, space, substance, matter, corporeality . . . Not only metaphysical terms, but also moral attributes are completely absent." One wonders, however, if La Condamine had sufficient knowledge of the language to make such judgments. Humboldt, who traveled extensively in present-day Venezuela and Colombia, prepared dictionaries of the Chaymas language that consisted of only about 140 words (he does give words for fever, hammock, boy, girl, bridegroom, fire, sun, moon, and phrases such as "he likes to kill" or "there is honey in my hut").[27]

Humboldt was keenly aware of the problems of communication and the power relations involved in privileging one language over another. He commended the Amerindians for their facility in learning new languages, especially Spanish, and remarked that because the Cassiquiare, Guahibo, Poignave and several other peoples inhabiting the missions did not understand each other they were forced to converse in Spanish—the language of the mission but also that of the occupying civil power. Humboldt admired the Jesuits for attempting to make Quechua the universal language of South America. Humboldt knew that most Amerindians did not understand specific Quechua words, but he supposed that they were familiar with its structure and grammatical form. He held this proposal much wiser than the one made by a provincial council in Mexico that different native American peoples communicate with one another in Latin.[28]

The problem of language in the West Indies was not merely one between persons of far-flung cultures. In the course of his travels to Saint Vincent, Jean-Baptiste Leblond happened upon an English doctor, Mr. Johnston, age thirty. He knew not a word of French, and Leblond not a word of English. "I wished to speak to him in Latin," Leblond wrote, "but we could not understand each other because of the differences in pronunciation." Henceforth, they communicated in written Latin and from then on understood each other perfectly. Leblond stayed with Johnston two years, helping him run the hospital and pharmacy and learning the local island medicine.[29]

Communication across cultures was sometimes ameliorated by the native populations in the Caribbean, who served as active linguists. The Caribs, for example, created "a jargon" through which they dealt with the French in Saint Domingue, which was a mixture of "Spanish, French, and Caraïbe pell-mell [all] at the same time."[30] African slaves in Surinam created a language that served as common currency there called "Negro-English," composed of Dutch, French, Spanish, Portuguese, and mostly English.

The hybrid peoples of the Caribbean—new biological entities formed through contact—also served to ameliorate bicultural communications. Pierre Campet, a physician working in Guiana in the 1760s, was anxious to collect Amerindian information concerning tetanus, a disease that attacked African newborns throughout the Caribbean and dashed European hopes of breeding creole slave populations. Campet, whose 1767 "Traité du tétanos" detailed his treatment of twenty-five cases of tetanus primarily among slaves held by the French there, had been assured by the "Indians" of both sexes that their own newborns never suffered from this terrible disease. Anxious to discover their secret—which he suspected lay in a certain thick balm applied to the umbilical cord—Campet found "an old Indian and his wife" whom he wished to question on this matter but whose language he did not speak. Among a company of Amerindians newly ar-

Figure 2.3. Typical European fortification *(chasteau)* in the Caribbean. This one was in Saint Christopher (Saint Kitts) Island and was occupied by General Philippe de Lonvilliers, chevalier de Poincy, governor of the French Antilles from 1647–1660, for whom the *Poinciana pulcherrima* was named. Laid just behind the castle is its kitchen and pleasure garden, still visible among the ruins today. Inside the wall on the left is a small chapel. The tower outside the wall (on the bottom left) is the armory. Note the slave huts ("Ville d'Angole" on the right) also outside the fortification wall. De Poincy kept six hundred and sometimes more slaves on his various plantations.

rived from the hinterlands, he found a "mulatto" *(mulâtre)* who could translate (the man's mother tongue was no doubt a mixture of creole and Indian languages; his father tongue—like his father—was probably French).[31] Earlier conquest and colonial mixing facilitated, in this case, Campet's medical query.

But problems of communication did not arise simply from lack of knowledge. Charles de Rochefort in 1658 placed the problems of language in the context of war and conquest. "Some of the French have observed," he wrote, "that the Caribbeans have an aversion to the English tongue; nay their loathing is so great that some affirm they cannot endure to hear it spoken because they look upon the English as their enemies." He noted that the Caribs had, in fact, assimilated many Spanish words to their own language, but that this was done at a time when there were friendly relations between the two nations. De Rochefort noted further that the Caribs shied away from teaching any European their language "out of a fear that their [own] war secrets might be discovered."[32] Botanical knowledge exchange may have foundered on similar shoals.

"Noise" in the biocontact zone also resulted from the inflexible theoretical frameworks that made Europeans unable to absorb radically new information. Medical historians have shown that Europeans tended to understand New World medicinal plants predominantly within Galenic explanations of disease and drug classifications. Europeans' understanding of Carib, Taino, Arawak, or transplanted African medicinal herbs was limited to what made sense to them within a humoral context of drugs that would heat or cool the constitution. European naturalists thus tended to collect only specimens and specific facts about those specimens rather than worldviews, schemas of usage, or alternative ways of ordering and understanding the world. They stockpiled specimens in cabinets, put them behind glass in museums, and accumulated them in botanical gardens and herbaria. They collected the bounty of the natural world, but sent "narratively stripped" specimens into Europe to be classified by a Linnaeus or a Jussieu, supporting once again the notion that "travelers never leave home, but merely extend the limits of their world by taking their concerns and apparatus for interpreting the world along with them."[33]

One clash in particular surrounded European encounters with obeah in Jamaica and voodoo in Saint Domingue. Europeans, who increasingly took pride in their empirical methods, tended to ignore, deride, and ridicule the ritual and spiritual aspects prominent in slave medicines. Obeah and voodoo both worked by combining various herbal concoctions with some sort of spirit possession or trance; these were invoked to cure or to induce physical as well as social ills.³⁴ It is especially frustrating when approaching subjects that Europeans reported so negatively that historians have no direct access to these eighteenth-century West Indian practices.

It is an understatement to say that British physicians did not understand obeah. James Thomson, a native of Jamaica schooled in Edinburgh, considered many slave illnesses the result of the expensive "charms" and unreasonable "prejudices" provided by obeah men or women. "A medical man" hardly stands a chance, he wrote, in the management of disorders where obeah is involved. Rather than attempt to understand the "intimate union of medicine and magic" joined in these practices, Thomson attempted to "destroy the hold of the obeah over the patient through reasoning"—but, more often than not, he failed. "It is quite a mistake to conceive that Christianity has abolished its influence," he concluded.³⁵ Even physicians such as Charles Spooner, who appreciated the abilities of the obeah people and used their cures with "great success," limited his research to the herbs they employed and shied away from incorporating the spiritual aspects of obeah into his practices.

Benjamin Moseley, surgeon-general of Jamaica in the late eighteenth century, left a lengthy description of the manner of employing hair, teeth, hearts of bird and livers of mice in the "science of Obi" (he believed obeah and gambling the only instances in which immigrants from "Negro land" combined ideas at all). Although he considered obeah akin to quackery, he was perplexed by its irreversible power. If a slave was "bewitched," he or she surely died; European medicine was as helpless against its spells as colonial laws were to suppress it. According to Moseley, African males—world-class masters in these arts—"could instruct even Frier [sic] [Roger] Bacon and frighten Thomas Aquinas." He judged their skill in consuming humans or beasts in "lingering illnesses" to far outstrip Amerindians' proficiency in poisoning arrows. And, Moseley opined, the knowledge of the obeah-men was superior to that of the Obi-women: men dealt in matters of life and death, whereas women dealt only in the realm of passions and merely tormented inconstant or jealous lovers.³⁶

By the end of the eighteenth century, attitudes in Jamaica had hardened and efforts were made to outlaw obeah. In 1760, the Assembly passed an act making the practice of "Obeah or Witchcraft" a felony; in 1789 the Assembly ruled further that any slave who "pretended" to any supernatural power, "in order to affect the health or lives of others, or promote the purposes of rebellion, shall upon conviction thereof suffer death."³⁷ Other islands followed suit in the first part of the nineteenth century. Despite the Europeans' sense of superiority in medical matters, one should recall that their cures at this time consisted primarily in bleeding, purging, blistering, vomiting, and sweating patients, and often did more harm than good.

In addition to problems involving language, conceptual frameworks, and physical hardships in biocontact zones, Europeans, Amerindians, and slaves each had their own economic interests and cultural aims—all of which further curtailed transcultural exchange. Humboldt complained that his Amerindian guides were interested in trees only as timber for building canoes, and as a result paid little or no attention to their leaves, flowers, or fruit. Exasperated, he exploded: "like the botanists of antiquity they deny what they had not taken the trouble to observe. They are tired of our questions, and have exhausted our patience in turn."[38]

Encounters between Europeans and Amerindians or African slaves were not pure meetings between equals but antagonistic struggles full of exaggerations on both sides. Humboldt derided his "Indian" pilot, who spoke to him in Spanish, for exaggerating the dangers of water-serpents and tigers. "Such conversations are matters of course, when you travel at night with the natives," he wrote. "By intimidating the European travelers, the Indians believe that they shall render themselves more necessary, and gain the confidence of the stranger." La Condamine some years earlier had suggested that the marvelous virtues that Europeans attributed to New World cures had been "greatly exaggerated by both prejudice and ignorance."[39]

Even had conditions been ideal, the enormity of the task of understanding the medical qualities of New World plants was mind-boggling. From a point of near exhaustion in Quito, La Condamine wrote that to detail all of the plants of the Amazon basin would require "years of toil from the most indefatigable botanist . . . and draughtsman . . . I speak here merely of the labor which a minute delineation of all these plants, and their reduction into classes, genera, and species, would require, but if to this were superadded an examination of the virtues ascribed to them by the natives of the country, certainly the most interesting part of a study of this nature, how daunting would be the task."[40]

But conditions in biocontact zones were hardly ideal. Humboldt attributed the exaggerations of the Amerindians to "the deceptions which everywhere arise from the relations between persons of unequal fortune and civilization." The excesses of colonial rule brought a Carib rebellion on Saint Vincent and slave uprisings in Saint Domingue and elsewhere in the 1790s. Fear of poisonings led planters to outlaw slave medicine in English colonies in the 1730s and in the French islands in the 1760s. Even as early as the 1680s, Sloane reported the dangers of herbalizing in islands where "run away Negroes lye in Ambush to kill the Whites who come within their reach."[41]

Secrets and Monopolies

Europeans were often curious about Amerindian and slave medicines and eager to learn, but the indigenes and slaves were less eager to divulge this knowledge to their new masters. Along with miraculous cures came the silence of secrets. Bourgeois in Saint Domingue characteristically remarked that though "the Negroes treat themselves successfully in a large number of illnesses . . . most of them, especially the most skilled, guard the secret of their remedies." A Dutch physician, Philippe Fermin, confirmed that the "Negroes and Negresses in Surinam know the virtues of plants and offer cures that put to shame physicians coming from Europe . . . but," he continued, "I could never persuade them to instruct me."[42] James Grainger, on Saint Christopher's island, found a "maroon negro" who cured lepers using an ingenious cure, but Grainger could not "discover the secret of his art."[43]

The guarding of secrets from colonial aggressors occurred worldwide. In his influential 1577 *Joyfull Newes Out of the Newe Founde Worlde*, Nicolás Monardes told the story of a company of Spanish soldiers patrolling in Peru who were curious about bezoars, gastric concretions used at that time in Europe as antidotes to poison, scorpion stings, worms, melancholy, and plague. The soldiers asked "certain Indians" hired to serve them about the stones, but because the Indians considered the Spanish their enemy, they refused to reveal their secrets. After a while, a twelve-year-old Indian boy, sensing that the soldiers truly wished to know about these stones, showed them that they came from the "stomachs of beasts." Immediately, the boy was killed by his compatriots "for the advice that he gave." Little had changed a century later when Alonso de Ovalle, a Jesuit writing from Chile, reported that "here are many plants of great virtue in physic known only to the Indians called Machis, who . . . are their doctors. Yet they conceal these plants, particularly from the Spaniards. If they do reveal the knowledge of one or two of them, it is a great mark of their friendship." In eighteenth-century Peru, La Condamine commented that "the naturals," as he called the Incas, had guarded the secret of *Cinchona* from the Spanish for some 140 years (some said for more than 200 years).[44]

Naturalists in the West Indies devised various methods for wresting secrets from unwilling informants. Bourgeois in Saint Domingue attempted to win slaves' confidence with friendship. Failing this, he offered money "to be instructed in the details of all that they know" but again without success. Fermin in Surinam, anxious to save the colony the cost of foreign drugs and the malfeasance of ill-intentioned slaves, attempted to learn from the "black slaves" their knowledge of plants, "but these people are so jealous of their knowledge that all that I could do," he wrote, "be it with money or kindness [*caresses*], was of no use."[45]

Europeans also used a number of different tricks when dealing with other Europeans in the West Indies. Nicolas-Joseph Thiery de Menonville tried flattery on the Spanish while at the same time maneuvering to steal their secrets. Landing in Cuba, he let it be known that he was a botanist and had come to collect herbs. Upon hearing this, the people of the country asked if the French did not have plants in their own country. Thiery de Menonville acknowledged that France and its colonies were not deficient in that respect but, playing to "Spanish vanity," added "that the herbs of Havana have the reputation of possessing superior virtues." Europeans also threatened and coerced reluctant informants. Sloane tells how Europeans learned about contra yerva, the potent antidote to poisoned arrows. The story was told to him by an English physician named Smallwood, who had been wounded by an Amerindian's poisoned arrow while fleeing the Spanish in Guatemala. Not having much time, he took one of his own Indians prisoner, tied him to a post, and threatened to wound him with one of the venomous arrows if he did not disclose the antidote. Fearing for his life, the Indian (of an unnamed people) chewed some contra yerva and placed it into the doctor's wound, which healed soon thereafter.[46]

Perhaps the most egregious example of attempts to purchase, beg, or steal medical secrets in the eighteenth century comes from another part of the world—the British establishment in India, where "black doctors" were commonly employed to collect simples for European physicians. Edward Ives, an English physician working there in the 1770s, was much taken with "a poor Portugueze widow woman" who cured venereal disease that even he had pronounced incurable. Ives instructed the East India Company's surgeons to buy the secret, offering the widow a very "considerable premium." She refused, saying that this secret would sustain her family financially in the present as well as in future generations. Unsuccessful with his purse, Ives instructed company surgeons to spy on her, "to watch her every movement." With continuous surveillance, they were able to see her gather a woody shrub known as the "Milky Hedge." Ives conjectured that it must be the juice of this shrub that she used to such effect.[47]

The effort to secure secrets against enemies or competitors was not unique to the vanquished in the colonies. Europeans, of course, had many secrets of their own. The historian William Eamon has discussed medieval and early modern literary texts that purported to reveal "secrets of nature." These texts supposedly held esoteric teachings associated with nature's occult forces. Many of these books did indeed divulge recipes, formulas, and "experiments" associated with various arts and crafts, such as instructions for making quenching waters to harden iron and steel, recipes for mixing dyes and pigments, cooking recipes, and practical alchemical formulas used by jewelers and tinsmiths.[48]

More commonly, trade secrets protected profitable knowledge all across Europe. In the medical domain, physicians and apothecaries often protected their remedies by keeping their recipes secret until they could sell them for a good price. In a celebrated case, the apothecary Robert Talbor of Essex (1639–1681) garnered fame and wealth from his "marvelous secret" that cured fevers. His *remède de l'Anglais* secured him a knighthood in England and an annual pension of 2,000 livres from Louis XIV, enough to live like a rich nobleman.[49]

The great trading companies of the early modern period guarded their investments through scrupulously protected monopolies. Thunberg, traveling with the Dutch East India Company, told how that company held a monopoly on the spice trade and on opium in this period: "If anyone is caught smuggling," he warned, "it always costs him his life, or at least he is branded with a red hot iron and imprisoned for life." Although many naturalists caught passage on company ships, the Dutch East and West India Companies cautioned them not to reveal too many of their findings in their publications. The French Compagnie des Indes also admonished authors to limit published information. The French company blocked British efforts to buy Michel Adanson's papers concerning the natural history of Senegal. After the British took the West African port of Saint Louis in 1758, few French academic papers dealing with this part of the world were published. Adam Smith pointed out that a monopoly granted to a trading company had the same effect "as a secret in trade or manufactures." And historian Francisco Guerra noted that "medical science has been the heritage of the human race, but the drug trade has been throughout history the object of economic monopolies" and (thereby) substantial profits. Companies, of course, still guard secrets today. Through patents, U.S.

pharmaceutical companies seek to recoup the average $802 million typically required to develop a new drug.[50] Commercial exchange and desire for profits both by individuals and joint-stock companies countermanded emerging practices of free scientific exchange in the seventeenth and eighteenth centuries as today.

Drug Prospecting at Home

Beginning in the late seventeenth century and throughout the eighteenth century, academic physicians prospected for drugs inside Europe itself, using techniques similar to those employed by their colleagues in the colonies. From Sweden, Linnaeus remarked, "it is the folk whom we must thank for the most efficacious medicines, which they . . . keep secret." In England, Joseph Banks's interest in botany was kindled when, as a youth in the 1750s, he watched women gathering "simples" for sale to druggists. In encounters strikingly similar to those in the West Indies, European medical men cajoled their countrymen and women into disclosing the secrets of their indigenous cures. Sometimes persuasion, other times the power of the purse yielded the secret of some purported magically efficacious ingredient. As with colonial plants, physicians began testing on the basis of ethnobotanical clues and then later (sometimes) published the results. Thomas Sydenham provided a popular rationale for these new practices: any "good citizen," he wrote, in possession of a secret cure was duty-bound "to reveal to the world in general so great a blessing to his race." Medical experiments were to benefit the public good (not only physicians' purses) for, as Sydenham continued, "honors and riches are less in the eyes of good men than virtue and wisdom." Not everyone would have agreed.[51]

As in the colonies, much bioprospecting at home was sponsored by the crown. The introduction of an "admirable essence" known as "styptick" into England in the 1670s followed common procedures. An unidentified "curious person" traveling in Paris heard about a Mr. Denis's secret formula for a "Styptick" that miraculously stopped bleeding "without any need of binding up the wound." The efficacy of Denis's wonder balm was demonstrated many times, first upon dogs "of whom was cut the crural and carotid arteries and the thigh itself." After applying the styptic, all bleeding soon stopped, and the wound healed "without any scar, suppuration, or cicatrices." Thereafter, "trials" were made on humans, whose arteries were opened, or whose hands or faces had been cut "for the purpose of the experiment." The styptic succeeded as well with them as it had with the dogs.[52]

After successful trials in France, the astringent was sent to Sergeant Richard Wiseman in England. Wiseman had occasion immediately to employ the wonder drug: a patient upon whom he had operated that day was brought back to him with uncontrolled bleeding, "wetting almost a whole sheet" during the coach ride. The offending vessel lay so deep in the neck that it was hard to reach. Wiseman "dipped two pladgets in the Liquour aforesaid and thrust them into the two orifices whence the blood came." The blood was immediately stopped, and the neck dressed without any considerable bandage. Wiseman employed the styptic that same day to stop the bleeding in a young woman who had undergone a mastectomy. Again, it worked successfully, "the blood continu'd stanched, and the Mouth of the arteries remain'd closed."

These successful trials brought Mr. Denis himself to London, where he demonstrated his substance in a session at the Royal Society using a dog and two calves. At this point the king purchased the secret of the formula and had "a Quantity" prepared in his own laboratory. The English brew worked well, stopping the bleeding of wounds cut in the legs of three calves "to the Admiration of all the Spectators." Further tests were made on two patients at Saint Thomas's Hospital: each patient—a woman and a sailor—had had a leg amputated. Rather than the usual excruciating cauterization of the remaining limb, the French styptic was used, leaving the patients "free from pain." Thus was the testimony of the physicians and surgeons sent by the king to witness the operations. "All did acknowledge," the report read, that the patients could not "look more fair and ruddy."

Within the year, the substance known as the "Royal Styptick Liquor" was ready for wartime use. In skirmishes with the Dutch, the styptic was used experimentally in the field by surgeons in the service of the Earl of Offory, Sir Edward Spragg, Sir John Berry, and others, "with admirable success."[53]

Money was often at stake as quacks and legitimate healers alike attempted to make their fortune. Sloane walked a fine line between respecting the monetary security bound up in secrets and advancing the emerging ethic of making effective cures available to the public for the "welfare of mankind." Toward the end of his life Sloane published *An Account of a Most Efficacious Medicine for Soreness . . . of the Eyes* (a secret remedy he had purchased many years earlier); the fifty-four page booklets he sold for six pence each. In England, as in Jamaica earlier in his career, Sloane was ever watchful for "real cures" wherever they might be found. Sitting at the center of England's medical empire, he received unsolicited "receipts" from all over the world. Other recipes he actively sought out. Such was the case with Dr. Luke Rugeley's medicine for ophthalmia. Avoiding direct contact with the doctor, Sloane attempted to procure the secret of his eye balm from a "very understanding apothecary" who knew both Rugeley and Sloane—but the apothecary either did not know it or was not talking. After Rugeley's death, Sloane somehow got hold of all his books and manuscripts, including his *Materia medica,* but still "all in vain." Finally, a "person" who had made the medicine for Rugeley sold Sloane the secret of the eye liniment on the condition that Sloane would not divulge it until after the unnamed person's death (because he or she was still making a living by it). After seeing the recipe, which consisted of tutty (a zinc oxide), lapis hematite, aloes, and "prepared pearl" mixed in viper's grease or other fat, Sloane realized that it had originally belonged to Matthew Lister, a member of the College of Physicians, who, Sloane presumed, had communicated it to Dr. Thomas Rugeley, Luke Rugeley's father.[54]

Sloane wrote his booklet in the standard form of the day, supplying the recipe for the eye balm; reporting on his own experiments and experiences, the findings of other physicians, and proof of effectiveness of the balm from as far away as the East Indies; and discussing the use of the balm in treatment. As was (and often still is) the case, success in medical treatment was probably over-reported and failure under-reported (Chapter 4): Sloane's balm was said to be so effective that it cured an impossible 500 out of 500 cases! He cautioned that a patient required careful preparation, which consisted, among other things, in "bleeding and blistering the neck behind the ears in order to draw off the humors from the eyes."[55]

By 1745, Sloane felt that a physician of his standing had to account for holding such a useful remedy secret for so many years. He laid out the issues, as he saw them, in some detail. He should not be accused of withholding valuable information from the public. In this case, he had "religiously kept 'till now" his vow of secrecy that he had been obliged to swear in order to acquire the recipe. Neither should he be charged with making a "mystery of [his] practice." Unlike some other physicians of "good morals and great reputation," he had not willfully "concealed" or "monopolized" genuinely helpful medicines.[56] Sloane's apology makes clear that by mid-century, if not earlier, an ethic of public usefulness was supplanting a physician's right to secrecy in medical cures.

Physicians cast their nets widely in the search for effective cures at home. Published instructions encouraged travelers inside and outside Europe to question and learn from people of all stations and sexes—from statesmen, scholars, and artists as well as from craftsmen, sailors, merchants, peasants, and "wise women." Women in the sixteenth and seventeenth centuries were still widely acknowledged healers. Upper-class ladies were routinely educated in medicine; Thomas More's daughters in England, for example, were educated in religion, the classics, and practical medicine—the distilling of waters and other chemical extracts along with the use of minerals, herbs, flowers, and plants. Diaries and books on housewifery reveal that women who administered large households routinely dressed wounds, administered medicines, distributed herbs from their gardens, and attended childbirth. Several of these women left books of family recipes. Among the lower classes, women served as unlicensed healers of all sorts. In 1560, surgeon Thomas Gale estimated that sixty women were actively practicing medicine in London (in a population of some 70,000). Of the 714 unlicensed practitioners prosecuted by the College of Physicians of London between 1550 and 1640, more than 15 percent were women. These women treated males and females alike and often, like Joanna Stephens, discussed below, specialized in a particular kind of cure. Female practitioners generally purchased their drugs from apothecary shops and were distinct from the herbwomen who supplied those shops.[57]

The eighteenth century was an important swing century for women in medicine and science. The freer structures of early modern European science had allowed women to practice midwifery, supply medical cures, and practice a variety of sciences, including astronomy and physics. During the course of the eighteenth century, women healers' requests for admission to university—increasingly the route required for work in medicine and science—were denied. Nonetheless, physicians were anxious to gather the knowledge that many women held. This included their traditional cures. In the sixteenth century, Carolus Clusius, who worked in Vienna and Leiden, praised country "women root cutters" (*rhizotomae mulierculae*), who supplied him with information about the medical properties of plants and the names of indigenous varieties. The fashionable seventeenth-century London physician Thomas Sydenham declared, "I know an old woman in Covent Garden who understands botany better [than any academic]."[58]

The process of collecting information from women "root cutters," old women, or a particularly successful woman healer was strikingly similar to that of prospecting abroad. The major strategy was to buy the cure and often the government put up the money. One woman who did well for herself was the spinster Mrs. Joanna Stephens, daughter of a gentleman of good estate and family in Berkshire, England. She was paid £5,000 on March 17, 1739, by the king's exchequer for her drug—an eggshell and soap mixture reported to dissolve bladder stones. Her cure for this "painful distemper" was highly prized because the only other option was surgery, known as "cutting for stones." The London surgeon William Cheselden was able to extract stones in less than one minute (a procedure he perfected on dead bodies); the operation was nonetheless excruciating and dangerous.[59]

According to the records from the case, Stephens had a knowledge of medicines and was skilled in preparing them, though we are not told how she was trained. In her narrative, she reported that "about twenty years ago" she "accidentally" came upon a recipe for "the stone" consisting of eggshells dried in an oven, which she administered to several persons. Over the years she made "trials," experimenting with burning the egg shells and with adding various quantities of soap. The fame of her cures grew, culminating in her successful treatment of Edward Carteret, the postmaster general. By 1737, her cure was so celebrated that the speedy publication of its secrets was judged "of great importance to mankind." At this time, her supporters attempted ("with her consent") to raise from the general public £5,000 "as a reward for her discovering the medicines," but this failed, and a proposal for payment went instead to the House of Commons.[60]

In early 1739, Stephens herself petitioned the House of Commons for payment. Her petition was made into a bill, which passed both Houses and won royal assent on June 14th of the same year. Before payment was made, however, her medicines were to be submitted to the "examination of proper judges." A board of trustees, appointed by Parliament exclusively for this purpose, had the remedy prepared and administered to four men (ranging in age from 55 to 79) who had long suffered from stones. About one year later, the men were examined by various surgeons, physicians, and apothecaries, each of whom verified that each man had had a stone before the medicine was administered and that no stone was now to be found. These witnesses appeared before twenty-two men of a Parliamentary committee that included the Lord Archbishop of Canterbury, the Lord High Chancellor of Great Britain, various dukes, reverends, doctors, and surgeons. All but two (who still had doubts) certified the efficacy of Stephens's medicament.[61]

Stephens received her rich reward and "communicated her method of preparing and giving her medicines" along with "every minute caution and circumstance" that she had observed in her years of working with it. It was now assumed that her cure could be even more successfully applied by "persons learned in the science of physick."[62]

A similar case occurred in France. A Swiss surgeon's widow, a Madame Noufer, acquired great celebrity by employing a secret remedy to expel tapeworms. The secret was thought of such importance that some of the principal physicians in Paris received royal approval to make a complete trial of its efficacy. When satisfied that the drug worked as advertised, the recipe was purchased by the French king and published by his order.[63]

Most of the women whose cures were eventually adopted and published in the various European *Pharmacopoeia* remained nameless, as was true also of most of the West Indian indigenes and slaves who offered cures. The story of the development of digitalis provides a celebrated example of a medicament still in use today that came originally from an unnamed "old woman." The Birmingham physician William Withering told that he learned from "an old woman" that the leaves of *Digitalis purpurea* were useful for treating dropsy (edema, an abnormal accumulation of watery fluid in the body). According to Withering's account, "in the year 1775, my opinion was asked concerning a family receipt" that was kept secret by the woman, "who had sometimes made cures after the more regular practitioners had failed." "This medicine," he continued, "was composed of twenty or more different herbs; but it was not very difficult to one conversant in these subjects to perceive, that the active herb could be no other than the Foxglove." Withering then embarked on a decade of testing in order to find the most effective dosages (too much of the herb induced vomiting and violent purging). It is worth noting that he experimented first on the sick poor, whom he treated gratis one hour per day (he treated between 2,000 and 3,000 of these patients annually). "I found the Foxglove to be a very powerful diuretic," he recorded, "but then, and for a considerable time afterwards, I gave it in doses very much too large, and urged its continuance too long." Over the years he perfected his dosage and learned how to ensure that the powder he administered was of uniform potency. He also experimented with mixing in small doses of opium to avoid excessive purging. Only after he had developed a safe and pleasant form of the drug did he prescribe it for his paying patients. Since Withering's day, more than thirty different cardiac glycosides, including digitoxin, digoxin, and digitoxigenin, have been isolated from *Digitalis purpurea*.[64]

As with informants in the West Indies, we have little access to the women's own reactions to their encounters with academic European naturalists. The historian Lisbet Koerner has highlighted an article in a 1769 Stockholm magazine purporting to give voice to "wise women." It was not uncommon in the eighteenth century for articles such as this to be written by men under female pseudonyms. Nonetheless, the wise women's assertions echo complaints that appear in other women's writings from this period. The women noted their "joy and pleasure" when a physician, standing by a sick child's bed, ruled that "here no other help can be had than that of finding an experienced wise woman." They complained that physicians "exert themselves to both smell and taste our pouches, creams and bandages," attempting to divine the secrets of the medicines. The women, like so many at the time, ended by asking to be admitted to professional training, in this case to Stockholm Medical College, "for we are after all considered as highly as the gentlemen [physicians] in the homes into which we are called."[65]

Swedish (male) physicians offered to purchase the wise women's secrets, but, like other folk healers, these women usually would not sell them because their livelihood, and often that of their descendants, depended on monopolizing the cure. Mr. Ward, for example, who developed Ward's pills, would not divulge the recipe "because it would be worth so much yearly to him and his successors."[66] Thus a conflict arose between irregular healers (and Sloane classed some London physicians in this category), who depended on their secrets for their livings, and more academically trained physicians, whose research was conducted for the "benefit of humankind"—or at least such was the ideology. Patents, developed in subsequent centuries to safeguard university and company investment in drug development, tempered this ethic.

Not everyone extolled the virtues of the herb women or folk healers. In efforts to tighten their own monopoly on healing, the Royal College of Physicians in London discussed in 1583 how to prevent "itinerant and inexperienced old women from practising medicine." For many years the college continued prosecuting herb women whom they described as "old," "poor," "little," "ignorant," "bold," "blind," and "stubborn." In 1632 Thomas Johnson, a London apothecary, inveighed against the "greedy and dirty" old herb women of the Aldersgate and Broad Street green markets who allegedly "thrust" any old root on druggists and doctors. A century later, Patrick Blair, a fellow of the Royal Society, similarly attacked the "pittiful filly Herb-women" who "cheat" the poor and ignorant apothecaries by "imposing" upon them any herb they pleased, without control. But the herb women did not budge; records of the Fleet and other markets show women selling "physick herbs" throughout the first half of the eighteenth century. What is remarkable in Johnson's and Blair's statements is that herb women were primary suppliers to many apothecary shops and that prescriptions by college physicians ultimately depended on these women's knowledge of roots.[67]

Brokers of International Knowledge

Here I want to digress for a moment from my focus on bioprospecting in the West Indies and Europe to take an example of the transfer of a medical

Figure 2.4. Lady Mary Wortley Montagu in Turkish dress. Montagu was instrumental in introducing the smallpox inoculation into western Europe in the early eighteenth century. The artist has drawn her with eyelashes, although hers had been taken by smallpox.

technology from Central Asia into England, and eventually into Europe and its overseas colonies. What interests me about the story of the introduction of smallpox inoculation into Europe is the role played by Lady Mary Wortley Montagu—who, in this instance, might be styled an international broker of women's knowledges.[68] What is of particular interest is who is credited with having introduced the new knowledge, and who is not.

Inoculation was introduced into Europe at the beginning of the eighteenth century with the upsurge in smallpox epidemics that devastated people of all classes, including the royal families of England and France (Queen Mary died in the epidemic of 1694 and the Dauphin of France in 1711). Reports of the practice of opening the "pustules of one who has the Small Pox ripe upon them" and "engrafting the virulent matter" into another person came from China, the Barbary coast of Africa, and, most impressively, from Constantinople.[69] It was here that the practice of transplanting smallpox by inoculation caught the attention of well-educated and influential Europeans.

Since the eighteenth century, historians have fought over who should wear the laurels for introducing smallpox inoculation into the West, whether it should be the learned graduates of Padua university, Emanuele Timoni and Jacob Pylarini, whose reports on smallpox inoculation were published in the *Philosophical Transactions*, Lady Mary Wortley Montagu, who created interest in the operation in England through the successful inoculation of her own son, or Sloane, the royal physician, and Charles Maitland, the royal surgeon, who oversaw the first inoculations in England.[70]

Genevieve Miller, a historian who wrote a comprehensive account of this chapter in history, credited Sloane for having overseen the scientific aspects of the transfer of this procedure into England. Having returned from the West Indies interested in exotic medicines, Sloane requested information about the safety of the procedure from Timoni, a Greek physician, who published the first scientific paper in Europe on the procedure in 1714. Miller found accounts of Lady Mary's contribution in this process "exaggerated," and was keen to put her "in her place," as Miller declared in an article of the same title.[71]

Miller's attempt to deflate Lady Mary's reputation in this regard responded to eighteenth-century accounts—by Voltaire, La Condamine, and Maitland—that, perhaps in deference to her class standing, celebrated Lady Mary's role in the introduction of this lifesaving procedure. La Condamine, for example, characterized Montagu as "a tender mother" concerned for the safety of her children and celebrated her influence with the Princess of Wales, a woman of "genius superior to that of the rest of her sex," who set a powerful example for all of Britain by having her own family inoculated (1722), "by which means her royal highness preserved not their lives only, but the beauty, of some of the most amiable princesses in Europe."[72]

Charles Maitland, surgeon to the Montagu family while they were stationed in Andrianople, where Lady Mary's husband Edward served as ambassador for the English Levant Company, also credited his former patroness, Lady Mary, for having brought the inoculation for smallpox to Europe. Lady Mary herself had been tragically struck by smallpox (her eyelashes were gone and her face deeply pitted) and her brother had died of the disease. In 1717 both Lady Mary and Maitland began their inquiries, and "thoroughly convinced" that it was safe, Montagu resolved to have her only son, aged about six years, inoculated. Delighted with the success of the operation, she asked that her newborn daughter also be inoculated against "the fatal distemper." Maitland refused, and convinced her to wait until the girl was older and after they had returned to England—hoping, as he wrote, "to set the first and great example to England, of the perfect safety of this practice, and especially to persons of the first rank and quality."[73]

All of these discovery narratives focus on learned Europeans as conduits for information traveling from Constantinople into Europe. Rarely mentioned in this regard are the old Greek women who kept the practice of inoculation alive in the Mediterranean. One of these shadowy figures reputedly inoculated 4,000 men, women, and children during the 1706 epidemic in Constantinople and the outlying areas. (Some accounts have her inoculating 6,000 people.) Known to us only as a "Greek woman from Morea," she performed the technique primarily on persons of the lower classes, but, as epidemic threatened, inoculation came into vogue also among noble families, and especially those of the English, Dutch, and French merchants residing in the area. Many years later, La Condamine praised this woman's success and attributed it to her precision and care in the operation.[74]

This Greek woman must be the same "old woman"—with her needle and nutshell (used to cover the incision)—whom Lady Mary asked to inoculate her son. The old woman did not invent "sewing the pox" but rather kept ancient knowledges alive in her practices. The origin of engrafting against smallpox is not known, but, according to La Condamine's

1754 *Histoire de l'inoculation,* it was practiced from ancient times by many peoples around the world. From China, Father Dentrecolles related that smallpox was "sewn" by thrusting cotton impregnated with the matter of the dried pustules reduced into powder into the nose of children. From Circassia (on the northeast shore of the Black Sea) and Georgia, the technique was said to have been "practiced from time immemorial" by peoples anxious not to blemish their reputed beauty. It should be remembered that Georgian beauty—especially among females—was of such mythic proportions as to inspire both high prices for women from this region sold into Turkish harems and the coining of the racial term "Caucasian," devised by the great Johann Blumenbach in the late eighteenth century to honor the Caucasus as the cradle of humanity and the people of that region as the most perfect and thus original humans. From the similarity in practices across Central Asia and Africa (the use of three pins tied together to make punctures in the skin), La Condamine conjectured that the practice was taken from Circassia into Egypt by slaves sold into Cairo's militia. From there it was thought to have traveled into Tripoli, Tunis, Algeria, and the interior of Africa. From Africa, slaves carried the knowledge across the Atlantic to the Americas, several eventually imparting it to their masters. Apparently unrelated, "buying the smallpox" (for two or three pence) was said to have grown up independently in Wales. Here "schoolboys" reportedly gave each other smallpox by rubbing pustules from a sick boy's arm on a well boy's until blood appeared.[75]

In addition to the inoculatress of Andrianople, La Condamine noted other Greek women who practiced the art—one from Philippopolis, the other from Thessalonica. According to La Condamine, these "inoculatresses" mixed good skill with "quackery" and "superstition." The Thessalonian woman claimed that her knowledge came from the Holy Virgin, and consequently she made her incisions in the form of a cross in eight different places on the body, including the forehead, cheeks, chin, both wrists, and feet.

Many physicians in England and New England objected to the prominent role women played in inoculation. The ardent anti-inoculist William Wagstaffe, a physician at Saint Bartholomew's Hospital in London, was aghast that "an Experiment practiced only by a few Ignorant Women, amongst an illiterate and unthinking People shou'd on a sudden, and upon a slender Experience, so far obtain in one of the Politest Nations in the World, as to be receiv'd into the Royal Palace." In New England, Zabdiel Boylston also recommended that this important procedure not be left to "old women and nurses" but be taken over and managed by "good physicians and surgeons." Only if physicians were not available were women to be allowed to inoculate.[76]

If old Greek women had their superstitions, so did European physicians. Physicians taught that in addition to being in good health, patients had to be "prepared" with numerous bleedings and purges before undergoing inoculation. Lady Mary, who disliked medical men, opposed their desire to monopolize (for personal gain) such a "useful invention" as engrafting. She feared her resolve to introduce inoculation into England in its simplest form would place her at "war" with the medical profession. Perhaps for this reason Maitland, her surgeon, emphasized that his experiments with inoculation were intended to benefit the "Publick Good" and not his own pocketbook; he did not wish, he remarked, to retain a monopoly on the procedure by keeping it secret. Secrecy was, however, incidental to Maitland's ability to profit from the new procedure. As royal surgeon, he received the generous sum of 1,000 pounds sterling for inoculating Prince Frederick in Hannover.[77]

Historians of science often present the creation of new knowledge as an unencumbered search for truth characterized by open communication of the best results. As we have seen, however, conditions were less than ideal in European biocontact zones at home or abroad. Mortality was high; informants were unwilling. The search for potential lifesaving new drugs was mired in relations of conquest, commerce, and slavery. European colonial expansion depended upon and fueled the search for new knowledge in the realm of tropical medicines. At the same time, colonialism bred dynamics of conquest and exploitation that impeded the development of this knowledge.

It is difficult to resolve who should be honored with having introduced inoculation into Europe. Lady Mary was one of the few women who served as a conduit for a practice that eventually became part of standard medicine all across Europe and its colonies. As we shall see in the next chapters, despite her vivid reports concerning the West Indian use of abortifacients, Merian was unsuccessful in introducing them into Europe.

3

Exotic Abortifacients

> When I look to gather fruit, [I] find nothing but the savin-tree, too frequent in our orchards, and there planted by all conjectures, to destroy fruit rather.
>
> THOMAS MIDDLETON, 1624

> Such miserable creatures lose the natural desire to have children. Mothers, going against all natural instinct, kill their children to protect them from cruelty. Unhappy slaves listen less to the cry of nature than to their hatred for those who oppress them.
>
> ANONYMOUS, 1795

For Europeans in the seventeenth century, importing exotics from Europe's West Indian colonies was big business. Already in the 1680s, sugar, letterwood (so named for its black spots that resemble hieroglyphics), cotton, tobacco, indigo, pimento, various gums and resins, and Amerindian commodities such as hammocks and cassava promised substantial profits.[1] By 1688, Surinam was exporting 7 million pounds of sugar annually. A century later, the French colony of Saint Domingue produced sugar worth 227 million livres per year.

Commerce in medicines was a trickle by comparison, yet it, too, was significant. In the 1760s, trade in Saint Domingue medicines such as guaiacum (along with the mahogany and logwood listed in the same category) amounted to 14,620 livres.[2] Were herbs used to concoct exotic contraceptives and abortifacients among the precious cargoes brought into Europe? Was trade in such plants lucrative or desirable? In a sense, savin—one of Europe's leading abortifacients—was an exotic, brought northward into Europe from the Mediterranean basin. Very early on it was cultivated in gardens all over Europe, readily available and inexpensive. Perhaps there was no need to import new anti-fertility drugs from Europe's tropical colonies? But Europeans continued to search everywhere for the best medicines the world had to offer, and many people, then as now, were simply taken with exotics. To what extent did naturalists search for new anti-fertility agents?

106 | Plants and Empire

Figure 3.1. A sugar mill circa 1660. The cane fields (#5) and mill powered by oxen (#1) are shown in the background, stoves and boilers used to boil the juice (#2) are in the foreground. Sugarcane stalks are first squeezed through rollers (#1). The juice then runs down a trough to a settling vat and on to four boiling pans, heated by four wood ovens (#2). The concentrated syrup is then ladled into conical forms (#3), which are later dried (not shown). A European overseer keeps the slaves moving with his stick. A slave hut (#10) is pictured in the lower right corner.

Merian's Peacock Flower

My attention was first drawn to the topic of abortion by Maria Sibylla Merian's vivid report of slave women (in this case both Amerindian and African) aborting their offspring, which appeared in a book on the metamorphosis of caterpillars. Merian immediately placed abortion within the context of colonial struggles, identifying the killing of slave progeny as a form of political resistance. According to Merian, slave women killed the children in their wombs for the same reasons that many of them hanged themselves from trees or ingested deadly poisons—to find release from the insufferable cruelty of New World slave masters.

Did naturalists introduce into Europe any of the exotic abortifacients they encountered in the Caribbean? To address this question I have chosen to trace the history of Merian's peacock flower, known today by its Latin/Linnaean name as the *Poinciana pulcherrima* (L) or *Caesalpinia pulcherrima* ([L] Sw). The French call this plant the *fleur de paradis;* English speakers know it as the Red Bird of Paradise, Dwarf Poinciana, Flower Fence, and Barbados Pride (it is the national flower of Barbados and adorns that country's coat of arms along with the bearded fig tree). The peacock flower grows profusely in Florida, Central and South America, India, and Africa, where in all these places it is still sometimes recognized as an emmenagogue (an agent that induces the menses) and an abortifacient. In some cases it is the flowers, in others the seeds or bark that are considered the effective part. Like most plants, the peacock flower has many uses. In Guatemala and Panama, for example, the leaves are used to poison fish and the seeds employed to execute criminals. The plant is also used as a remedy for sore throat, lung disease, fever, eye and liver complaints, constipation, and skin rashes, and for making black dyes and ink (allegedly "the most beautiful black ink in the world").[3] The brilliant beauty of its flowers has also made it a popular ornamental.

I have focused on Merian's peacock flower because of her moving remarks and also because the plant's abortive qualities were discovered independently in the West Indian colonial territories of the Netherlands (by Merian), England (by Sloane), and France (among others by Michel-Étienne Descourtilz, who confirmed Sloane's and Merian's findings with firsthand observations of his own). Sloane, working in Jamaica some years before Merian's travels in Surinam, discussed a plant he called the "flour fence of Barbados, wild sena, or Spanish carnations" and identified it later in his published work as the same plant as Merian's peacock flower. A perusal of his specimen in his herbarium, the largest extant collection of plant specimens from the pre-Linnaean period (now housed in the Natural History Museum of London), indicates that it is indeed the same plant. The history of the peacock flower, like that of other plants in this period, is fraught with ambiguities: it is not clear that the name always refers to the same plant or that varieties have similar medicinal virtues. Specimens labeled with this name, collected from all parts of the world and held in the Laboratoire de Phanérogamie in Paris, vary in their appearance and characteristics.[4]

The very first European notice of the peacock flower was recorded by General Philippe de Lonvilliers, chevalier de Poincy, governor of the French Antilles from 1647–1660 (see Chapter 5). Poincy, not a naturalist but a military man concerned with the health of his troops, was taken with the plant's ability to cure fevers and dosed himself "to good effect." It was not until much later that Descourtilz, a French doctor sent by the government to Saint Domingue in 1799, highlighted the use of the *poincillade* as an abortifacient. This beautiful thorny shrub, he wrote, is cultivated in some gardens in Europe but grows naturally in the Antilles. He reiterated other French physicians' reports that it was useful in treating lung ailments and fevers. After detailing its chemical properties and medical preparations, he added that a strong dose of the flower (not the seed, as reported by Merian) could be employed to induce the menses, but that it must be used with extreme caution: "ill-intentioned Negresses," he added, "use it to destroy the fruits of their guilty loves."[5]

What is of interest is that Merian and Sloane each independently collected this plant as an abortifacient. Although Sloane cited Merian in his *Voyage*, he did not learn about the abortive qualities of this plant from her (or vice versa); her *Metamorphosis* is cited only in an appendix which included things that Sloane learned after the body of his *Voyage* had been written. Sloane and Merian never had a scholarly exchange, although one assumes that Sloane saw the copy of her *Metamorphosis* purchased in 1705 by James Petiver, a fellow of the Royal Society of London when Sloane was president. Although the flower had been brought into Europe from the East Indies long before either Merian or Sloane traveled to the West Indies, Merian's is the first European report of its abortive qualities (see Chapter 4) that I have found.

Although Merian, Sloane, and Descourtilz all mentioned abortifacients, they placed the peacock flower in very different social contexts. Merian and Descourtilz both located it within the colonial struggle; she emphasized the importance of this plant for the physical and spiritual survival of West Indian slave women, while he stressed the "ill intentions" of the "Negresses" who used them (see below). I examine first Sloane's experience in Jamaica with the plant he called the flour fence.

John Stedman gave firsthand accounts of the extreme brutality slaves in the Caribbean endured. In Surinam in the 1770s, he observed a "revolted negroe" hung alive upon a gibbet with an iron hook stuck through his ribs, two others chained to stakes and burned to death by slow fire, six women broken alive upon the rack, and two slave girls decapitated. Slaves in French holdings were treated no better. The *Code noir* of 1685, celebrated at the time for its "humanity," required that fugitive slaves at large for a month have their ears cut off and be branded on one shoulder with the *fleur-de-lys*. If slaves escaped a second time, they were hamstrung and their other shoulder was branded. A third offense brought execution. Sloane described how slaves in Jamaica were burned for running away, "by nailing them down on the ground with crooked sticks on every limb, and then applying the fire by degrees from feet and hands, burning them gradually up to the head, whereby their pains are extravagant." For crimes of a lesser nature, a foot might be chopped in half with an axe. For negligence, slaves were whipped by the overseers "with lance-wood switches" after which pepper, salt, or even melted wax were poured into the open wounds "to make them smart." John Dalling, a governor of Jamaica in the 1770s, estimated that "the Spaniards treat their slaves better than we do; we treat them better than the French; and the French treat them better than the Dutch."[6]

Although Sloane was well aware that slaves "cut their own throats" to escape such treatment, he did not see his "flour fence" in this light. Sloane wrote rather dryly, "it provokes the Menstrua extremely, causes Abortion, etc. and does whatever Savin and powerful Emmenagogues will do." Sloane placed his discussion of abortive qualities of his flour fence in the context not of the colonial sufferings but of the growing conflict between doctors and women seeking assistance in abortion. In this instance, Sloane carried fully-formed notions concerning abortion with him to Jamaica; his attitudes toward abortion mirrored those of the majority of his male medical colleagues in Europe. Concerning his service as physician to the governor in Jamaica, he wrote:

> In case women, whom I suspected to be with Child, presented themselves ill, coming in the name of others, sometimes bringing their

own water, dissembling pains in their heads, sides, obstructions, etc. therby cunningly, as they think, designing to make the physician cause abortion by the medicines he may order for their cure. In such a case I used either to put them off with no medicines at all, or tell them Nature in time might relieve them without remedies, or I put them off with medicines that will signifie nothing either one way or other, till I be furthered satisfied about their malady.

Sloane finished his passage on abortion with a strict warning: "if women know how dangerous a thing it is to cause abortion, they would never attempt it . . . One may as easily expect to shake off unripe Fruit from a tree, without injury or violence to the Tree, as endeavor to procure Abortion without injury or violence to the Mother." Sloane did not discuss the social or political status of the women he treated in Jamaica in this regard, whether they were English, creole, or slave. Rather he accused "dissembling" women in general of seeking abortions from unsuspecting doctors. His attitudes were shared by many European physicians at this time. The German physician Johann Storch also reported "tricking" a pregnant woman, whom he suspected to be seeking an abortion, by prescribing only a mild laxative. Some physicians claimed that women even endured inoculation against smallpox, hoping that the operation would cause an abortion. Warnings to midwives, physicians, and apothecaries about giving unmarried women medicines that might induce abortion date to at least the sixteenth century.[7]

European physicians who discussed abortion in the seventeenth and eighteenth centuries emphasized its dangers. It is possible that this correctly reflected their experience, since (male) physicians were generally called to attend women only when things went wrong. Sloane himself noted that when an abortion was absolutely necessary, he preferred "the hand" to herbal preparations. Following this ancient method dating back to the first century A.D. and perhaps further, a physician positioned a woman on her back across a bed where she was to be held down by three women with her knees pushed up to her chest (as advised by the seventeenth-century Parisian master-surgeon, François Mauriceau, who wrote extensively on birthing and female maladies). The doctor, sitting on a stool, anointed his hand with oil, fresh butter, or unsalted lard, and "gently" introduced his fingers, one by one, through the cervix and into the uterus until the whole hand slid inside. Herman Boerhaave suggested giving the woman opium to relax those parts. Once the physician's hand was in the womb, he broke the membranes, took hold of the feet of the fetus, and "pulled it away." Next he separated the placenta from the womb with his fingers and extracted it. Some physicians suggested that in the first few weeks of pregnancy, one finger, bent like a "blunt hook," would suffice to draw the embryo from the womb. European physicians also suggested other nonherbal methods to induce abortion, including excessive bloodletting, various douches, vigorous jumping and horseback riding, and applying pressure to the main artery in the thigh (a technique known as the Hamilton method).[8]

Given his disapproving attitudes toward abortion, how might Sloane have procured information in Jamaica about abortifacients? Merian informed her readers that she learned about the abortive qualities of the peacock flower directly from the enslaved women of Surinam. When I began my research, I had conjectured that abortion and contraceptives were women's business and that Merian's report on the abortive qualities of her peacock flower was as unique as her presence in the field. John Riddle, an historian of pharmacy, confirmed my notions in his fine two-volume history of contraceptives and abortifacients in ancient and early modern Europe. Likewise, Edward Shorter, an historian of medicine, argued in his history of women's health care that birth control was women's knowledge. And it is true that birthing in the Caribbean, as in Europe, was generally a female affair. A slave woman typically gave birth in her own hut with the assistance of a slave midwife, perhaps a nurse, and several of her kin or friends. After 1780, large Caribbean plantations offered infirmaries or "hothouses," as many were called, with lying-in rooms for slaves. These hospitals were typically run by a female slave, in the French islands called a *hôpitalière*, usually an older woman no longer able to work the fields. The *hôpitalière* generally took care of birthing and was assisted by several younger nurses (mostly female), a cook, and sometimes an additional midwife, either slave or free. As an observer noted in the 1790s, those attending slave births on plantations were "Negresses." Only if a birth were extremely difficult was a costly surgeon called. Throughout the seventeenth and eighteenth century, local white physicians or surgeons visited the plantations under their care once or twice a week to supervise slave nurses. Planters' wives (when present) might oversee preparations of medicines for the plantation.[9]

Not only were European physicians removed from slave birthing, several physicians serving in the Caribbean mentioned that they did not deal with the diseases of women and children at all. Thomas Dancer, a Jamaican physician, wrote in *The Medical Assistant; or Jamaica Practice of Physic: Designed Chiefly for the Use of Families and Plantations* (1801) that modesty did not allow "the sex" (that is, women) to seek advice from physicians in the West Indies, the majority of whom were "young bachelors." He encouraged women (and particularly the matrons among them) to learn "how to manage themselves in their various situations." Philippe Fermin, working in Surinam in the mid 1700s, further confirmed that physicians and surgeons had "little to do with women . . . who hardly complain of anything but headaches or some constipation." Apart from a terrible suppression of the menses ("which can almost never be cured"), according to Fermin, "the sex" in Surinam did not seek help from physicians.[10]

European doctors in the islands were far removed from slave birthing and so were European midwives, who generally lived in towns. Until the end of the eighteenth century, few European midwives set up practices in the colonies. In Saint Domingue, for example, of the 102 French medical personnel (physicians, surgeons, apothecaries, and the like) working there between 1704 and 1803, only five were midwives and these were typically under the supervision of a male royal physician and *accoucheur* for the colony. All five practiced in Cap Français. One of these midwives, a Demoiselle Renouts, cared for women in her home (an expensive proposition) and also taught "Negresses" from the plantations the art of delivering pregnant women. Charles Arthaud reported, however, that even where French midwives were available, "white" women often preferred the services of "women of color." In some islands, such as Barbados, European (Quaker) midwives were more plentiful. They were at times called to attend slave births, but usually only when a slave woman was not available.[11]

Clearly much information concerning the use of contraceptives and abortifacients passed from woman to woman, neighbor to neighbor, midwife to client. But things were more complicated than they seem. Riddle's own examples show that some men—some learned, some not—were privy to these secrets. In some instances, it was the male partner who provided a woman with a contraceptive herb—but he might hold that information secret so that she could not betray him with another man. In other recorded instances, women went to apothecaries, barbers, or even their priest lovers for contraceptives. Mauriceau noted that "the ancients, Avicenna and Aëtius, taught us many remedies to induce abortion when it is judged necessary." And as we have seen through the examples of Sloane and Descourtilz, European physicians had firsthand experience with abortion.

Certainly Sloane collected reliable information about the abortive qualities of his "flour fence" while in Jamaica. He also treated numerous women during his stay. In his *Voyage,* he discussed cases involving some thirty-eight English women and four African women with various complaints. He also looked in on sick nuns when summoned en route by their abbess on Madeira, the Portuguese-held islands off the coast of northwest Africa. So although many women knew about abortifacients and often employed them, many men were also familiar with these herbs.[12]

I do not want to make too much of the differences among the attitudes of Sloane, Descourtilz, and Merian toward abortion. Merian, to my knowledge, discussed only one abortifacient. Her chief interest was insects, and she described plants primarily as they related to insects (her passage concerning the peacock flower is devoted to the caterpillars that live on the plant's leaves). Whether women "do science differently" is currently a topic of heated debate in feminist theory; distinctions should not be drawn too sharply between individual men and women scientists, however.[13] Many European women—plantation owners' or governors' wives, for example—had little interest in their newly adopted countries, and most came and went without collecting any information from the indigenous populations or cultivating any special sympathies toward the women of the region.

Abortion in Europe

Before exploring further what Europeans knew about abortifacients in the West Indies, I will explore attitudes toward abortion in Europe in the seventeenth and eighteenth centuries. Did Europeans have their own abortifacients? Or was abortion such a "heinous crime" and "abominable sin" that collecting such medicines was out of the question? Did European women need new and exotic anti-fertility drugs from abroad, or did they already possess effective abortifacients?

Today the word abortion generally means ridding a woman of an unwanted pregnancy. When this is done for medical reasons to save the life of the mother, we call it "therapeutic." English speakers today generally reserve "miscarriage" for the spontaneous loss of a pregnancy. These distinctions were not made in early modern Europe, however. The words miscarriage and abortion were used interchangeably for any loss of the embryo, any birth before its time, and could apply to expelling a living or dead fetus. Now as then, "abortion" was also widely used to refer to an arrested growth or fruitless endeavor—as noted in the *Oxford English Dictionary* (*OED*), "the breasts of male beings are aborted teats."[14] The 1694 *Dictionnaire de l'Académie Françoise* also defined *avortement* as an imperfect development or a miscarried plan, as in "the most beautiful trees always produce some abortion," or references to a literary text as "only an abortion." Today we continue to use these terms figuratively, as in a "miscarriage of justice" or an "aborted mission."

Only in the course of the eighteenth century did the terms abortion (*abortus, avortement, Abtreibung*) and miscarriage (*aborsus, fausse-couche, Fehlgeburt*) become more clearly distinguished. Urbain de Vandenesse, author of the short article on abortion in Diderot and d'Alembert's *Encyclopédie*, noted that doctors and surgeons reserved *avortement* for animals and used *fausse-couche* (literally a false confinement to bed) to refer to women, but that physicians did not bother themselves with these distinctions. Mauriceau, writing nearly a century earlier, had rejected those who wished to reserve "abortion" to refer only to animals. He applied the term abortion—accidental and induced—to women because for him miscarriage (*fausse-couche*) referred to the expulsion of a *faux-germe* or a "mole," a mass not yet human issuing from the womb. De Vandenesse suggested that before the second month of gestation anything issuing from the uterus was merely a *fausse-conception* or *faux-germe* and not really a human creation of any kind at all; but after the second month, any interruption of pregnancy was known as an *avortement*. An abortion could also be known as a *fausse-couche forcé*. Robert James, author of the *English Medicinal Dictionary* (1743), remarked that there was no "foundation" for the distinction between "abortion" (*abortus*) and "miscarriage" (*aborsus*), because these two terms referred to the same phenomenon. Some thirty years later, the man-midwife and first editor of the *Encyclopedia Britannica*, William Smellie, recorded the technical distinctions made by medical personnel of his day: "Miscarriage" happened before the tenth day of pregnancy and expelled nothing more than a liquid conception; from the tenth day to the third month an interrupted pregnancy was known as an "expulsion"; between the third and seventh month a woman was said to have suffered an "abortion"; and any time after that the woman was "in labor." Smellie also noted more generally that a woman whose pregnancy was interrupted anytime from conception until the ninth month could be said to have "miscarried."[15]

Not until the nineteenth century did "abortion" and "miscarriage" take on their current meanings. The 1835 edition of the *Dictionnaire de l'Académie Française* listed *avortement*—knowingly expelling a fetus through the effect of "some beverage"—as a "criminal abortion" (*avortement criminel*). Terminology criminalizing abortion arose most visibly in France after 1790. Increasingly, the term "miscarriage" was restricted to an accidental loss of the fetus. Germans coined special vocabulary for these terms relatively late (relying before that time on Latin): *Fehlgeburt* (miscarriage) was not commonly used until the nineteenth century. Zedler's massive *Lexikon* (1732) did not include this term but listed instead *Missgebärung* (misbirth), *frühzeitige Gebärung* (premature birth), and *unzeitige unrichtige Geburt* (untimely, improper birth). Nor did Zedler index this discussion under the German *Abtreibung* but retained the Latin-derived *abortiren* and Latin *abortus*, though he used the term *Abtreibung* in the text. *Abtreiben* in the early eighteenth century meant many different things and was used chiefly in forestry (meaning to clear the forest of wood) and in mining (meaning to drive out impurities such as lead from silver or some other precious metal). For the brothers Grimm in the early nineteenth century, the first meaning of *Abtreiben* was "to drive the cows from their summer grazing in the high Alps to winter quarters." In medicine, *Abtreiben* referred to driving anything (through the use of medicines) from the body, including stones, worms, or "the fruit" of a pregnancy.[16]

Today, one of the first questions that people ask in discussions concerning abortion—even abortion in the past—is "what was the law?" This is a modern question, and arguably even a distinctly North American question, that obscures earlier practices and ways of thinking. By the second half of the nineteenth century, it was easy to say what the legal status of abortion was in most parts of Europe. In Germany, for example, Section 218 of the German penal code (*Reichsstrafgesetzbuch*, 1871) punished abortion with five to ten years' imprisonment for both the mother and the person who assisted in the abortion. Austria required similar punishment of the father also, if he took part in "the crime." In the course of the nineteenth century, most European states—including France, England, Denmark, Italy, Belgium, and the Netherlands—introduced centralized laws governing "criminal abortion" (Chapter 4).

The situation was different in earlier centuries, however. No legal consensus governed early modern European practices of "abortion" and the use of anti-fertility agents. Abortion in the modern sense—meaning the induced expulsion of a living conceptus—was rarely condoned by church or state. Many towns and local areas did have their own laws and conventions, but many practices regulated in towns went unregulated in the countryside. Even the entry on abortion in the *OED* notes that the practice of abortion was one to which few persons in antiquity attached any "deep feeling of condemnation." In ancient Greece—a source of knowledge valued by early modern physicians and jurists—Hippocrates advised against teaching how to induce abortion, and his famous Hippocratic Oath spoke against providing women with abortive remedies, yet Plato and Aristotle both favored limiting population by means of contraception and abortion. In ancient Greece and Rome, abortion was generally "resorted to without scruple" and was punished only in cases in which a married woman, motivated by resentment, practiced it to "defraud her husband of the comforts of children."[17]

Around 1600, several legal traditions co-existed in Europe: Roman, Salic, Germanic, Canon, and diverse "common laws." Almost everywhere the topic was broached, laws declared the purposeful abortion of a living fetus to be a crime punishable by death. In Germany, Article 133 of Charles V's imperial legal code (1532) made aborting a "living child" a crime punishable by death (by the sword for men, by drowning for women). Church law was more lenient, teaching that "it is lawful to procure abortion before ensoulment of the fetus, lest a girl, detected pregnant, be killed or defamed. It seems probable that the fetus (as long as it is in the uterus) lacks a rational soul and begins first to have one when it is born; and consequently it must be said that no abortion is homicide." English common law left the matter of abortion to ecclesiastical courts until approximately the seventeenth century. Common law courts made abortion an indictable offense only after "quickening" (when the fetus was felt to move by the mother) and sometimes only as a manslaughter, not murder; or, as the great commentator Sir Edward Coke put it, "this is a great misprision, but no murder." Sir William Blackstone's eighteenth-century codification of English common law also noted that abortion was punished only if the woman was "quick with child." France's approach was slightly different. Henry II's Edict of 1556/7 made infanticide (the death of a child already born) a capital crime and regulated both infanticide and abortion via prohibitions against hidden pregnancies. From 1556 to 1810, all expectant mothers were required to register their pregnancy with the *officiers de santé*; those who hid their pregnancies were suspect.[18]

The trump card for women seeking abortions in this period was that church, medical, and local judicial authorities tended to agree that a woman was *not* pregnant—not truly with child—until "quickening" or "ensoulment" took place, usually considered to occur near the midpoint of gestation, late in the fourth or early in the fifth month of pregnancy (or, according to Aristotle, forty days after conception for a male child and ninety days for a female child). Within elite European legal traditions, then, deliberately employing abortive herbs before quickening was not considered a crime. A fetus that had not quickened was not considered a person, but simply a part of the mother's own body (*ein Theil mütterlicher Eingeweide*). Germany's imperial code was exceptional in punishing (as a felon) a woman who expelled a child unformed or "not yet living," and even here punishment was discretionary.[19]

A few medical authorities held that "ensoulment" took place at conception. Mauriceau in Paris believed that the child was formed at the moment of conception. The ardent anti-abortionist and dean of the Parisian medical faculty, Guy Patin, employed Antoni van Leeuwenhoek's homunculus theory (holding that offspring exist preformed in the sperm) to argue the same thing. (Interestingly, Patin did not employ the counterargument, later favored by Carl Linnaeus, that all future offspring preexist in the ova.) This notion that conception represented the essential spark of life often led to disagreement between physicians and women over the definition of ensoulment and hence abortion. Mauriceau reported the case of a woman who, in 1682, with the assistance of a "wicked midwife," had attempted to induce an abortion. Responding to her physician's objections, the woman told him that "she would not have done it, if she had not thought, with the child neither shaped nor quickened, there could be no great harm in procuring a miscarriage." The physician then launched a diatribe against abortion, arguing that such a sentiment was "ill founded," and that it was as "pernicious, as the action she had endeavored to commit was wicked." This "false persuasion," he continued, "though of long standing . . . has encouraged an abundance of profligate women to procure themselves a discharge of the embryo after conception and an abortion in the first months of their pregnancy." For him, it seemed "very true, that from the first day, and immediately after conception, the soul is actually introduced into the little speck of matter." This idea of life beginning with conception was a minority opinion in Europe in the seventeenth and eighteenth centuries. The majority still held to the traditional notion that the child was not "ensouled" until quickening.[20]

The problem for those holding the minority opinion was that in a time prior to reliable pregnancy testing in the early twentieth century, doctors, jurists, and churchmen had to rely on a mother to say when she was pregnant. As Barbara Duden has pointed out, only the pregnant woman could feel her child first quicken.[21] The practical problem for would-be prosecutors in this era was that there were few "sure" signs of pregnancy, and a woman could not be prosecuted for abortion if she could not be proven pregnant.

Physicians noted, for example, that "stoppage of the menses is an unclear sign of pregnancy" since menstruation could be obstructed by other causes, such as ill health. Many physicians in this period did not even subscribe to an essential link between pregnancy and the cessation of menstrual blood. John Freind in his elaborate *Emmenologia* (1729) wrote that "not a drop of the mother's blood is carried to the fetus; because there is no anastomosis between the uterine and umbilical vessels." These physicians believed that the fetus was nourished not by blood but by a "milky juice" which it received by mouth. They further denied that the "final cause" of the "menstruous blood" was to nourish the fetus "because the quantity evacuated in the space of nine months seems too small to be sufficient to sustain it." Hence, even prominent physicians found it almost impossible to distinguish between an obstructed menses and an early pregnancy.[22]

As late as the early twentieth century, medical lawyers labeled the following conditions "uncertain" and thus unusable in a court of law: the cessation of the monthly flow, morning sickness, darkening of the areola around the nipple, enlargement of the breasts, increased size of the abdomen, and the growth of the womb. Only fetal limbs palpated through the abdomen by a physician or the pulse of the fetal heart heard by means of a stethoscope (the fetal pulse being quicker and not synchronous with the maternal pulse) were considered "sure" signs of pregnancy.[23]

These and other ambiguities meant that prior to the development of pregnancy tests in the twentieth century women enjoyed considerable freedom to judge for themselves when quickening took place—that is, when they truly were pregnant. John Riddle has argued that this gave women a wide window of opportunity to "invoke the menses" without restriction. Taking a menstrual regulator or emmenagogue (generally an herbal concoction to "bring on the menses") was not necessarily considered "inducing abortion." This disconnect—both in the minds of women and the law—between these concoctions and abortion was such that they were not referred to as "abortifacients" until sometime in the nineteenth century. A woman ingesting an emmenagogue might not have been sure whether she was inducing a late period or provoking what would be known today as an early-term abortion—and she had little reason or ability to distinguish between the two. One anti-abortionist lamented in 1812 that the "law's silence concerning abortion" had resulted from the fact that it was nearly impossible to prove the crime of abortion without a "confession" from the mother, a confession that was almost never obtained.[24]

Why in early modern Europe would a woman seek an abortion? In many instances, it was to protect her life when giving birth threatened to be difficult—when, for example, her hips were too narrow or bad health or other circumstances prevented her from carrying a child to term safely. The article on *fausse-couche* in the *Encyclopédie* also asserted that married women aborted children to protect family property, and that unmarried women aborted to be rid of the fruit of fornication. Recall that in many settings, the discovery of an adulterous pregnancy threatened a woman's life as much as the prospects of a drug-induced abortion. Physicians from the eighteenth century also suggested that women were willing to undergo the rigors of abortion to preserve their youth and beauty, and to escape the cares and duties of family life.[25]

Laws do not tell us much about actual past practices surrounding abortion and the use of abortifacients, and historians are far from agreeing about the frequency of abortion in early modern Europe. Norman Himes and John Noonan claim that anti-fertility herbs simply did not work, and that nothing much was available to pre–twentieth-century women for birth control. Louis Lewin, a pharmacologist who wrote extensively in the 1920s about abortifacients, and Larissa Leibrock-Plehn, an historian of pharmacy, both suggest, contrary to Himes and Noonan, that the strengthening of European laws against abortion in the early modern period indicated extensive use of abortifacients. Riddle shares this view and has argued against Angus McLaren's view that widespread use of contraceptives began only at the end of the eighteenth century (McLaren's argument is based on demographic evidence that he interpreted as showing stable fertility rates from the sixteenth to the eighteenth centuries). Along with Edward Shorter, Riddle contends that what was once a rich collection of knowledge about anti-fertility agents became "lost" over the course of the early modern period, and that while women continued to practice abortion behind closed doors, the legal crackdown on abortion suppressed knowledge of abortifacients. Gunnar Heinsohn and Otto Steiger, both sociologists, have claimed in their much-criticized *Vernichtung der weisen Frauen* that women's knowledge of contraceptives and abortifacients was destroyed in the virulent witch hunts, and that knowledge of abortifacients burned with the witches (many of whom were midwives). Ulinka Rublack has documented the widespread use of abortives in seventeenth-century Germany, and has vigorously (and correctly) opposed the notion that only women knew about and used anti-fertility agents.[26]

One reason for these disagreements is that abortion was disreputable and almost always performed quietly and behind closed doors, if at all. Another is that practices differed significantly across Europe. Still another reason is the confusion over the term "abortion," noted above. Disagreement has also arisen from the fact that scholars support their arguments with different types of documents—herbals, midwifery manuals, physicians' case books, trial records, literature and poetry—that often yield divergent results.

Riddle and Leibrock-Plehn, for example, have scrutinized herbals for clues about abortion. These compendia of botanicals were used by herbalists, physicians, apothecaries, and the literate public to identify plants for specific uses. These books usually provided Latin and common names, the uses of plants, geographical areas where a particular plant could be expected to grow, and, importantly, illustrations to insure that herbalists collected the correct kind of plant for their medicines. Many European herbals, learned or popular, merely translated, often word for word, the work of Dioscorides (the bible of European pharmacology until the seventeenth century), Avicenna, Galen, Hippocrates, Pliny, Theophrastus, or Constantinus Africanus. Riddle and Leibrock-Plehn have shown that these books contain a number of brews and pessaries that make use of abortifacients such as savin, pennyroyal, myrtle, wallflower, bitter lupin, rue ("an enemy to generation"), barrenwort and birthwort, death carrot, juniper, squirting cucumber, and other substances, including beeswax. Though herbals often identified abortifacients, information concerning their preparation and dosage was often left vague. John Gerard's *The Herbal or Generall Historie of Plantes* (1597) provides a typical entry: "the leaves of Savin boiled in wine and drunke provoke urine, bringeth downe the menses with force, draweth away the after-birth, expelleth the dead child, and killeth the quicke." Effective use of plants required knowledge of the parts of the plant appropriate for use (root, sap, bark, flowers, seeds, or fruits), the proper time for harvesting (spring, summer, autumn), when to administer the drug in relation to the menstrual cycle and coitus, in what amounts, with what frequency, and so on. Herbals rarely included such details, and are best viewed as a codification of certain natural knowledge of plants, but not a source providing potential recipes for their medical use.[27]

We might assume that in contradistinction to herbals, most of which were written by learned men, midwifery books would be a good source of information about practices concerning contraception and abortion. Precisely because women's health lay primarily in the hands of midwives, it would make sense that midwives, if anyone, would have cultivated a rich practical knowledge of abortifacients. The extent to which this is true is difficult to evaluate. Midwifery was an art learned orally through apprenticeship. Books by midwives by and large did not appear until the seventeenth century, and many of these were written as polemics against the appearance of the newfangled "man-midwives."[28]

Historians again disagree about the role of midwives in procuring abortion, though there is some evidence that midwives' work required them to have reliable drugs and procedures for therapeutic abortions. A medieval midwifery manuscript provided such a recipe. The author gave priority to the mother's life over that of the fetus, as was common in this period; and it is clear that the author understood "abortion" to mean after quickening—in this case, to aid in a difficult labor: "For when the woman is feeble and the child cannot come out, then it is better that the child be killed than the mother of the child also die." This remedy, also given to abort a dead fetus and to promote afterbirth and menstruation, contained a number of likely effective abortifacients:

> Take half a pound of iris roots and half an ounce of savin, boil them in white wine, and add to half an ounce of powder of ground ivy, 1 ounce of honey; with 1 measure of the boiled proceeds, take 1 drachm of bull's gall, and make a pessary, and give a pill of 2 drachms of myrrh with this liquid. Take 1 ounce each of bishop's weed, woodruff, parsley seed, balm, caraway, dill, iris, artemisia, and 3 pounds of white wine. Boil them; then chop up 1 ounce of savory, hyssop, woodruff, and dittany in equal amounts, dilute with 4 drachms of the liquid, etc.

Other women's recipe books, such as Lady Alathea Talbot's, included formulae (eleven in all) for draughts to induce the menses. The active ingredients in these brews, such as rue, would certainly also have induced an early-term abortion.[29]

It is clear, then, that at least some midwives possessed knowledge of effective abortifacients. But midwives also often referred—publicly—to abortive brews as "damnable remedies . . . wicked in a high degree." By the late fourteenth century, midwifery in most of Europe had become a public office; licensed midwives were responsible to the church and increasingly to the state to report hidden pregnancies and names of fathers, and more generally to keep order in reproductive matters. Midwives' oaths from the sixteenth century onward prohibited them from deliberately causing abortions. The Parisian Oath of 1587 was typical in forbidding midwives from "ordering or administering any potion, or other kind of medicine to a woman, whether married or unmarried, in order to procure the abortion of their child, on pain of death."[30] Books on midwifery are remarkably silent on the questions of abortion and abortifacients. The great seventeenth-century midwife to the French queen, Louise Bourgeois, did not openly instruct midwives in abortion. Her *Recueil de secrets* included nine recipes for menstrual regulators. Although some of these certainly could have induced abortion, none was labeled as an abortifacient. In her advice to her daughter, then embarking on her own career as a midwife, Bourgeois cautioned that she should never deliver a woman in her own home because it would be assumed that she was performing an abortion. Male midwives, such as the seventeenth-century Englishman Nicholas Culpeper, warned midwives against giving menstrual regulators to women with child, "lest you turn Murder." "Willful murder," he warned, "seldom goes unpunished in this world, never in that to come." Jane Sharp's *Midwives Book* (1671) specified things a woman who wanted to conceive or to carry her child to term should avoid (this may or may not be coded language for how to induce abortion, as Riddle has suggested). Justine Siegemund, a seventeenth-century German midwife, included nothing on abortives in her well-known text. Nor did the audacious Angélique Marguerite Le Boursier du Coudray, the "king's midwife," mention induced abortion of an unwanted pregnancy in her 1777 treatise on birthing. She did, however, describe a number of herbal concoctions for aborting a dead fetus and also how to use "the hand," this latter a method she associated with physicians and surgeons. Her justification for usurping their monopoly on this technique was that female midwives often treated women in the countryside where no male doctors were available.[31]

Prescriptive oaths and texts, however, do not reveal actual practices. Already in the sixteenth century, midwives had developed elaborate systems whereby an expectant mother—sometimes taken blindfolded—could be hidden away in total secrecy in the midwife's house until her child's birth. The midwife then disposed of the child, either handing it over to a wet nurse or (after about 1740) to a foundling hospital. Physicians and midwives in eighteenth-century Göttingen also delivered women secretly for the price of room and board at the newly established lying-in hospital there. These paying patients were not offered abortions but rather secrecy and anonymity.[32]

Secrecy and silence on the part of midwives is understandable. Throughout the seventeenth and eighteenth century, authorities turned a blind eye to uneventful abortions and the sequestering of pregnant women—in fact, they rarely knew much about these practices. A failed attempt, however, could result in execution. In France, where fashionable upper-class women were known to wear contraceptive sponges tied to their belts, we encounter the writings of the rabid anti-abortionist Guy Patin, who claimed that in the year 1660 not fewer than 600 Parisian women confessed to their priests that they had "suffocated their children in their wombs." These women were not prosecuted, nor were the herbalists from whom they had procured or learned about the drugs. The death of a highborn woman, such as Mademoiselle de Guerchi, however, caused a sensation in Paris. She was denied burial in Saint Eustache, and the midwife Marie le Roux, wife of Jacques Constain, was prosecuted and executed (her two male surgeon assistants were tried but not convicted). Madame Constain was not a hack or amateur abortionist (she actually denied giving de Guerchi anything) but rather an esteemed *jurée-matrone* of Paris, one of the city's sworn midwives. Her misfortune was to have caused the death of a woman much loved by the French court in a failed abortion attempt. The trial proceedings did not focus on details, such as whether the fetus had quickened or not, but rather on whether the unhappy midwife was to be burned alive or merely hanged. She was condemned to be hanged and left on public view, her flesh to be torn and eaten by birds and other vermin until it was "no more." Not stolen by anatomists hungry for corpses to dissect, her body was still on display ten days later when the twenty-eight-year-old Louis XIV returned to Paris from the Pyrenees.[33]

Trial records, like those recording the le Roux case, offer a glimpse into medical practices in the past. Ulinka Rublack has discussed court records from late seventeenth-century Germany, where women were tried for ingesting potions of laurel, pennyroyal, savin, and a host of other potent herbs administered by mothers, friends, herbsellers, physicians, and lovers.³⁴ We have to be cautious in generalizing from such sources, however, since they are by definition texts about things gone awry. Court records teach us about extraordinary cases—when something has gone wrong—but may not accurately reflect ordinary life.

One source that historians have not exploited when discussing abortifacients in early modern Europe are *Pharmacopoeia* and physicians' *Materia medica*, the former being the official compendia of drugs that secured uniformity in remedies, the latter being published lectures for medical students that discussed the uses and efficacy of those drugs. Europe's top abortifacients—savin, rue, and pennyroyal—were listed in these *Pharmacopoeia* as soon as they began to appear in the sixteenth century. Entries on abortifacients generally admonished care in administering them because of their strength.

Pharmacopoeia tended merely to list drugs, but physicians' *Materia medica* discussed their virtues in some detail. Paul Hermann, a professor of botany in Leiden in the 1680s, was explicit about the use of drugs in relation to abortion: some herbs worked to prevent abortion or miscarriage; others provoked the menses, induced abortion and labor, and helped to expel the fetus and afterbirth; still others "killed" the fetus. This latter required preparing an infusion of Spanish flies in Rhenish wine with one scruple of flies to four ounces of wine. Savin, he continued, had the same power, "as is too much known to common creatures" (meaning women of the lower classes). Although Hermann aired his disapproval, he provided two recipes for savin in abortive doses: Take savin and motherwort, express the juice, and drink two drams of it for a dose; or, take a handful of the powder of savin or the fresh herb itself and castor to one dram. This was to be put in a linen cloth and made into a pessary. When thrust into the vagina, it "draws forth or expels" the child. Hermann also recommended savin for purging children of worms. Hermann further noted that "vile women" used radix asari or asarabacca root to occasion "miscarriage," and again he offered the recipe (castor, one scruple; borax, half a scruple; roots of asarabacca, half a scruple made into a powder or bolus). Carl Linnaeus, a physician and botanist, listed five abortives and a surprising fifty-three emmenagogues in his 1749 *Materia medica*. The abortives were Elaterium, Colocynthis, Aristoloch, Muscus erect., and Sabina (savin).³⁵

These many sources taken together—herbals, midwifery manuals, trial records, *Pharmacopoeia*, and *Materia medica*—reveal that physicians, midwives, and women themselves had an extensive knowledge of herbs that could induce abortion. As we shall see in Chapter 4, physicians' case histories also reveal both physicians' and patients' knowledge and use of abortives.

Finally, popular imaginative literature also sheds some light on early modern women's use of abortifacients. In his fourteenth-century *Decameron*, Giovanni Boccaccio portrayed two nuns discussing the "sweets of the world" and how a woman can enjoy a man without unwanted consequences. If pregnancy ensued, "there are a thousand means," they proclaimed, "by which to take care of the matter so that no one is the wiser, if we ourselves say nothing about it." Three hundred years later, in seventeenth-century England, Ben Jonson's "Mistress Haughty" declared that women who wished to preserve their youth and beauty had many "excellent receipts" for preventing conception. The scandalous Marquis de Sade likewise suggested in his *La Philosophie dans le boudoir* (1795) that a good wife "who delights herself in adultery" should be ready to induce abortion if her contraceptive devices fail; he recommended savin as a reliable abortifacient.[36]

Back across the English Channel, Mary Wollstonecraft, in her novel *Maria, or The Wrongs of Woman* (1798), portrayed the trials and tribulations not of an upper-class woman pleasuring in surreptitious liaisons but of a serving girl, seduced by her master and thrown out of the house by his jealous wife. With little money and few friends, the suicidal girl decided to end her pregnancy or her life in abortion:

> I hurried back to my hold and, rage giving place to despair, sought for the potion that was to procure abortion, and swallowed it, with a wish that it might destroy me, at the same time that it stopped the sensation of new-born life, which I felt with indescribable emotion. My head turned round, my heart grew sick, and in the horrors of approaching dissolution, mental anguish was swallowed up. The effect of the medicine was violent, and I was confined to my bed for several days; but, youth and a strong constitution prevailing, I once more crawled out, to ask myself the cruel question, "Wither I should go?" I had but two shillings left in my pocket, the rest had been expended, by a poor woman who slept in the same room, to pay for my lodging, and purchase the necessaries of which she partook.

Wollstonecraft does not tell us what the serving girl took. Of the many abortifacients available in early modern Europe, as I have mentioned, savin (*Juniperus sabina* in the Linnaean) was among those most widely used.

The abortive properties of savin are attributed to its essential oil: sabinyl acetate. An eighteenth-century German doctor noted that this colorless or yellowish oil was most often distilled and used in pills because infusions and decoctions were generally upsetting (*übel*). The oil stimulates the smooth muscles of the uterus and can provoke strong contractions.[37]

Savin (*Sabina, Sabine, Savenbaum, Sage-* or *Sadebaum*), European women's abortifacient of choice, was mentioned by Cato in ancient Rome. Dioscorides knew it as an abortifacient, as did Galen, Avicenna, and Constantinus Africanus. Native to the south of Europe and the Levant (savin grows from Spain to the Caucasus and even high in the mountains of Austria and Switzerland), it was supposedly carried from Italy by monks and nuns to the south and middle of Germany, France, and other parts of northern Europe where it was planted in their kitchen gardens. By the 1560s it was known and cultivated in England and Scotland: Mark Jameson, a physician and deputy rector of the University of Glasgow, planted a garden of plants for gynecological complaints that included savin and several other abortifacients. Said to be so named for saving a young woman from shame, savin was also known popularly as "Kindermord" (child murder), "Jungfernpalme" (a woman who used this plant could, despite all, stand as a virgin at the marriage altar), "abortion tree," "lucky herb," "tree of life," "bastard killer," and "plant of the damned." Linnaeus, a conservative Lutheran, remarked that the savin tree was used by "women who are whores, and though they think their sin is secret, God sees it." In addition to its anti-fertility virtues, savin was used as an emmenagogue, a diuretic, and for "sanguine movements."[38]

Throughout the eighteenth century, physicians judged savin "the most efficacious drug in the *Materia medica* for producing a determination to the uterus." In 1738, doctors in Halle, near Leipzig in Germany, wrote opinions on the question: "whether savin, when used internally, infallibly and unavoidably kills and aborts the fetus?" After consulting medical "reason, experience and authority," the doctors answered with a resounding "yes." Though then as today most contraceptives were used by women, in the ancient world Pliny the Elder recommended rubbing crushed juniper berries on the male part before coition to prevent conception.[39]

As time went on, too-ready access to savin trees began causing alarm in some quarters. In botanical gardens in Munich, Zurich, and Thüringen, savin trees were fenced off or planted in inconspicuous places. In the late eighteenth century, a woman of "questionable reputation" came under suspicion when, despite her many-year affair with a "lusty" hunter, she never became pregnant. Finally, the savin tree in the nearby castle garden was "ripped out" (the report does not say by whom) and, almost immediately, the woman became pregnant. An investigation reported that "regularly" each month at the time of menstruation she had prepared a drink from the leaves of the savin tree.[40]

In Berlin in the late 1790s, the savin trees in the public Tiergarten were felled because visitors showed too great an interest in them. Around this time a German professor remarked: "when I travel in the countryside or pass by a small village garden in which I see a savin tree or bush, . . . I suspect it is the garden of a barber-surgeon (Barbierer) or of the town midwife." In Austria toward the end of the eighteenth century, the cultivation and sale of this tree was stringently forbidden. As late as 1935 in the flush of Nazi pronatalist frenzies an herbalist in Germany wrote that because savin was still used as an abortifacient "its cultivation in public areas . . . is forbidden." The author continued that in many botanical gardens savin trees were fenced off from the public. Indeed, in some gardens, those fences sometimes bore suspicious signs of surreptitious entry.[41]

What can we conclude, then, about how widely abortion was practiced in early modern Europe? It is, of course, impossible to say with any precision. Patin in the seventeenth century argued that the numbers were appallingly high, pointing to the case of Cathérine Voison, midwife at the court of Louis XIV, who was hanged after being accused of carrying out some 2,500 abortions. Patin also mentioned the 600 abortions in the year 1660, nine of which resulted in criminal proceedings against the midwives. Patin's numbers are surely too high. One has to remember, though, that despite laws to the contrary, European principalities also had high incidents of infanticide and abandoned children; abortion was only one method for disposing of unwanted offspring. The *État de médecine, chirurgie et pharmacie en Europe pour l'année 1776* reported that of the 19,353 registered births of that year, 3,152 boys and 3,181 girls (or nearly one-third of the babies born) were abandoned. Louis de Jaucourt, writing in Diderot and d'Alembert's *Encyclopédie,* noted that states that supported foundling hospitals significantly reduced their incidence of abortion and infanticide.[42]

It is also possible that abortion in Europe was practiced more often by the upper classes. Discussing fertility rates in his 1776 *Wealth of Nations,* Adam Smith wrote that while a "half-starved Highland woman" frequently bears more than twenty children, "a pampered fine lady is often incapable of bearing any, and is generally exhausted by two or three."

"Luxury in the fair sex," he continued, "while it inflames perhaps the passion for enjoyment, seems always to weaken and frequently to destroy altogether, the powers of generation."[43] Perhaps it was something more specific than "luxury" that was curbing the powers of generation in these sexually active women.

Although we cannot say how widely savin and other abortifacients were employed in this period, evidence provided by herbals, *Pharmacopoeia*, *Materia medica*, trial records, and classical fiction, among other things, suggests that many European women regulated their fertility. Jaucourt was probably right in saying that the number who braved the perils of abortion in the eighteenth century was "considerable" and the methods "numerous."[44] I turn now to the West Indies and abortion practices there. To what extent did attitudes in Europe affect abortion practices and policy in these colonies? And to what extent did such practices and policies influence naturalists' interest in collecting exotic abortifacients for sale in Europe?

Abortion in the West Indies

Abortion was more publicly discussed in the colonies than in Europe, in part because abortion in the colonies concerned principally slave (not European) women and was debated often in terms of property (not morals). As abolitionists threatened to shut down the slave trade from about 1770 onward, slave women's suppressed fertility became a topic of growing concern among doctors, legislators, and planters. Edward Bancroft in his *Essay on the Natural History of Guiana, in South America* (1769) volunteered that on the Wild Coast "this unnatural practice is very frequent, and of the highest detriment to the planters, whose opulence must otherwise be immense."[45] Abortion, especially among slave populations, was not a matter of private conscience or "family planning," as we might think of it today; it was rather a part of the colonial struggle of victors against vanquished and a matter of economy and of state.

Before turning to abortion among slaves, to what extent were contraception and abortion practiced by the native populations in the Caribbean and South America? There is good evidence that the indigenous populations of the Caribbean—the Tainos, Caribs, and Arawaks—knew and made extensive use of abortive herbs well before European contact. The first European accounts from the New World described how Taino women aborted in the face of extreme circumstances. Bartolomé de Las Casas, who sailed with the conquistador Nicolás de Ovando to the New World in 1502, perhaps exaggerated how the horrors of Spanish cruelty—the fierce attack dogs, the swords used to disembowel or to hack off arms, legs, noses, and women's breasts—caused desperate Taino mothers to drown their infants. Nonetheless, he chose to frame Taino response in terms of abortion: other women, when they felt they were pregnant, he continued, "took herbs to abort, so their fruit was expelled stillborn." An Italian adventurer, Girolamo Benzoni, who traveled to the New World in 1541, recorded how the Spanish extinguished the Taino in Hispaniola, causing such grief that the indigenes sought relief in suicide, infanticide, and abortion: "Many, giving up hope, went into the woods and hanged themselves from trees, having first killed their children; . . . the women, with the juices of some plants, interrupted their pregnancies, so as not to give birth." Whether from experience at home or in the New World, these Europeans interpreted abortion—the willful ending of a pregnancy—as one response to desperate circumstances.[46]

But use of abortives (and contraceptives) was apparently also a part of Amerindian everyday life. During his travels in the New World from 1799 to 1804, Alexander von Humboldt reported extensively on the Indians (he named the Macos and Salivas) living along the "Oroonoko" (Orinoco) River, which flows through modern-day Venezuela and Colombia. The first European to describe many of the plants of this region, Humboldt deplored the young wives who did not wish to become mothers and their "guilty practice . . . of preventing pregnancy by the use of deleterious herbs." He deplored further the use of "drinks that cause abortion." Humboldt reported being astonished that "these drinks do not destroy health." As was typical of many of the learned men of his day, Humboldt had assumed that abortion ended in death. To his surprise, however, he found that the Indian women he observed were still able to bear children after using these herbs.[47]

We know very little about when or how natives of the extended Caribbean developed abortive techniques. Humboldt reported that native women aborted so as to time their pregnancies precisely, some women thinking it best to preserve their "freshness and beauty" when young and to delay childbirth until late in life in order to devote themselves to domestic and agricultural labors. Others, he wrote, preferred to become mothers when very young, thinking this the best way to "fortify their health" and "attain a happier old age." Further north in Virginia, Thomas Jefferson reported that Indian women "learned the practice of procuring abortion by the use of some vegetable" because they attended their men in war and hunting, and because childbirthing was inconvenient for them.[48]

Although European males on the whole did not approve of abortion, they reported it among Amerindians as noteworthy. Abortion was often observed among Amerindian populations throughout this period and still figures in the ethnographic record today. Women in Brazil, Colombia, Mexico, Peru, and Dominica (where indigenous populations still exist) continue to use a variety of abortifacients—including some originally brought from Europe (such as rue), others from Africa (yam beans, for example), and some apparently indigenous (gully-root).

In the eighteenth century, however, the real struggle over abortion in the West Indies surrounded slave women. Europeans were well aware that African women in the Caribbean used "specifics," drugs designed to treat a particular ailment, to induce abortion. The naturalist priest Jean-Baptiste Labat, traveling in the French Antilles in the late seventeenth century, claimed that "Negresses" were adroit in the use of "simples" that ended pregnancy with "surprising effectiveness."[49]

Even though abortion was more widely discussed in the colonies than in Europe, here too it is difficult to know much about the particulars of abortive practices. Women who underwent the trials of abortion did so quietly and rarely committed details concerning their experience to writing. Nowhere is information available on the number of abortions actually carried out on plantations. Abortion in the colonies as in Europe was largely female-initiated and beyond physicians' control. Reports, sometimes to legislative bodies, about the methods and frequency of abortive practices were prepared primarily by European physicians who lived and worked for many years on plantations. Historians also have notes and observations written by European voyaging naturalists, such as Merian, Sloane, and Humboldt, who were curious about plants and their uses in the region, and by miscellaneous European travelers to the area. Working with plantation records from the year 1795 kept by the owner of the Worthy Park Estate in Jamaica who, like many plantation owners in this period, was exceedingly concerned about slave fertility, the historian Michael Craton found that the rate of miscarriage (spontaneous and induced) for that year was 1 in every 4.6 births. Had all stillbirths, miscarriages, and abortions in that year been eradicated, he estimated that slave fertility rates would have increased 23 to 28 percent, a rate still insufficient to "grow the population" as the owner wished.[50] Records such as the ones Craton used are rare.

It must be kept in mind that European physicians in the colonies were deeply involved in colonial rule. As noted above, physicians were employed by planters to oversee plantation health care (although slaves' health was typically managed day to day by a female slave too old to work the fields). Physicians for their part healed when they could; they also policed slaves. In 1789, the Jamaican Assembly attempted to have surgeons register all instances of induced abortion. The act required "the surgeon of every plantation . . . to give in, on oath, to the justices and vestry, an annual account of the decrease and increase of the slaves of such plantation, with the causes of such decrease."[51]

Disciplining slaves, of course, was a horrific occupation, and one that often fell to the plantation doctor. John Stedman in Surinam in the 1770s pointed an accusatory finger at a Mr. Greber, a surgeon, who cut off one leg each of nine Negro slaves as a punishment for running away from their work at a plantation. Four of them died after the operation, and a fifth killed himself by plucking the bandages from the wound, and willfully bleeding to death during the night.[52]

Merian's poignant report of the use of the peacock flower cited in the introduction to this book placed abortion squarely within the colonial struggle, identifying it as a form of slave resistance. Contemporary observers and present-day historians have identified many forms of resistance. One of those was largely male-led armed insurrection. Stedman, for example, drew a vivid portrait of guerilla warfare in Surinam, where in the eighteenth century 5,000 Dutch employed 1,500 mercenary soldiers in the hopes of keeping in check the 75,000 slaves within the colony. The soldiers had their hands full, being also required to fight against the "maroons," or slaves escaped into the hinterlands, who were said to burn plantations, slice open the bellies of their former mistresses large with child, and poison entire plantations—Europeans, slaves, cows, and horses—with invisible substances sometimes carried under a single fingernail.[53]

Other observers emphasized the daily resistance of slaves, who shammed sicknesses, feigned inability to do simple tasks, were disruptively insolent, disobedient, and quarrelsome. Some even reported slaves committing suicide to spite their masters and find deliverance from their suffering in this world. Slaves were known to swallow their own tongues, to eat dirt, and even to leap into cauldrons of boiling sugar, "at one blow depriving the tyrant of his crop and his servant." Abortion, then, was only one type of resistance among many, and a number of contemporary observers saw it in such terms. As historian Barbara Bush has emphasized, in an economy in which planters sought to breed "Negroes" in like fashion as horses and cattle, refusal to breed became a political act.[54]

Slave women's desperate willingness to risk abortion responded to two aspects of a distinctive West Indian sexual economy: the pressure on slave women to "breed" themselves to enhance plantation property and wealth; and the pressure on slave women to provide sexual services to planters, soldiers, and sailors in the islands (also of course to their own husbands and lovers in an economy in which slave men far outnumbered slave women). By aborting, slave women were well aware that they saved their children from a life of servitude since, no matter who the father, the child inherited its mother's legal standing. As early as 1671, the Dominican priest Jean-Baptiste du Tertre quoted a Guadeloupe slave woman who refused to marry even when her master agreed to purchase any man she fancied: "I am miserable enough without giving birth to children who may live a life more pitiable than mine." Nearly a century later the political tensions had intensified dramatically, and Edward Long reported again that slaves "refuse to marry in order to avoid generating a race of beings enslaved to such [brutal] masters."[55]

The colonial sexual economy driving slave abortion was fueled by the notion among European men that black women were free for the taking; it was also fueled by the extreme youth of the colonial populations, by the lawlessness of life in the colonies, and, importantly, by the fact that colonial populations—both European and African, free and slave—were fiercely male. European settlements in the Caribbean from their beginnings in 1494 were heavily populated with men. In efforts to establish colonies in the New World, Christopher Columbus had brought 1,500 souls with him on his second voyage—an expedition of seventeen ships carrying livestock, seeds, plants, a doctor, a mapmaker, several clerics—but not a single woman. Columbus's men exploited the Indians, demanding food, shelter, labor, and also sexual services from the women. The first Spanish women crossed the Atlantic in 1497, but only in a trickle. The shortage of Spanish women in the colonies coupled with a lack of laws or prejudices against intermarriage led Spanish men to take Amerindian wives, so that by 1514 one in three married men were cohabiting with indigenous women.[56]

Dutch, English, and French settlements in the West Indies of the seventeenth century were quasi-military operations in which private soldiers planted and reaped under the command of their officers—an arrangement that kept the proportion of men inhabiting this region high. As these settlements matured, the majority of the planters, soldiers, sailors, merchants, naturalists, physicians, surgeons, and slaves occupying the islands continued to be men. Figures for the ratio of males to females are not exact and vary from colony to colony, but everywhere men dominated: in 1613 only two European women were to be found in all of Saint Christopher, for example. In Martinique in the 1670s and 1680s, the ratio of European men to women was almost three to one (in 1671, 2,200 men versus 730 women), despite the fact that female orphans and women of "ill repute" were rounded up from the streets of Paris and shipped out to the colony in 1680 and 1682. In Guadeloupe the ratio was 2.5 men for each woman in 1671. These ratios began to equalize by the mid eighteenth century in Martinique (120 men per 100 women) and in Guadeloupe (which reached parity), where settlement was encouraged. Saint Domingue, with its high absentee populations, remained heavily male with eight men for each woman in 1681, and still two to one in 1700. Du Tertre described the desperate search for wives in Caribbean islands in 1667: "the scarcity of women compels the inhabitants to marry the first who come; . . . the first thing demanded of ship captains is whether they have any women. No one asks questions about their birth, virtue, or beauty; two days after they have arrived, they are married, without anyone knowing anything about them." Indeed the boatloads of Europeans sailing from France across the whole of the eighteenth century remained 85–90 percent male; women constituted only about one-third of the European population in Jamaica, for example, as late as the 1780s. Europeans, of course, were a minority of the population in the West Indies (with populations varying from about 30 percent in Guadeloupe to 8 percent in Jamaica in 1720).[57]

Males also dominated slave cargoes. It is true that in the seventeenth century males and females were sometimes imported in equal numbers, but, as the slave trade increased, at least two-thirds of all slaves carried from Africa were adult males. Everywhere in the Caribbean, the preference for males increased as sugar monoculture spread. West Indian planters favored youthful males and paid high prices for them. Even as late as the 1790s, slave cargoes consisted of between 150 and 180 males per 100 females. There were a few exceptions to this male predominance. Towns harbored larger numbers of women. In the countryside, too, female slaveholders tended to keep more female slaves, sometimes breeding them for sale. Toward the end of the eighteenth century, with the number of creoles increasing and with political upheavals threatening to end the slave trade, sex ratios began to equalize. French colonies such as Martinique and Guadeloupe experienced a greater degree of creolization than did French Guiana and Saint Domingue. So, in Guadeloupe, for example, the numbers of male and female plantation slaves had equalized by the 1780s, but in Saint Domingue sexual parity was not achieved until after 1791. Barbados was an exception, with male and female slave populations reaching parity in the late seventeenth century. In Jamaica this did not happen until 1817, the end coming only after the ban on slave imports.[58]

In addition to being predominantly male, colonial populations were also generally young. Sailors in the Caribbean died, on average, at age twenty-six; other Europeans in the colonies at age thirty-eight. The French islands were typical, with a majority of colonists aged twenty to thirty years. Not only were the vast majority of men young, they were also single. Reverend James Ramsay remarked concerning the British West Indies that plantation owners preferred unmarried overseers because they were cheaper than men with families, even though an overseer's wife might take care of the sick, deliver slave babies, and cook for the plantation. "Planters," Ramsay asserted, "have determined it to be better to employ perhaps a dissipated, careless, unfeeling young man, or a groveling lascivious, old bachelor (each with his half score of black or mulatto, pilfering harlots, who at their will, select for him, from among the slaves, the objects of his favour or hatred) rather than allow a married woman to be entertained on the plantation."[59]

In colonies where the numbers of European women more nearly equaled those of European men, contemporaries perceived that greater moral order prevailed and that slaves more often replenished their own numbers. Barbados serves as a good example. It was one of the few among the British West Indian islands to attract large numbers of European women. Historian John Ward has argued that the high numbers of European women there enforced moral "decency," so that their menfolk did not engage in profligate intercourse with the slaves under their command so much as in Jamaica and elsewhere in the Caribbean. Barbados was also one of the few islands, after 1780, to "grow" its own slaves.[60]

Many of these relatively young men—planters, slaves, and the many soldiers required to defend Europe's colonial interests—demanded sexual services. John Stedman commented extensively on the commerce in sex required of young black and mulatto girls in this regard. His diary, ringing perhaps with the bravado of a young lieutenant, detailed the number of women offered to him. He was hardly off the ship from Amsterdam when "a Negro woman offer[ed] me the use of her daughter, while here, for a certain sum. We didn't agree about the price." The next morning he was "astonished" to see an elderly Negro woman enter his room and present him her daughter to be "what she pleased to call my wife." Even his own beloved Joanna, a slave girl whom he eulogized as the "fairest creature on earth" and intended to take to England as his lawful wife, came to him through a bargain struck with her mother. Slave girls routinely provided sexual services as part of prearranged commercial transactions. In Surinam, this practice was raised to a quasi-official system of concubinage called the "Surinam marriage," an arrangement Stedman noted that the "sedate European matrons" would deplore. To contract a Surinam marriage, a European man paid an agreed-upon price to a member of a slave woman's family for her services as both his housekeeper and concubine during his residency in the colony. The "marriage" was generally concluded through a secular ceremony and continued until the man died or returned to Europe. European men rarely returned to Europe with these mistresses, partly because many already had wives back in Europe.[61]

A certain lawlessness also characterized the West Indian sexual economy. Stedman found the European male planter absolutely dissolute, going to bed late and passing the night in the arms of one or other of his "sable sultanas (for he always keeps a seraglio)." Masters, whether they had an European wife in residence or not, kept slave women for this purpose and offered them freely to male guests. Europeans jealously guarded these women for their own use: in Jamaica, should a male slave dare to trespass on the black mistress of the master or overseer, he faced castration or worse. Observers from as far away as the Cape of Good Hope confirmed these practices. Carl Thunberg noted that the Dutch soldiers there "ruin themselves" by connections with black women and that the Dutch East Indies lodges were full of children born of female slaves with European fathers. Historians today estimate that one in twenty-five infants born to slave women were fathered by whites on Saint Domingue's sugar estates in the 1790s.[62]

Efforts had been made in the early years of colonial settlement to curtail European men's intercourse with female slaves. The Dominican missionary Jean-Baptiste Labat, working in Martinique in the 1690s, reported that the Sisters of Charity compelled slave mothers to name their mulatto children's (white) fathers, who would then be taken before a tribunal and fined. A particular priest became particularly skilled in extracting confessions from female slaves, even though masters carefully instructed their paramours in the art of deflecting such questions. One slave turned tables

Figure 3.2. Dissolute Surinam planter in morning dress, by the poet William Blake. Note the elaborate headgear. He is served by one of his slaves, who is perhaps also his mistress. This is one of thirteen images Blake did for John Stedman's book about his experiences in Surinam.

on the clever priest, naming him the father of her child. He was dumbfounded, but had to admit that, in fact, he had overnighted at her master's plantation nine months earlier. Witnesses to the scene could not decide which was more outrageous: the pretended naïveté of the "Negress," the embarrassment of the outwardly pious priest, or the shaken dignity of the judge, who had to dismiss the case and send the Negress home with her master until further information was available. Well-counseled slaves often simply named as the white fathers of their children sailors whose ship had already left port, or soldiers whom they had encountered in the street but whose names they had never known. Midwives (most of them also slaves) were often part of the scheme, and many hid the color of mulatto babies when they brought them for baptism. It was commonly agreed that before the tenth day after birth all babies were white, and that only the genitalia and fingernails betrayed a baby's true complexion.[63]

Although officially illegal, prostitution—of all sorts—was common throughout the Caribbean. Plantation masters often leased out their own slave mistresses in order to obtain extra cash. European women in town—especially widows—also owned taverns and inns where itinerant males could find "washerwomen," "seamstresses," or "housekeepers" on short notice. Social custom and pocketbooks also dictated that masters not marry their slave lovers, even if they wished to. Father Labat reported knowing only two white men who married "Negresses" during his thirteen years in Martinique. Even these men soon left their African wives because of ridicule. Plantation profits discouraged masters from marrying slaves; the *Code noir* required that a master who married his slave mistress set her free, along with any children that might come from their union.[64]

Free sexual intercourse with slaves was less tolerated for European women. Early in the seventeenth century, when white women first emigrated to the colonies as indentured servants, some married black men. With the collapse of the system of indenture, fewer European women emigrated to the colonies, and over the course of the seventeenth century, relations between European women and non-European men became taboo. When a white woman did have relations with a male slave, the consequences could be severe. In Martinique in 1698, for example, a father found his daughter pregnant by one of his slaves. She was to be dispatched to Guadeloupe or Grenada to birth the child secretly; the slave was to be deported to Spain and sold. Shortly before the girl was to be sent away, a Polish man, newly arrived in the colony, offered to marry her and recognize the child as his own. He declared himself the (unlikely) father in the church registry and that is as much as we know about this case.[65]

The Dutch were no more lenient in this matter. In the 1720s, when two women of European origins in Paramaribo gave birth to children of mixed race, they were banished from the city. Ten years later, the daughter of a Jewish planter in Surinam confessed to having had sexual relations with an Amerindian; she, too, was expelled from the colony. Writing in the 1770s, Stedman lamented that "should it ever be known that a female European kept a carnal intercourse with a slave of whatever denomination, the first is detested, and the last loses his life without mercy."[66]

Free black women seeking to marry European men did not fare much better. Elisabeth Samson in Surinam, a fifty-year-old free woman of color and wealthy plantation owner, announced her intention to marry the thirty-year-old white organist of the Dutch Reformed Church—to gasps all around. Townspeople found it "repugnant" and "disgraceful" for a white to want, "perhaps from misdirected lust," to marry a black. The historian Cornelis Goslinga reported that, during the dispute over this particular marriage, the white groom died.[67] The wealthy Samson easily found another prospective husband.

Men of science—physicians, botanists, and naturalists—were by no means disinterested observers in this sexual economy. European biopirates who passed through these areas often presumed that indigenous and slave women, like other bioresources, were there for the taking. Nicolas-Joseph Thiery de Menonville, traveling in New Spain in the 1770s, fancied an Indian woman whose family had given him shelter for the night. Struck by her "perfect beauty," he looked for "deficiencies" in her but could find none. Talking with her, he learned she was married, which "rendered her all the more interesting." Finally, he reached into his pocket for a gold coin—but drew himself up short, not because he feared ruining the woman's virtue and well-being, but rather because he worried that, should he yield to these "enervations of voluptuousness," he would compromise his mission to win the coveted cochineal dye for France.[68]

By the late eighteenth century Thiery de Menonville's encounter with an Amerindian woman was atypical for the Caribbean islands; the majority of the women with whom European men had sexual relations were African. Indigenous women sometimes held a privileged position in the colonial sexual economy. In the 1680s, the Surinamese governor Aerssen van Sommelsdijck, whose wife had refused to make the arduous journey to the "Wild Coast," took a beautiful Carib girl as his concubine. Amerindian girls were often much in demand, but this particular one was the daughter of a chief and taken to cement political allegiances between the two peoples. Some years later Edward Bancroft described how some of the most prominent Dutch families in Surinam descended from these marriages. In this way, Bancroft recorded, the Dutch acquired control over the Amerindians so that their governor, and not the indigenous chiefs, could broker agreements concerning trade, war, and peace in the colonies.[69]

Other voyaging naturalists also succumbed to the going sexual economy during their lengthy stays in the colonies. Jean-Baptiste-Christophe Fusée-Aublet, the French botanist working for the Compagnie des Indes on the Isle de France, was said to have married a Sénégalese woman, Armelle, whose freedom he purchased from the company. Other sources report that he freed his slave (which is likely, since he wrote passionately against slavery) and married a woman of color with whom he had numerous children. Still other sources say that he "abandoned his science for debauchery" and left 300 children in the countries to which he traveled—this would have required that he sire some thirty children per year. While unlikely, this latter report is not completely inconceivable when one considers Thomas Thistlewood, the promiscuous overseer in Jamaica, who left a thirty-seven-volume diary detailing his daily sexual exploits with multiple partners. Aublet's alleged debauchery, however, was most likely a fabrication. Aublet, it seems, purchased Armelle and wished her to "share his bed," but she refused until she became deathly ill, and Aublet nursed her with great devotion. Won over by his affection, she eventually had three children with him, one of whom survived to adulthood.[70]

Jean-Baptiste Leblond, another French biopirate who was later elected (by the white population only) first deputy of the Assembly in Guiana in 1790, also fathered two children of mixed race. Leblond sent his older son to France to be educated; this son later became secretary to the president of the Republic of Haiti. The younger son was born after Leblond had left Guiana; he never saw his father or bore his name.[71]

The West Indies, then, were wild, lawless places, populated by a good number of fortune-seeking males. Many European observers saw abortion as an outgrowth of this overheated sexual economy. Yet when abortions were discovered, blame was attributed not to the cruelties slaves endured or to the lawlessness of West Indian sexual customs but rather to what was seen as the natural licentiousness of Africans. Sir Hans Sloane, for example, reported two abortifacients during his short stay in Jamaica: his "flour fence" and the "Caraguata-acanga" (*Bromelia pinguin*). This latter grows, he noted, "very plentifully in the Caribes and Jamaica. It is very diuretick, and brings down the Catamenia very powerfully, even in too great a quantity if the dose be not moderate. It causes abortion in Women with Child, of which Whores being not ignorant make frequent use of it to make away their children."[72]

Edward Long, writing in Jamaica a half-century later, also saw abortion in these terms. "The women here are, in general, common prostitutes; and many of them take specifics to cause abortion in order that they may continue their trade without loss of time or hindrance of business." Long also assumed, as was common among learned European men at this time, that such numerous and "promiscuous" embraces necessarily hindered, or destroyed, future possibilities for conception. Janet Schaw, who traveled with her brother and kinsfolk from Scotland to Antigua in the 1770s, similarly denounced the "young black wenches" who, in her words, "lay themselves out for white lovers." Schaw remarked that in order to prevent a child from interrupting their pleasure "they have certain herbs and medicines, that free them from such an incumbrance." Governor Edward Trelawny of Jamaica added that these "Wenches" lie with "both Colours, and do not know which the Child may prove of" and, to avoid difficulties, procure abortions.[73]

Like abortion, it is hard to know if prostitution was as widespread as Long and others implied. Sloane, Long, and the other European critics of African mores do not distinguish between prostitution required of slaves by their masters and prostitution initiated by the women themselves (usually for lack of other ways of procuring a livelihood). Prostitution—whether forced or chosen—was seen to result from a natural moral deficit in Africans and not from the desperate circumstances of their lives. One historian has estimated from figures given in the 1770s for Saint Domingue that one out every ten free (not slave) women of color and one out of every twenty women of European origins were prostitutes, but there are too many ambiguities in such categorizations to take these estimates seriously.[74]

Bancroft, working on plantations on the "Wild Coast," again dismissed the idea that abortion among slave women resulted from hard work, poor food, and even the extreme corporeal cruelty. Coarse food and hard labor, he reminded his readers, "are ever accompanied with the blessings of increased health and vigour." Bancroft added his voice to the litany of complaints against "young wenches" who were said not to wish to lose their income from prostitution to the inconveniences of pregnancy. He wrote in

1769: "The 'true cause' of [slaves'] decrease results from the intercourse of the Whites with the young wenches, who derive no inconsiderable emolument therefrom; and as child-bearing would put an end to this commerce, they solicitously use every precaution to avoid conception; and if these prove ineffectual, they ever procure repeated abortions, which incapacitate them from child-bearing in a more advanced age, when they are abandoned by the Whites."[75]

Descourtilz, working in the thick of the Haitian revolution, devoted the final volume of his elaborately illustrated *Flore pittoresque et médicale des Antilles* to emmenagogues, or "hysteriques" (as they were often called), many of which he learned about from his "mulatress" informant (Chapter 2). Plants in this class had a strong, penetrating, and disagreeable odor, and served to provoke the menstrual flux and cure illnesses (jaundice, migraines, "vapors," convulsions, and uterine spasms) caused by its suppression. Five of the nineteen emmenagogues discussed induced abortion when administered in higher dosages. Although Descourtilz, like his European confrères, saw abortion among slaves as issuing from the West Indian sexual economy, in which women aborted babies conceived in prostitution or rape, he also saw it as resulting from the West Indian political economy, in which women, worn down by the cruelties of slavery, aborted to spite their masters. The plants he discussed included the *Poinciana pulcherrima*. They also included the *Aristolochia bilobata*, which, he wrote, was regarded by physicians in the Antilles as pernicious because, as was known already to Hippocrates, Galen, and Dioscorides, it could provoke abortion. (Although this particular variety of birthwort is specific to Saint Domingue, other varieties were well known to Europeans.) Another emmenagogue in Descourtilz's pharmacopoeia was the *Trichilie à trois folioles*, or *arbre à mauvais'gen* (literally, "tree of bad people"); the slave women in the colonies, he wrote, too often use this root to "destroy their fruit" and "wreak vengeance" upon their masters. The plant was so violent that he counseled against its internal use. "And the women who have been miserable enough to resort to its use are punished by it; most of them lose their life in atrocious pain and uterine hemorrhages that nothing can stop." According to Descourtilz, the plant also provoked violent vomiting.[76]

The *Veronica frutescens*, also used by slave women and known to the Caribs as the *cougari*, was, Descourtilz wrote, prescribed by "wise physicians with trembling and grave circumspection." Its virtues, he reported, were sadly too well known. It was used by "criminal matrons and by the

Negresses who are guilty of infanticide." Finally, the *Eryngium foetidum*, Descourtilz wrote, had already been described by his predecessor, Pouppé-Desportes, who denounced those depraved persons who employed dangerous emmenagogues "with the criminal intention of taking the life of the very creature they are meant to protect."[77]

Given the political and economic dimensions of abortion, it would be interesting to know whether slaves owned by free persons of color aborted at the same rate as those owned by whites. Because records of induced abortions were not kept, this question cannot be answered. It should probably be noted, however, that we have no reason to believe that the treatment of slaves by freed persons of color was any kinder than that by whites. Free women of color regularly branded their slaves, sometimes in bold letters across their chests, just like their white counterparts. European women in the colonies were no kinder. Stedman in Surinam told of a plantation mistress, a Mrs. Stolker, who, disturbed by the cries of a Negro baby, held it underwater until it was drowned. According to Stedman, women of her ilk detested the quadroon and "Negro" girls for taking European men for their husbands and "persecute[d] them with the greatest bitterness and most barbarous tyranny." A Doctor Jackson who went out to Jamaica in 1774 could not say enough about the cruelties of creole women (he did not specify, but probably of European descent). He reported them flogging slave drivers with their "own hands" if the drivers did not punish slaves harshly enough. Some slaves were tied down and stretched between four stakes on the ground. To accommodate a pregnant slave, "a hole was dug to receive the belly." Owners who killed slaves by such maltreatment were considered "incautious," but they were rarely punished—they were thought chiefly to have sustained a loss of property.[78]

Abortion and the Slave Trade

Overwhelming evidence thus exists that women in the colonies, especially slave women and free women of color, practiced abortion in this period, though at what rate and for what reasons is less clear. Concerns about abortion intensified toward the end of the eighteenth century when European states threatened to cut off the planters' supply of slaves from Africa. Until this time, planters had used slave women chiefly as "work units" by day and "sexual servants" by night but only rarely as "breeders." According to Michel-René Hilliard d'Auberteuil, planters calculated that it cost more to breed slaves than to purchase them. The price of a slave woman's labor lost for eighteen months (three at the end of the pregnancy and fifteen while she worked half-time nursing the baby) was estimated at 300 livres. A slave baby at fifteen months was valued at only 150 livres; at age ten, however, the child could fetch 1,500 livres, and at age fifteen, a full 2,000 livres. Whether this represented a profit in the long run depended on how much a planter spent feeding and clothing the child for fifteen years. From about the 1760s on, planters began to recognize that although it was more costly to breed slaves in the islands, "creole Negroes" were much less often sick than imported "salt-water Negroes," and planters began urging plantation physicians to improve breeding practices.[79]

This was a very different situation from that in the southern part of the United States. African slaves in the Caribbean did not replenish themselves, even after planters implemented reforms aimed at improving conditions for pregnant women. According to Stedman, the mortality rate in Surinam was about 5 percent annually. This meant, by his calculation, that "the complete number of Negro slaves, consisting of 50,000 healthy people, goes extinct once every twenty years." Aublet also noted that slaves in Saint Domingue reproduced only a "very small part" of their population. He figured that 20,000 slaves had to be imported each year to maintain a population of 200,000. Reports from the British West Indies put the loss of slaves to dysentery alone at one-fifth of the total slave population. By modern calculations, Caribbean slave populations would have disappeared every century or so unless steadily resupplied from Africa. Natural increase was not achieved in Jamaica and several other parts of the Caribbean until the middle of the nineteenth century.[80]

Historians today attribute the low rate of natural increase among West Indian slave populations to many things. Some emphasize the high death rate: life expectancy for slaves in French colonies was between twenty-nine and thirty-four years, compared to forty-six years for Europeans in this period in France. Other historians stress low female fertility rates among slaves. One Jamaican planter estimated that in 1794 and 1795, only half of his 240 resident female slaves at Worthy Park ever became pregnant. Of these, approximately a quarter (thirty-five) suffered miscarriages. More appalling still was that of the eighty-nine children born, only nineteen survived past infancy. Fifteen of these nineteen would be the only child that a particular female slave ever produced. Among the female slaves at Worthy Park in the eighteenth century, only two women produced two surviving children and only one woman produced three. In nearby Saint Domingue, fertility levels were among the lowest in the Americas; on sugar plantations, fewer that half of adult females ever gave birth.[81]

Amenorrhea (suppression of menstruation) and sterility among female slaves were caused by disease, hard labor, poor living conditions, and long periods of suckling that suppressed ovulation. But abortion also played a role. Robert Thomas, who practiced surgery from 1777 to 1785 in the islands of Saint Christopher and Nevis, listed "frequent abortions" as the second major cause of the decrease of slave populations on sugar plantations ("free and early intercourse" among slaves he signaled as the first cause; "epidemical diseases," alcoholism, and long periods of suckling followed as lesser causes). A Dr. Collins, who styled himself a "professional planter," found "frequent abortions" to be the fourth most important cause of the decrease among slaves, following the unhealthy climate, the lesser number of imported females, sterility among females, and preceding high rates of infant mortality. Some historians have also pointed out that Africans may have selectively sold infertile women into slavery. The extreme brutality of plantation overseers, combined with the disruption and separation of families, must also have given slaves little desire to bear children. Whether driven by high death rates, abortion, or low fertility rates, conditions were such that planters were continually forced to purchase new slaves from Africa.[82]

As criticism of the slave trade sharpened in the explosive atmosphere of the American (1776), French (1789), and Haitian (1791) revolutions, planters and government councils began to explore the questions of slave fertility and abortion practices. Population and its increase were considered matters of state, and colonial governments increasingly took action (see Conclusion). In 1764 the French king sent a new governor to Saint Domingue to study the causes of depopulation among the slave population. At the same time, French colonial officials attempted to improve the art of midwifery, establishing schools in the principal cities of the colonies in the 1770s where a professor of medicine trained and certified midwives. In British colonies, there was also discussion of establishing lying-in hospitals.[83]

Government councils also investigated the causes of abortion. The Jamaican physician James Thomson returned again to the topic of promiscuity, suggesting that female infertility derived from slave women's early and unbounded indulgence in venereal pleasures. "The parts are left in so morbid and relaxed a state," he wrote, "as to be unfit for impregnation." Many of these young females, he charged, "endeavor to procure abortion by every means in their power, in which they are too often assisted by the knavery of others. The effect of these repeated miscarriages operates dreadfully on the tender frame of the mother, and not unusually terminates in death, or incapability of future impregnation." If these women did finally "settle down," Thomson wrote, they produced "weakly and diseased off-spring that perish in a short time, or prove incapable of propagating their own race." It was also sometimes said that their "loose" sexual practices caused slave women to cease menstruating at an unusually young age. Edward Long added that the medicines slave women took to treat their repeated bouts of venereal disease often killed the fetus and sterilized both the woman and her sexual partner. Long further traced "Negresses'" frequent obstructions (of the menses) to their too-frequent bathing in cool water. Plantation slave midwives were also held accountable; their unskilled management of births was said to destroy women's abilities to reproduce.[84]

But critics of slavery also echoed the theme that Merian had sounded a century earlier: slave women refused to reproduce in efforts to revenge their masters and to save their offspring from the horrors of bondage. Father Nicolson, a Dominican priest, attributed the cause of abortion among slaves to their "barbarous" masters. "One sees," he wrote, "Negresses who abort themselves so that their masters will not profit from their condition." While he did not approve of these "homicidal mothers," he pitied them their vengeance in this act. To the planters he cried: "Inhumane hearts! These atrocious crimes recoil upon you." The French physician Pierre Campet wrote similarly in 1802 that slave women's miserable state of servitude extinguished from their hearts the maternal tenderness observed even among animals; their state of degradation made raising children "repugnant" to them. Hilliard d'Auberteuil also blamed slave women's frequent abortions on the tyranny of their masters: If "Negresses" often abort, it is almost always the fault of their master whose "excess of tyranny has suffocated their maternal sentiment." By the late eighteenth century, even those not critical of slavery, like Dr. Collins in Jamaica, agreed that abortion resulted from slave women's refusal to birth "a being, like herself, [into] the rigours of eternal servitude."[85]

The anonymous author of *Histoire des désastres de Saint-Domingue,* too, denounced the cruelty of planters' smothering in these "miserable creatures" the natural desire to bear children. "Mothers," the author moaned, "go against all natural instinct and kill their children to protect them from cruelty." These "unhappy slaves," the author continued, "listen less to the cry of nature than to their hatred for those who oppress them." There were many examples, the author reported, of women who commit hideous crimes (abortion and infanticide) in order to avoid enriching those who oppress them. In a note to the text, this author added that suicide among slaves in Saint Domingue was common because many of them believed (as Merian had reported of the slaves in Surinam a century earlier) that upon death they return to the country of their birth. "But abortion and infanticide are even more common among slaves than suicide." These crimes, this author added, have their source sometimes in the fear of giving birth and sometimes in the desire to have nothing impede their promiscuous love affairs, but the principal source is almost always the unhappiness and hatred inspired by a detested master. The "Negresses" have "numerous secrets" for doing away with the "seed of their maternity."[86]

Slave women also engaged in infanticide, though direct reports are rare. Jean-Barthélemy Dazille, royal physician in Saint Domingue during the upheaval of revolution, recounted how one mother sacrificed her two children in order to "steal them away from slavery." Like so many others, she, too, believed that slaves were transported back to their African homes after death where their rank, fortune, parents, and friends were restored to them. Dazille warned physicians to be constantly on their guard against infanticide. He reported one particularly horrific case—the discovery of thirty-one dead infants at an otherwise thriving plantation in Saint Domingue, "one of our richest colonies." The infants had all died within the first nine days of life, and tetanus was suspected. The surgeon responsible for the plantation found the thirty-second child well on the tenth day of life and sent his compliments to the plantation owner. The next day the owner replied that the child had died in the night of tetanus. The surgeon, finding this improbable, had the body exhumed. He discovered that the infant had been suffocated by much "vegetable matter" stuffed into its throat. The "abominable" mother said that she was no more culpable than the other Negresses who had killed their infants in the same manner. Descourtilz, also witness to revolution in Saint Domingue, reported an even more horrendous crime: A midwife by the name of Adradan (slaves did not yet have last names) confessed to killing seventy newborns with her own hands. "See if I deserve death!" she challenged, "It is a shameful custom to raise children into slavery. My position as a midwife delivers them newly born into my hands . . . I plunge a pin into the brain through the fontanel, this brings on the deadly lock jaw that plagues this colony and whose cause you now know."[87]

The consequences for producing a dead infant, whether stillborn or euthanized, were severe: the attending midwife (usually a slave) was

Exotic Abortifacients | 147

Figure 3.3. A romanticized image of a female slave being punished. Note the pedestal, the anguished posture, the classical breasts, and the depiction of African women carrying burdens on their heads.

whipped as was the mother, who might also be placed in an iron collar, where she remained until she became pregnant again. Colonial law further required that all "Negresses" who became pregnant declare their pregnancy to a midwife, who was in turn to report the fact to a surgeon, who registered it. A "Negress" declared pregnant and known to have provoked a "miscarriage" (a common euphemism for abortion) was whipped and made to wear the iron collar. In 1785 Girod-Chantrans witnessed the chains and iron collars that afflicted Negresses suspected of inducing abortion; some women were obliged to wear these "day and night until they had given their master an infant."[88]

The British House of Commons held hearings on the slave trade in the 1780s and 1790s, and debated the question of slave abortion and infanticide. A Doctor Jackson, testifying before a select committee, did not find "Negro" mothers deficient in affection for their children, but described how slave women practiced abortion and infanticide to spare their children the "hard usage" and "cruel treatment" they could expect from a life in slavery. Most physicians, however, closed ranks around the planters and did not hold them responsible for declining slave populations. Stephen Fuller, agent to the island of Jamaica, reported to the Assembly of Jamaica in 1789 that the decrease in slaves was not "comprehended" in Great Britain. It did not arise, as charged by abolitionists, from ill treatment or insufficient care, but from three primary causes: the disproportion between the sexes in the annual importations from Africa (according to his count, five males transported for every three females); the high mortality among new Negroes imported into Jamaica from Africa; and the frequent deaths among the Negro infants born in Jamaica (one quarter of whom perished within fourteen days of birth). Three physicians, all trained in Great Britain, supported Fuller, testifying that deaths among infants in Jamaica were due to tetanus, want of cleanliness, insufficient linen, poor housing, and the custom of employing too few wet nurses to care for too many infants, not corporal punishments. The well-known physician John Quier provided extensive testimony, noting that "Negro women, whether slave or free," simply did not breed as frequently as the women of the laboring poor in Great Britain. This he ascribed chiefly to the "promiscuous intercourse" that slave women supposedly indulged in. Although he did admit that abortion was "rather frequent among them," he ascribed this to their natural promiscuity rather than to any ill treatment or excessive labor. He testified further that among the four to five thousand slaves under his care in Jamaica, he had never met with any cases of abortion that he could attribute to ill treatment or hard labor. Moderate labor, he continued, was in fact beneficial to pregnant women and the best means of preserving general health.[89]

Historians today have no way of saying how many women killed their unborn children in the West Indies. That abortion was practiced is not in doubt, however. A keen observer like Médéric-Louis-Elie Moreau de Saint-Méry believed that slave midwives routinely induced abortions. (Moreau de Saint-Méry himself sold syringes used to dilate the cervix and bring on abortion from his bookstore in Philadelphia, where he fled in 1801 during the Haitian revolution.) Historians suspect that reports of slave abortions, like reports of slave poisonings (of their masters, plantation slaves, and livestock), have been exaggerated. Indeed, Thomson in Jamaica found that when planters began offering incentives for slaves to become pregnant—a dollar, a silver medal, perhaps a scarlet girdle to be worn on feast days and holidays, and a relaxed work schedule—these allurements encouraged some slave women to feign pregnancies. When the pregnancy did not yield a child, these women produced "some bloody discharge in evidence." To end such practices Thomson advised appointing a "faithful" midwife who would not "connive" with them, and implementing "exemplary" punishment for the guilty.[90]

No one will ever be able to say how often women aborted their children in efforts to resist the desperate miseries of slavery. Slave women, no doubt, killed their unborn children for many different reasons, and not just from a desire to save their offspring from a world of hard labor, poor nutrition, frequent rape, disease, and moral and psychological dejection. Slave women were known to have used their sexuality as a political weapon in various ways: on the eve of the revolution in Saint Domingue, for example, some prostituted themselves to French soldiers in exchange for bullets and gunpowder.[91] But there are many forms of resistance, and many forms of submission, and it is not always possible to discern among them. Slave women may have submitted to conditions they could not change, but they clearly also used a number of means within their power to control their own fertility, confounding their masters' efforts to have them reduced to breedable beasts of burden.

4

The Fate of the Peacock Flower in Europe

> Some years ago, while at the University of Edinburgh . . . a few of us associated for the purpose of making experiments on various medicines.
>
> JAMES THOMSON, 1820

Europeans recreated a bit of Europe wherever they landed, planting gardens with carrots, beets, and peas in Barbados, raising chickens, goats, cows, and horses in Saint Domingue. Interestingly enough, Edward Long in the 1770s listed several known European abortifacients—pennyroyal (*Mentha pulegium,* native to Europe and Asia) and rue (*Ruta graveolens,* native to southern Europe and northern Africa)—growing in English kitchen gardens in Jamaica. Whether these plants were originally brought to Jamaica as anti-fertility agents is not known; like most herbs and spices, they enjoyed many uses. In addition to being employed as contraceptives and abortives, pennyroyal was said to relieve giddiness and headaches, and was used to purge the lungs of "gross humors."[1]

Did abortifacients travel in the opposite direction? Did the peacock flower (*Poinciana pulcherrima*), well known to Europeans in the West Indies, voyage to Europe? Was it planted there in the kitchen and botanical gardens of Paris, London, and Leiden? Did its abortive qualities become known in Europe, and did it pass into European *Pharmacopoeia* as an approved drug? Was it used by European women the way they used sugar, tea, chocolate, quinine, jalap, and ipecacuanha?

When analyzing whether a contraceptive such as the peacock flower moved into Europe, we need to distinguish between movement of *knowledge* and movement of the *plant itself.* One thing is clear: The peacock flower—the plant itself—moved easily into Europe. From about 1666 onward, the flowering bush was brought repeatedly into Europe from both the East and West Indies. It was cultivated in the major botanical gardens all across Europe, as we know from records of the catalogues of plants grown in these gardens. The first documented specimen in Europe that I have found grew in the Jardin du Roi in Paris in 1666; this specimen came from the Americas, and was most likely sent to Paris by General Philippe de Lonvilliers, chevalier de Poincy, governor of the French Antilles, or Jean-Baptiste du Tertre, the Dominican missionary and naturalist in French West Indian territories whom we met in earlier chapters. (The list of plants growing in the King's Garden does not indicate the source of the specimen.) By 1682, the peacock flower also grew in the Hortus Medicus in Amsterdam and the Horto Academico in Leiden from seeds sent by Jakob Breyne from the East Indies. The *Poinciana* also flourished in the Chelsea Physic Garden outside London, and in Carl Linnaeus' gardens in Uppsala, and may even have grown in the Royal Botanical Garden (Real Jardín Botánico) in Madrid after its founding in 1755. The Spanish connection is probable, given that the plant was described by Francisco Hernández already in the sixteenth century.[2]

The elegant peacock flower was well known to gardeners; its flaming yellows and reds made it a favorite ornamental. Philip Miller at the Chelsea Physic Garden provided precise details concerning cultivation of the peacock flower in his gardening manual. With proper management, he wrote with remarkable hubris, the *Poinciana* will grow much taller in England than in Barbados, although the stems will not be larger than "a man's finger, which is occasioned by their being drawn up by the glasses of the stove [a stove house or heated greenhouse, a term coined in the seventeenth century but rarely used even in the eighteenth century]. I have had some of these plants near eighteen feet high in the Chelsea Garden, which have produced their beautiful flowers some years."[3] For Miller, though, it was exclusively an ornamental.

Although the *Poinciana* was grown in Europe, the knowledge of its use as an abortifacient did not take root there. Merian's *Metamorphosis insectorum Surinamensium,* including her report of the abortive qualities of the peacock flower, was published in 1705. Caspar Commelin, director of the Hortus Medicus and professor of botany in Amsterdam, prepared bibliographical notes for this book and would have been intimately familiar with its content. Despite this, Herman Boerhaave, a leading authority on Europe's *materia medica* and professor of botany at Leiden, just down the road from Amsterdam, reported "no known virtues" of this plant in 1727. Boerhaave knew abortifacients well; he warned that "ecboliques avortifs" (abortive agents) be used "with reserve" because they endanger not only the fetus but also the life of the mother. It is also significant that the *Poinciana* was not listed as growing in the Parisian apothecaries' garden, which supplied plants to French physicians. And, although Charles

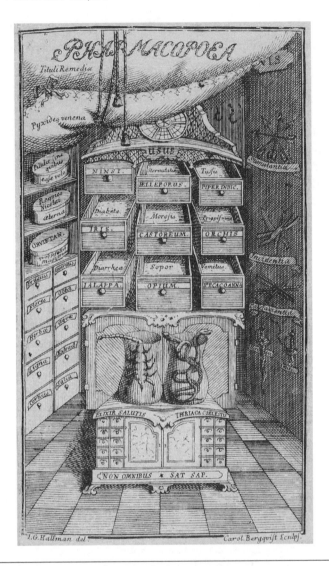

Figure 4.1. Linnaeus' "Pharmacopoea"—meaning both the printed book describing drugs and their uses, and the medicaments approved for use in apothecary shops. The drawers of the apothecary's cabinet are marked with drugs, several—jalap and ipecacuanha—from the New World. A plant that successfully entered into European drug therapies was generally tested according to the standards of the day and eventually became listed as an official simple in one of the *Pharmacopoeia* published in major cities around Europe.

Alston grew the *Poinciana* in his gardens of medicinal plants in Edinburgh, like his colleagues across Europe he did not discuss the *Poinciana* as an abortifacient.[4]

My argument in this chapter is that although the peacock flower was taken many times into Europe, the knowledge of its use as an abortifacient did not transfer to Europe. European physicians had easy access to reports of its use as an abortifacient in the West Indies, but none of these doctors promoted remedies made from the peacock flower for this purpose in their practices. It is telling that in all my searches, I have found only two recipes for the use of the peacock flower: one as a cure for belly-ache, the other a cure for fever.[5] Nothing suggests that knowledge of the abortive virtues of the *Poinciana* was actively suppressed, yet this knowledge only rarely made its way into Europe. And where it did, it did not flourish.

How can a "nontransfer" of knowledge be documented? I have wrestled with the problem of what constitutes proof that something of significance did not occur. One way to look at the nontransfer of knowledge surrounding the *Poinciana* is to compare its introduction (or nonintroduction) into Europe with that of other medicinal plants at the time. If physicians had introduced the peacock flower into Europe as an abortifacient, what pathways of testing and certification would it have followed? By comparing the fate of Merian's peacock flower to that of other newly identified drugs from the New World, we can see to what extent it did or did not participate in the culture of drug experimentation in Europe of its day. This chapter, then, lays out the eighteenth-century culture of drug testing and looks in particular at the extent to which the *Poinciana pulcherrima* and other West Indian abortifacients were tested according to standard eighteenth-century practices.

Animal Testing

Physicians in the eighteenth century complained that the "experimental knowledge" of medicines was defective and that the virtues ascribed to many plants were often based wholly on the authority of Dioscorides (or some other agent) rather than empirical experience. The Jamaican James Thomson, who was part of a new experimental program at the University of Edinburgh in the late eighteenth century, added that in some cases, the "most opposite qualities" were ascribed to the very same plant; in other cases, cures were attributed to substances that were known to possess no active properties whatsoever.[6]

Medical testing was not (is not) always something set apart from everyday medical practices. As the great experimentalist Claude Bernard wrote, "Physicians make therapeutic experiments daily on their patients, and surgeons perform vivisections routinely on their subjects." Since antiquity, physicians and healers of all sorts have tried new and untested cures in the regular care and management of patients, and especially in desperate situations. By the eighteenth century, however, as Andreas-Holger Maehle has shown, drugs were more often being tested according to a set of procedures agreed upon by the European medical community at large. Physicians at the time labeled these procedures "trials" or "experiments," *expériences* or *pareilles épreuves*, and even "controlled experiments" *(regeln Versuche)*.[7]

The first step in eighteenth-century drug development was to identify a potentially useful substance. As we saw in Chapters 1 and 2, drug discovery was big business in this period; trading companies, governments, and private individuals invested large sums of money in search of new and potentially profitable drugs, perhaps equivalent in value to the "Peruvian bark," or new spices or dyes that were often tested also for culinary and industrial use. After identifying a potentially useful medicament, an experimenter examined its color, smell, and taste in order to understand its relation to other known drugs. Laboratory testing with fire or other substances, especially blood, was considered important to determine the chemical makeup (acidity and alkalinity) of the substance.

Testing for toxicity was a next step of crucial importance and done through the experimental use of animals. Anton von Störck, physician at the city hospital in Vienna and pioneer of experimental pharmacology, followed these procedures when developing his nonsurgical treatment for breast cancer. One can imagine Störck's excitement in attempting to develop a drug to cure breast cancer in a period when the only other option was mastectomy. Störck's experiments in the 1760s involved feeding his wonder drug—a hemlock extract—to a little dog. It would be "criminal," he remarked in his published report, to make the first trial of this substance on a human. He gave the dog a "scruple" (1.3 grams) of the drug in a piece of meat three times a day over the course of three days, and took it as a good sign that the dog remained healthy and eager for the food.[8]

Animal experimentation, of course, was nothing new. Since ancient times, experiments on animals had been carried out primarily to test poisons and their antidotes. Rhazes tested the effects of quicksilver on apes. Paracelsus experimented with ether-like substances on chickens. In the eighteenth century, the Italian abbot Felice Fontana employed 3,000 vipers, 4,000 sparrows, numerous pigeons, guinea-pigs, rabbits, cats, and dogs in his experiments with snake venom. Experimentalists in this period were especially keen to use "higher" animals to increase the likelihood that results could be generalized to humans. To this end even horses met their fate at the hands of experimentalists. Since the Renaissance, however, the dog had emerged as the experimental animal of choice. Apes, bears, and lions, according to Renaldo Columbo, Andreas Vesalius' successor at Padua, are internally more similar to humans, but they become angry on being cut, which makes vivisection difficult. Pigs were too fat and their squeals annoying.[9]

It is important to see that European methods of testing were used not only inside Europe but also in Europe's West Indian colonies. European physicians working in the colonies brought established methods for drug testing with them to the colonies and often tested new substances *in situ*. Philippe Fermin in Surinam, for example, gave one dram (the eighth part of an ounce) of juice expressed from the cassava root to a three-week-old dog. Two minutes later Fermin found the animal "turning from side to side and in terrible agony." Death followed in thirty-two minutes. Two young cats also died. Needless to say, Fermin did not taste the poisonous root himself.[10]

Even the mathematician Charles-Marie de La Condamine, wildly jumping fields of competence, experimented on a chicken with the fabled Amazonian curare while traveling in Guiana. "During my sojourn at Cayenne," he wrote, "I had the curiosity to test, whether the venom of the poisoned arrows, which I had preserved for about one year, still retained its potency, and whether sugar in reality was as secure an antidote as it was said to be." As was customary, the experiments were made in the presence of witnesses: the governor of the colony, several officers of the garrison, and a royal colonial physician. A first hen, slightly wounded by a small arrow blown through a hollow pipe, lived about a quarter of an hour. Another chicken, pricked only in the wing with an arrow, was seized with convulsions. Despite the bird being made to swallow sugar after it had "fainted," it expired. A third chicken, pricked with the same arrow and immediately administered sugar, showed "not the least inconvenience."[11]

Experiments on potentially useful substances were repeated in Europe. Upon his return, La Condamine repeated these experiments with the curare (composed of "more than thirty kinds of herbs or roots") in Leiden before Professors Mussenbrock, Van Swieten, and Albinus. But the experiment did not work. La Condamine attributed his ill success to the cold climate—the experiments being made in January—and to the fact that the poison had lost its force during the lengthy journey from the Amazon. "The experiment," he wrote, "was not repeated."[12]

Explorers like La Condamine often shared substances for testing with well-known chemists, physicians, and botanists who themselves never left Europe. Upon request, La Condamine "offered, with the best grace in the world" samples of the Amazonian poison to the experimentalist Hérissant in Paris, who immediately suffered two accidents. Unaware of the poison's strength, Hérissant prepared it in a small closet where his assistant was working. The lad soon became motionless. Hérissant, coming toward him to "reprimand him for his laziness," saw that he was very faint and remembered that La Condamine had warned that in preparing the poison, the native American peoples, the Ticunas and Lamas, "oblige a criminal old woman to take care of the boiling of this poison, . . . when she died, it was a sign, that the poison was sufficiently boiled." The boy was revived with a pint of good wine, a measure of sugar, and plenty of fresh air.[13]

Anxious to retest this effect, Hérissant tried the experiment on himself, using the standard reasoning that "it would be inhumane, not to say criminal, to test it on any other person but myself." After about an hour his legs bent under him and his arms became weak. Near collapse, he stumbled from the closet and, like the boy, was revived in the courtyard by wine and sugar. He then experimented with the poison on four dogs, eight rabbits, four cats, six horses, a bear (that René-Antoine Ferchault de Réaumur wanted to have killed for his cabinet of natural history), and every sort of hawk, pigeon, hen, blackbird, sparrow, duck, goose, magpie, innumerable worms, vipers, and insects. Hérissant found that only the mammals and birds succumbed to the poison. In his report, he emphasized that the animals had felt little or no pain before dying and that neither sugar nor seasalt served as a specific antidote to this quick-acting poison.[14]

Much more could be said about animal experimentation. As Johann Friedrich Gmelin emphasized in 1776, however, while experiment with animals was one of the few ways to test drugs that might prove useful to humans, "in the end, no other option remained except to experiment on the human body itself."[15]

Self-Experimentation

Experiments in animals were followed by testing in humans, first on fluids from the human body—preferably blood, bile, chyle, or phlegm. Reactions were observed to see if blood, for example, thickened or fouled. The next important step was typically self-experimentation. Julia Douthwaite, a literary critic, has described this developing tradition as a form of "autobiographical empiricism," whereby credible subjects scrutinized minute effects in their own bodies and conveyed these as reliable data to other scientists. Historian Stuart Strickland has described how natural philosophers used their own bodies as "calibrated instruments" epistemologically equivalent to the voltaic columns, Leyden jars, thermometers, and other devices cluttering their laboratories. The body of the experimenter provided unique information not available through the use of other instruments, yet ideally it also became an instrument simulating as nearly as possible inanimate objects' indifference to the "prepossessions" of errant humans.[16]

Medical research participated in this general culture of self-experimentation, but unlike the physicists, physicians only rarely used their bodies as unique instruments, as did Albrecht von Haller, who in the course of dying, recorded the effects of opium in the human body taken daily for two and a half years. More commonly, physicians considered their bodies representative of human bodies more generally and put themselves on the front lines of human experimentation by testing new treatments first on themselves. Physicians' bodies were considered "proficient" in a way that patients' bodies were not: a medical expert could presumably distinguish effects relevant to the experiment from other "subjective" states of his own body. The information was also "pure" in that it was gathered from healthy bodies. Autoexperimenters tested for toxicity not discovered in animals and for effects in the healthy human body (which, it was assumed, could handle a dangerous drug better than a body already weakened by illness).[17]

Self-experimentation also tended to exonerate physicians when they prescribed a medication—perhaps with fatal results—for others. The willingness to take a drug oneself, Boston physician Zabdiel Boylston noted, was a measure of a physician's "faith" in the drug.[18]

By the mid eighteenth century medical self-experimenters had built in certain safety procedures. Substances were first to be smelled, then touched to the skin, and finally tasted—first only with the tip of the tongue, and then, if appropriate, taken internally. These procedures can be seen in Störck's hemlock experiments. Encouraged by the benign effect of his extract on his dog, Störck took the next step by experimenting on himself and recorded that he took each morning and evening for eight days one "grain" of his hemlock extract (expressed from the needles of the tree) in a cup of tea. Perceiving no ill effects, he increased the dosage to two grains. Again perceiving no "ill or unusual" effects, he now felt justified "for the best reasons to try this on others."[19]

His experience with the hemlock root was less sanguine: Taking the juice from a freshly cut root, Störck rubbed two small drops on the tip of his tongue. His tongue immediately stiffened, swelling so painfully that he could not speak. "This untoward event frightened me," he wrote, "and gave me great apprehensions of the consequence." Remembering, however, that acid counteracted many poisons, he doggedly washed his tongue with the juice of a lemon, after which he was able to stammer. He repeated the lemon wash every fifteen minutes and, after two hours, was fully recovered. From this experience, he concluded that the strongest poison—and thus effective part—of the hemlock lay in the root and prepared pills from the dried powder of the root for use in his practice.

Although Störck seems to have been experimenting alone in his chamber, physicians taught that experimenters should not, in fact, work alone, for two different reasons. The first was that it was important for others to "witness"—to observe, learn from, and verify—the findings; the second was more immediately practical—namely, to aid in the event that the experimenter should fall unconscious (and also to continue to log the results of the trial).

Toward the end of the eighteenth century, self-experimentation became more systematic and organized. In efforts to overcome the idiosyncrasies of an individual experimenter's body (to calibrate his body against others), physicians and medical students tested potential drugs in groups. James Thomson reported that "some years ago, while at the University of Edinburgh . . . a few of us associated for the purpose of making experiments on various medicines, the active properties of which we had reason to question." According to his report, the doctors and medical students were each assigned specific drugs to test in his own healthy body. Each recorded in minute detail "the state of the pulse, vomiting, dizziness, and every other circumstance." When a particular drug seemed of sufficient importance, it was taken by different individuals at the same time, and the results were compared. Thomson continued, "[W]e then inferred that, generally speaking, the same results will follow in a morbid state, and combat successfully certain symptoms which we wish[ed] to obviate."[20] Here we see the beginning of testing on medical students and perhaps also the beginnings of a preference for testing in male bodies, since medical students were until the late twentieth century overwhelmingly male.

Although medical self-experimentation became a usual aspect of drug development in the eighteenth century, a sense of heroism—a willingness to sacrifice oneself for one's science—was also cultivated among physicians. Hérissant, testing La Condamine's curare mentioned above, looked upon himself as "a dead man" when a phial exploded in his hand and covered him with the poison.[21] Heroic self-experimenters at times also put their bodies at risk in self-abusive ways. In order to better understand gonorrhea, the surgeon John Hunter deliberately introduced the pus from an infected patient into three incisions in his own penis, thereby contracting both gonorrhea and syphilis.

Physicians, then, first tested the efficacy of drugs in humans in their own bodies. How, then, could they test drugs like emmenagogues or abortifacients, uniquely destined for women? The Edinburgh group, being at the university, was all male. To date, I have not found a single report of self-experimentation reported by a woman. Nor have I come across accounts of bodies of wives or female servants standing in for the physician's own body. It should be noted, however, that in the early modern period, women were in charge of household medicine, which required a keen knowledge of pharmacy. Women such as Ann Dacre and Alathea Talbot prepared medicines from their kitchen gardens and circulated—even published—large collections of household recipes; no doubt they experimented on their own bodies. Root or herb women must also have doctored themselves when ill and observed the effects. Midwives, responsible for preparing the medicines they administered, did test them in the course of their practice. The seventeenth-century midwife to the French queen, Louise Bourgeois, for example, admonished midwives to make no "trial of any new remedy or recipe, either on poor or rich, if thou be not assured of the quality and operation thereof," indicating that midwives kept careful watch over the efficacy of their cures. In another instance, an "anonymous lady" reported that she persisted in "trials" of a broom-seed remedy for dropsy in "numerous cases," including a gentleman and, significantly, a woman with child. It is possible that this lady first took the remedy herself.[22]

Human Subjects

Historians generally recognize that human experimentation—the testing of medicines and medical procedures in human bodies—for scientific purposes is very old, dating back to ancient Persia and Egypt. Maehle,

Laurence Brockliss, and Colin Jones all have emphasized that modern medical testing arose in the eighteenth century. This was a period in which medical practitioners developed for themselves implicitly and explicitly agreed on procedures—what we today call protocols. Unlike today, however, eighteenth-century experimentation was not governed by formal codes of ethics. Neither church, state, nor colleges of physicians set out specific guidelines for testing, though the Parlement of Paris stopped the celebrated Jean Denis's human blood transfusion experiments after two men died.[23]

Eighteenth-century physicians emphasized the need for many, repeated trials with new drugs. Because a physician in private practice could test a new therapy on only a few patients, physicians recorded and exchanged observations in the form of "case histories." Increasingly uniform in style, case histories included a description of the patient, an account of his or her past patterns in sleep, diet, and exercise, description of therapies, their effects, and outcomes—"complete cure," relief, or death. Enlightenment physicians circulated detailed reports of observations and experiments in a hope that systematic collection of scattered data could lead to a new medical science beneficial to the greater public good. Medical books in this period bulged with case histories published for circulation to other physicians but sufficiently novelistic to intrigue the reading public more generally. Case histories also attracted patients—some even treated by mail—making the fame and fortune of enterprising physicians.[24]

Explicit laws equivalent to our modern codes regulating the use of human subjects did not exist, but physicians' conduct was highly regulated. Historian Paula Findlen, taking the example of Francesco Redi, has suggested that seventeenth-century experimentation, designed to reveal truths of nature, was politically, materially, and aesthetically produced by court culture. As personal physician to the grand dukes of Tuscany, Redi was primarily a medical man, heading the famed Medicean pharmacy. He was also a natural philosopher who directed—at the Duke's pleasure—experiments on the generation of insects, the decapitation of turtles, and much else at court.[25] Vestiges of this culture survived into the early eighteenth century. The introduction of Turkish and Chinese methods of inoculation against smallpox into England, for example, was encouraged by royal patronage. The engineer of this testing, Charles Maitland, like Redi, served powerful families, and he worked with the royal physician Sir Hans Sloane, who brokered delicate relationships between science and its patrons. The use of experimental populations—condemned criminals, orphans, soldiers, and sailors—also required royal sanction in both England

and France. Experiments, many of which were carried out first on wards of the state, were regulated by the Crown.

Even though the seventeenth and eighteenth centuries shared a culture of royal patronage, English royal intervention, for instance, differed significantly from that of the Italian dukes. Experimentation with inoculation did not take place at court in the presence of the patron but in prisons, orphanages, charitable hospitals, and private homes. Royal physicians stood in for absent monarchs. Nor was the end of an experiment a witty conversation or dazzling display of nature for the entertainment of the court, as in the case of Redi; the point was rather a test—often with life or death consequences—of medicines in the body.

Medical testing in the eighteenth century was subject to restraints imposed by royal patrons also because it was often tied to state policy. The pros and cons of life-saving smallpox inoculation, for example, were much discussed publicly in the eighteenth century, one side seeing it as leading to an undesirable explosion in numbers of the nation's working poor, the other side seeing it yielding a desired increase in population, and hence the wealth and power of the state. In France, inoculation was to be administered in the Paris hospital for abandoned children in order to conserve the "greatest number of citizens" for the state. In Saint Domingue, engrafting the smallpox was viewed as "favorable to the interests of population, the state, and families" not only because it decreased overall mortality but because it also conserved the beauty of young people, whose disfigurement from the natural smallpox might otherwise have required them to seek a life of celibacy. Charles Arthaud, permanent secretary of the Cercle des Philadelphes at Cap Français, taught that it conserved the tenderness of a husband for a wife, whose beauty remained unblemished by ravages of the smallpox.[26]

Not all drug testing, of course, was a matter of state. Experimental culture broadened in the eighteenth century. Redi published in the form of letters to his patrons, but eighteenth-century physicians reported experimental results to scientific colleagues. Maitland sent his elaborate journal, written in English and French, chronicling the Newgate prison experiment with inoculation to Sloane; the French version was for the benefit of colleagues on the Continent. John Quier's elaborate experiments in Jamaica (discussed below) were sent in letter form to physician Donald Monro in London and later published for broader distribution. Letters, reports, and case histories were published and republished by scientific societies for circulation among the profession and learned public.

Mary Fissell, an historian of medicine, has argued that medicine in this period had no unique professional ethical code—indeed needed none—because physicians' conduct was based on the decorum, propriety, and decency of traditional gentlemanly ethics.[27] At the same time, certain restrictions were placed on an individual physician's conduct, often in connection with a specific incident, such as the failure of blood transfusion, by professional or civic bodies that included the medical faculty in Paris.

Medical experimentation was also regulated by the marketplace. Baron Dimsdale reminded his colleagues that those acquainted with the first aphorism of Hippocrates—"to help, or at least to do no harm"—will be cautious where the object is no less than the life of an individual.[28] Certainly a physician's livelihood depended on his reputation: A careless doctor soon had no patients. Even vulnerable populations, such as Caribbean slaves, were protected to a certain extent by considerations of masters' pocketbooks.

One purpose of drug testing in human subjects was to determine proper dosages. We must remember that drugs are not absolutes. Early-modern physicians were acutely aware that what heals at one dosage can be a deadly poison at another. The English physician and botanist William Withering, known for developing digitalis for the treatment of congestive heart disease, discussed the difficulties of ensuring uniform dosage when gathering and preparing the foxglove *(Digitalis purpurea)* for his medicines. Aware that the potency of the plant's root changed radically with the season, he chose instead to use the leaves, gathering these always at the same time of year and "in the plant's flowering state."[29]

The mode of preparation also influenced dosage. Arguing that the properties of the *Digitalis purpurea* were similar to tobacco *(Nicotiana tabacum)* because they were "of the same natural order," Withering realized that the active properties of the foxglove might be damaged by the long boiling required to prepare a decoction. During his course of study, he switched from decoction to infusion (simply steeping the leaves in hot or cold water). To gain even greater uniformity in dosage, Withering finally powdered the leaves.

Discovering correct dosages for a wide range of patients required testing in populations that extended beyond the bodies of individual physicians. Who were to be the subjects of these experiments? In addition to testing new therapies and dosages on charity and private patients, many eighteenth-century physicians followed the ancient pathway of testing new cures on condemned criminals [*Missethätern*]. The use of prisoners in human experimentation can be traced to ancient Pergamum, where fear of poisoning led King Attalus III Philometor to have toxins and their antidotes tested on prisoners. In the late seventeenth and eighteenth centuries these practices continued. Denis Dodart, a doctor of the medical faculty of Paris, advocated in 1676 the use of the condemned for the most extreme drug trials. Christian Sigismund Wolff in a 1709 disputation at the University of Leipzig went so far as to advocate the ancient practice of vivisecting criminals on the grounds that society would benefit from the knowledge gained. He pointed out, however, that the supply of the unhappily condemned was not plentiful.[30]

In the 1750s, Pierre-Louis Moreau de Maupertuis bemoaned the fact that few "new and dangerous cures" had been adequately tested, and proposed that criminals be employed for this purpose, again, "for the good of society." He supported the notion that a criminal who survived should be pardoned, "his crime having been expiated by the utility of his action." Prisoners, Maupertuis cautioned, should be used only for such operations as the removal of kidney stones or uterine cancers, for which "neither nature nor the art" had provided a cure. Even with the condemned, Maupertuis, echoing Dodart, advocated that for "the sake of humanity" the physician should diminish as much as possible the pain and the peril of the procedure. Further, physicians should first perform operations on cadavers and then on animals before performing them on criminals.[31]

Maupertuis, however, went beyond most of his colleagues in this period and thrilled at the thought of using condemned persons for purely speculative research. "Perhaps we will discover," he enthused, "the marvelous union of the soul and the body if we dare to search in the brain of a living human." Intoning again the theme of public utility, he continued, "[O]ne person is nothing compared to the human species." Still no one in the eighteenth century advocated the kind of cruelty that Francis Bacon reported, in which a man convicted of treason was opened alive, and whose heart, when thrown into hot water, leaped to the perpendicular height of two feet.[32]

Not all physicians, however, considered criminal bodies reliable instruments for experimentation. Jean Denis, who performed the first blood transfusion on a human in 1667, rejected the solicitations of "diverse persons of much gravity and prudence" who urged him "to beg some condemned criminal, on whom to make the first trial." Denis feared that a condemned man might deem "transfusion a new kind of death" and that the fear of it "might cast him into faintings and other accidents" that would unfairly defame Denis's own "great experiment."[33]

Experimentation on criminals was, as Maupertuis noted, "often proposed but rarely practiced." Historians have emphasized that physicians tested the poor and displaced in their surgeries and hospitals. It was not uncommon in the eighteenth century, however, to secure wellborn subjects for experiments to popularize new remedies. To be sure, nostrums and innovative surgical procedures were well tested on people of "inferior" status before offered to royalty, but, as in the case of the introduction of smallpox inoculation into England and the use of anti-malarial Peruvian bark (Chapter 5) reveal, the willingness of nobility to submit to a procedure or medicament did much to popularize its use. Early in the process of transmitting smallpox inoculation from Turkey to England, Charles Maitland had delayed engrafting Lady Mary Wortley Montagu's daughter because he felt she was too young. He also hoped to use the occasion of her inoculation to set the first great example of its use in England. It was the girl's successful inoculation that sparked interest within the English royal family and led to the Newgate experiments (discussed below). Two royal princesses were inoculated in 1722 without mishap, which further enhanced the procedure's public reputation. More than one physician remarked that inoculation conserved the lives and also the beauty of some of the "most amiable Princesses in Europe."[34]

Royal bodies, while epistemologically weighty, were few in number. In the eighteenth century, wards and employees of the state—hospital patients, soldiers, sailors, and orphans—provided opportunities for wider testing of new medical cures. Hospitals had long been charitable organizations for the care of the poor. In the eighteenth century, new and larger hospitals were founded in response to the needs of both states and the medical profession. Institutions such as the Royal Infirmary in Edinburgh and the Hôpital Général in Paris were established across Europe with the purpose of decreasing welfare costs by returning wards of the state to gainful employment. Lying-in hospitals, also created in this period, were to serve the "Wives of poor industrious Tradesmen," of "indigent Soldiers and Sailors," as well as unmarried women in order to secure a growing and healthy population also for the benefit of the state. These new urban hospitals—in London, Edinburgh, and Vienna—were associated with universities in such a way that they could be used to train new doctors and to develop new therapies. The Dutch physician Gerard van Swieten in Vienna, for example, inaugurated a systematic program of clinical drug studies in 1754 (Störck was van Swieten's successor).[35]

Hospitals in this period served as laboratories for the development of medical techniques, pushing forward the revolution in experimentation. Large hospitalized populations—both civilian and military—allowed for rationalized teaching, controlled bedside trials, and development of medical statistics. The class distinctions between the sick poor and the educated doctors lent medical men an authority over these patients they did not have over their paying patients, many of whom were well born. Moreover, patients in all hospitals became accustomed to strict diets and regimens that allowed for more controlled testing than doctors could secure from trials with private patients. Military hospitals, in particular, provided large numbers of patients as well as corpses for dissection and autopsy.[36]

Many colonial drugs were tested in European hospitals. Louis XIV, for example, awarded Adrien Helvétius, a Dutch physician, the right to make *diverses épreuves* with the colonial drug ipecacuanha, the "Indian Vomiting Root," a specific against dysentery at the Hôpital Général and the Hôtel-Dieu in Paris. Donald Monro in London reported on experiments at Saint Bartholomew's hospital in 1779 with Peruvian bark that had been captured in the cargo of a Spanish warship. It was "much larger, thicker, and of a deeper reddish color than the bark in common use" and was found to be even more efficacious than the common bark. Drug experiments were also carried out in colonial hospitals. In Spanish territories, already in the sixteenth century Francisco Hernández had tested some of the 3,000 medicinal plants he collected for the Crown in hospitals in New Spain. Historian Antonio Lafuente has judged the studies of Mexican plants conducted by Mariano Mociño and Luis Montaña in the Royal Native Hospital and the San Andrés Hospital, starting in 1801, among the most innovative episodes of the new creole science. In British and French territories in the West Indies, medical instruction and experimentation took place primarily in military hospitals. John Hume, a royal naval surgeon for forty years (ten of which were served in Jamaica), made observations on yellow fever and intermittent fever in 250 men. When men died, he dissected them in order to characterize the course of the disease more carefully.[37]

Orphans were another group utilized for experimentation. Without family to intervene, these children were freely used. Experiments with orphans continued well into the nineteenth century, when subjects were often compensated for their cooperation. In Russia in the early nineteenth century, the empress dowager arranged for an orphan to receive the new Jenner vaccination against smallpox. The child was rewarded by being named "Vaccinoff" and educated at public expense. In 1803, the king of Spain, in order to overcome smallpox in his North and South American domains (a major source of depopulation and thus decline of revenues) sent Jenner's vaccination over the seas via a living chain of twenty-two orphan boys. The royal surgeon, Don Francisco Xavier Balmis, kept the vaccine alive by passing it successively from boy to boy during the journey. While passing along the coast of Peru, the surgeons vaccinated no fewer than 50,000 people. The boys were settled in Mexico City and educated at the king's expense.[38]

Testing for Sexual Difference

Before turning to the question of how exotic abortifacients were tested in Europe, it is important to ask whether women were regularly included in eighteenth-century medical testing. In 1993, women in the United States were guaranteed the right to inclusion in clinical trials by federal law. Immediately before this time, females had been underrepresented in drug testing on the grounds of expense (adding females required more control groups) and of possible danger to a fetus (for fertile women).[39] Interestingly, we find that females were a required element in eighteenth-century medical testing and that an analysis of sexual differences was considered important to the results. Women served as test subjects for drugs and procedures developed for nonreproductive health as well as for cures developed specifically for female maladies, such as breast cancer.

Galenic humoralism, enriched with Paracelsian iatrochemistry, dominated much eighteenth-century medicine. Humoralism taught that disease and medical therapies often progress differently according to the age, sex, and temperament of the patient. Ideally, a physician took into account these and other individual circumstances affecting a patient—including the climate, water, winds, their morals, nourishment, employment, type of clothing, and so forth—before prescribing treatment. As one well-known physician put it, a proper experiment required repeated "trials" on persons of "different ages, sexes, and constitutions, in different seasons of the year, and in different climates." Experiments in this period were deliberately designed to take these factors into account (in human, but not in animal experimentation).[40]

Nowhere is this more apparent than in the public, royally-sanctioned Newgate experiments in which six prisoners underwent the Turkish method of smallpox inoculation in 1721. This carefully designed experiment was presided over by two royal physicians, Sloane and Johann Steigerthal; the royal surgeon, Charles Maitland; and observed by some twenty-five persons, including "several eminent physicians," surgeons, and apothecaries from different countries. The question of the safety of this potentially lifesaving procedure was crucial. Before having her own children inoculated, Queen Caroline "begged the lives of six condemned criminals" to undergo the experiment of inoculation "to secure her children" and "for the common good." It is significant that the six prisoners included three women and three men matched as closely as possible for age:

>Anne Tompion, age 25   John Cawthery, age 25
>Elizabeth Harrison, age 19   Richard Evans, age 19
>Mary North, age 36   John Alcock, age 20

As one observer reported, "it was desired to know how the operation acted in persons of all Ages, Sexes, and different Temperaments." The experiments began at nine o'clock on the morning of August 9, 1721. A German observer reported, perhaps only to heighten the drama, that the criminals trembled when Maitland took out his lancet, fearing that they were to be bled to death.[41]

Scattered throughout physicians' casebooks—recording the treatment of both private and hospital patients—are remarks on the importance of testing in both sexes. Sexual differences were carefully scrutinized, for example, in Dr. Fabritius's experiments with intravenous injection of laxatives carried out at a Danzig hospital in the late seventeenth century. Fabritius's first subject was a "lusty robust soldier" dangerously infected with venereal disease and "suffering grievous protuberations of the bones in his arms." Two drams of a "Laxative Medicine" (not further identified) were injected by means of a siphon into "the median vein of the right arm." Four hours later the laxative began to work; the man had five good stools. As reported in the *Philosophical Transactions of the Royal Society of London,* the experiment was hugely successful: without any other remedy, the man was "completely cured."[42]

The procedure was tried on two further patients of the "other sex," this time to cure epileptic fits. One was a married woman, aged thirty-five, the other a serving-maid, aged twenty. The married woman had "gentle stools" almost immediately after the injection. By the next day her fits were much milder, and soon "altogether vanish'd." As for the maid, she had four stools the day of the injection and several the next. She was, however, an unruly patient and her disobedience led quickly to her demise. "By going into the air, taking cold, and not observing any diet," the report read, she "cast herself away." Her unruliness, not the procedure, was faulted for her death.[43]

Thomas Fowler's experiments with tobacco in the 1780s on 150 persons offer another example of tests designed to include a nearly equal mix of males and females. Fowler, physician to the infirmary in Stafford, England, found the diuretic effects of the alkaline-fixed salt of tobacco (from the New World), administered internally, effective against dropsy. The cure, however, depended on the correct dosage; too high a dose might cause vertigo, nausea, or excessive purging. In the course of his testing, Fowler found that when age was held constant, sex, indeed, became the most important variable. He adjusted the dosages for his nicotine-based diuretic (to be taken daily two hours before dinner and at bedtime) accordingly:

1st class, 21 cases (3 men, and 18 women) 35–60 drops.
2nd class, 57 cases (29 men and 28 women) 60–100 drops.
3rd class, 13 cases (9 men and 4 women) 100–150 drops.
4th class, 3 cases (3 men) 150–300 drops.

The extremes in this study were at one end a weak and nervous woman, Sarah Dudley, who tolerated but twenty drops of the infusion, and at the other an old man, Charles Nicols, who, accustomed to the use of tobacco, required 400 drops. Fowler also tested his concoction on children, although not children under age five because, he wrote, "they could not so well describe the effects of so active a medicine."[44]

Significantly, Philippe Pinel and the Society of Medicine in Paris recommended that basic divisions of sex and age structure hospital teaching wards: "I believe," Pinel wrote, "that nature suggests a distribution according to age and sex. Every age has, so to speak, its own way of life and sickness, and demands fundamentally different therapy for the same disease. This holds equally true for the two sexes." According to Pinel, wards were to be divided first by sex and further by age. For men, the subdivisions included: (1) boys up to puberty; (2) adults to about fifty years of age; and (3) men from the "climacteric" to senescence. For women, subdivisions were similar: (1) girls up to menstruation; (2) the whole period of fertility, that is, from onset to the end of menstruation; and (3) from menopause to what he called *"femina effeta."*[45]

In addition to being represented as test subjects in the development of drugs and procedures for nonreproductive health, women also came under close experimental scrutiny for uniquely female diseases, such as breast cancer. We are familiar with Fanny Burney's harrowing description of her mastectomy in 1811—an excruciating "torture" that lasted for twenty minutes without anesthesia (save for a glass of wine) or antiseptic.[46] Hers was successful; the vast majority was not.

After testing his anti-cancer hemlock drug on a small dog and then on himself, Störck in Vienna began to employ it experimentally on a number of humans and in particular on five women with breast cancers. Initially Störck's new cure met with dazzling success. Such was the case of the healthful, twenty-four-year-old woman who came to him in October 1758 with a lump the size of a goose's egg in her right breast. For this particular woman, Störck prescribed "every morning and evening, three pills, of two grains weight each." By January the woman was cured "and from that time," Störck wrote, "I never saw her again."[47]

Many of Störck's test subjects, however, were hopeless cases sent to him by Gerard van Swieten and other professors at the university in Vienna after they had offered all the help they could. Störck's case XI, a seventy-year-old fruit seller with a fetid, running cancer in her left breast, was typical. Van Swieten, Dietman, Glasser, and Jaus examined her in the surgery at the university and sent "this miserable object" to Störck in late June 1759. In his case book, Störck recorded, "the whole breast was of a brownish black color, and full of lumps. The ichor stank most abominably, even at a great distance; and the patient could neither eat nor sleep for the pain." Störck prescribed that she take four hemlock pills internally morning and evening. He also applied externally a plaster of hemlock leaves. One month later, the cancer was already subsiding, and Störck took her back to the physicians at the university who were "greatly surprised" at her extraordinary improvement. Störck then reported that van Swieten gave the patient money. We do not know if it was for her cooperation in testing the drug or to pay for Störck's pills, which he advised her to continue using.[48]

A check of the patient in August showed continued progress. In early September, however, while sitting in the street selling fruit, the woman felt "a very great cold" in every part of her body. She was overtaken with "excruciating pains" in her belly, followed by "violent and painful" purging. Störck ordered an immediate halt to her hemlock pills and prescribed medicines only to ease her pains. The next morning she was no better; she voided blood in her stools and fainted frequently. Störck, accompanied by a surgeon named Laber, went to her and offered internally and externally any medicines they believed would help, but in vain. The third day her face had the appearance of a corpse, and the fourth day she died.

Medical investigations, however, continued. Laber cut off her breast and carried it to the university physicians for examination. According to Störck, van Swieten and the other professors were impressed by the good effect of the remedy and distressed that "the successful progress of the experiment" had been frustrated by what they called "the accidental death" of the poor fruit merchant.

Störck's hemlock cure for cancer gained such a reputation that other physicians began testing it. In Ireland a Dr. Sherratt used Störck's hemlock pills, carefully prepared according to Störck's own instructions, in 300 cases but without success. The Montpellier physician Jean Astruc also tried Störck's pills, again without success. Richard Guy in London reported that Störck's cure "has been tried by the most judicious practitioners in most of the hospitals in Great Britain, as well as in private practice, to its utmost extent, without a single instance of success." Guy conjectured that differences in climate (England being colder than Vienna) or the soils in which hemlock grew might account for these failures. Several "Gentlemen of the Faculty" contacted Störck to make sure they had the correct plant. Still having no success, Guy and his colleagues procured directly from Störck "considerable quantities" of the extract (at the cost of seven guineas per pound) prepared by his own hand. After repeated failures, Guy judged Störck's celebrated elixir (and similar preparations of belladonna popular at the time) injurious to many persons, causing vertigo, coma, languid sweats, sickness of the stomach, and paralysis. He even ventured that some people had been "sacrificed to the Experiments." "It would," he judged, "be highly culpable to remain silent."[49]

Guy was atypical in calling a fellow physician to task. It was more common in this period to blame the patient or "accident" when experiments failed. Fatal inoculations of smallpox were explained away, for example, as the result of earlier illnesses, not inoculation itself: a gentleman would have been out of danger had not his blood been inflamed by "all manner of excesses"; a young woman would not have died had not the "accidents" of a laborious pregnancy exhausted her strength; a child would have escaped had not a malignant fever and the "purples" aggravated the illness.[50]

Like today, physicians in the eighteenth century preferred positive to negative results and, from the overwhelming number of patients who went home "perfectly cured" (as one reads time and time again in physicians' case histories), one assumes that doctors tended to underreport when experiments failed and also to choose their subjects with care so that trials of medicaments had the best possible chance of succeeding. Physicians often simply refused to treat patients "too far gone." Störck's critic, Richard Guy, turned away the wife of Mr. Megus, a wire-drawer in Barnaby Street, languishing with a miserable breast cancer. Though he refused her treatment, he was "solicitous" to discover the exact "quality" of the "cancerous humor" flowing from her breast down her arm. With her permission, he collected the flowing lymph "in teaspoonfuls," which he tested with several "menstruums."[51]

Many sorts of human subjects, then, were used for medical testing in the eighteenth century—both in Europe and its colonies. In addition to living patients, surgeons and midwives commonly perfected new techniques on corpses and manikins. The London surgeon William Cheselden perfected his famous "cutting for stones" (which he could extract in less than a minute from the bladder) on dead bodies. John Leake, who founded the Westminster Lying-In Hospital for Women in 1765, invented an apparatus "intended as an artificial Substitute, for the real Bodies of Women and Children," upon which students could acquire the "dexterity required, which can never be taught by Description."[52]

The Complications of Race

Colonial physicians of European origins were tightly tied into the transEuropean community of drug testing. Most colonial physicians trained in Europe: Jamaican physicians studied in Edinburgh or London; Saint Dominguan physicians trained in Paris and Montpellier; and physicians from both English and Dutch territories trained in Leiden. Many colonial physicians were also corresponding members of learned societies in Europe. They answered queries from their European colleagues concerning colonial drugs or medical procedures, and they often published the results of their own experiments in established European journals. The Saint Dominguan Cercle des Philadelphes, which included a number of physicians, began publishing the results of their research, but the society ceased to exist after about 1791. The editor of the first issue (1834) of the *Jamaica Physical Journal* remarked that "it is somewhat strange, if not a reproach, that, amongst the various publications which have, from time to time, issued from the Jamaica Press, no periodical has ever emanated from the Medical Community." The Jamaican College of Physicians and Surgeons was founded only in 1833, which meant that throughout the eighteenth century the focal point for organized medical research remained in Europe. It is possible, of course, that there were independent medical traditions forged among native Amerindian or African slave communities.

The Caribs and Arawaks, for instance, may have developed methods for testing the drugs that the Europeans were so eager to know about upon their arrival, but documents have not survived that describe such practices.[53]

Trained in Europe, colonial physicians like Thomson, who was part of the experimental group at the University of Edinburgh, implemented European-style drug testing in the colonies. Thomson returned from his studies in Edinburgh to Jamaica at the end of the eighteenth century and there published his *Treatise on the Diseases of Negroes . . . with Observations on the Country Remedies* in 1820. As he had been taught, he began his experiments with colonial plants such as Capsicum peppers or unroasted coffee "on my own person." Having read about coffee in a foreign medical journal, Thomson hoped he had found something that would work as well as "the bark" (Peruvian bark) as a cure for intermittent (malarial) fever but that would not make him and his patients nauseous. "Satisfied with the results" in his own healthy body, Thomson waited for an opportunity to use this new cure on some diseased subject, "who I little imagined would prove to be myself." It "worked well," so he also gave it to a young gentleman suffering from intermittent fever and also to a "Negro woman who had long been troubled with fever for whom neither the bark nor snakeroot offered respite." He made further "trials," as he called them, on himself and individual patients—of either European or African origins—with *Zanthoxylum* (prickly yellow wood), quassia, the barks of the lilac or hoop tree and the neeseberry bullet-tree. After these tests, he used these drugs regularly in his practice, noting that the doses given to "Negroes" must sometimes be larger than those given to white persons.[54]

While colonial physicians tended to follow European protocols in drug testing, some developed remarkable experiments using unique West Indian populations—plantation slaves. In many instances European physicians in the colonies did *not*—as might be expected—use slaves as guinea pigs. Slaves were considered valuable property of powerful plantation owners whom doctors were employed to serve. But there were also abuses: in at least one instance plantation slaves were used in ways that Europeans rarely were. Here we look in some detail at John Quier's experiments with smallpox inoculation on at least 850—and probably closer to a thousand—plantation slaves in Jamaica.[55] Quier's experiments, carried out in the late 1760s, were remarkable for their size. Many physicians considered tests on five or six patients significant; large tests might include up to 300 patients. Quier's experiments are also remarkable for the light they shed on abortion practices among slaves and what European doctors knew about them. Many historians have used Quier's work in their discussions of smallpox inoculation, but no one has looked in detail at what he did (or did not) know about abortions that took place under his watch.

Quier studied surgery in London and medicine in Leiden, and early in his career served as an assistant surgeon in military hospitals. For fifty-six years he practiced medicine in Jamaica, where he had between four and five thousand slaves from various plantations constantly under his care. For this work, he was probably paid at the going rate of five shillings per head per annum.[56]

Inoculation had been introduced into Europe in the 1720s and into the West Indies as early as 1727.[57] Quier's experiments took place after the College of Physicians in London had endorsed the procedure in 1755, but while heated debates on the subject still raged in France and elsewhere on the Continent. In face of epidemics throughout the island of Jamaica, Quier and others in the colonies were persuaded to undertake inoculation because of the "extraordinary accounts of success received from England." Physicians in Jamaica followed the rules for inoculation laid down by Dr. Thomas Dimsdale in his treatise, sent by "the agent for the island," even though they were calculated for a colder climate and thus considered possibly invalid for England's Caribbean colonies.

Quier was employed by plantation managers and would have inoculated plantation slaves whether he was engaged in medical research or not. We can see from his reports, however, that he took what he considered a rare opportunity to explore questions about inoculation still pressing within European medical circles. Because his patients were slaves, Quier was able to answer questions that doctors back in Europe could not, including whether one could safely inoculate certain groups: menstruating or pregnant women (not inoculated in Europe for fear of miscarriage); newborn infants (not engrafted in Europe for fear of death); a person already suffering from dropsy, yaws, or fever; and so forth. In order to answer these questions, Quier reported that he (sometimes at his own expense) "made numerous repetitions of the inoculation in the same patient"—a practice rarely carried out on European subjects in either Europe or the colonies. The information he gathered was sent in three detailed letters (of 1770, 1773, and 1774) to Dr. Donald Monro of Saint George's Hospital in London, read to the College of Physicians, and subsequently published in book form.[58]

One question under consideration at this time was the amount of medical or dietary preparation required before inoculation could be safely carried out. Adherence to Galenic medicine (and often their own pocketbooks) led physicians in Europe to believe that inoculation patients required lengthy preparation and after-treatment in special isolation houses. Quier went to work "trying various experiments" with preparations. He pointed out that persons of African origin could not bear the "frequent repetitions of strong purgatives" commonly used to prepare Europeans. Many of them were also already being treated for "venereal complaints" and thus could not withstand the mercury preparation prescribed for hot climates. For slaves, Quier recommended a mild compound of calomel and emetic tartar to be administered only one day before inoculation or at most in two doses over the course of three or four days. Perhaps his greatest finding was that slaves required no preparation at all; he was pleased with his results from a group of 300 slaves inoculated without preparation of any kind (which was, in fact, medically sound from today's point of view). This fit nicely with planters' needs. In the Caribbean, where entire plantations were inoculated at one time in order to avoid the spread of disease, planters were usually anxious to dispense with preparations and the loss of their workforce for any extended period of time. Slaves were expected to continue their daily labor in the field during the whole process of inoculation. Only those who "experienced high fever or who had their feet and hands so swollen with pustules that they could not work or hold utensils" were exempted from labor. In these cases, one to three days' rest was granted.[59]

Quier did a number of other experiments in which he was guided by the Galenic variables of age, sex, and temperament. He found that old people of both sexes could be "very successfully" inoculated, as long as they were hearty and hale. At the other extreme of life, Quier experimented with children and, in particular, infants (Dimsdale had made it clear that children under the age of two were not to be inoculated unless "at the pressing entreaties of their parents"). In Quier's third letter of April 27, 1774, he reported his inoculations of 120 children, 50 of whom were still "at the breast." These children first were given a mercury preparation to cleanse their bowels of worms. Quier found it worth remarking that he inoculated many nursing infants after giving the preparatory medicines to the mothers only.[60]

Pregnant women, however, were Quier's special interest, and he designed his trials to determine whether or not inoculation caused miscarriage. Physicians in this period did not recommend knowingly inoculating pregnant women, though some remarked that at times women came to them hoping in this way to induce abortion. Two dangers made this question a matter of concern for both Quier in Jamaica and his colleagues in London. First, women who contracted smallpox naturally generally miscarried; and second, pregnant women who were not inoculated during a mass engrafting were in danger of contracting a more severe form of the disease.[61]

In his first letter reporting the results of his trials to Monro back in London, Quier wrote that "pregnancy is no obstacle to this process [inoculation] during the first six or seven months of gestation; afterward I think there is danger of abortion; not so much from the violence of the disease, as from the necessary method of preparation." His colleagues in London questioned whether this conclusion, derived from women of African origins, was valid for European women, especially women "of fashion," given the "delicacy" of their constitutions. In his reply, Quier rebuffed his London colleagues' assumption concerning the hardiness of slave women and noted that the females of that class were excused from all kinds of labor for three or four weeks after lying-in and generally treated with great care during pregnancy. He judged his experiments on slave women valid, if not for upper-class European women, then at least for women of what he called "the rustic part." Slave women and "the rustics" of Europe (peasants and anyone whose livelihood depended on physical labor), he argued, would have similar experiences in pregnancy.[62]

Here Quier also repeated his claim that his inoculations of pregnant women had caused "not a single instance of abortion." In his third letter of 1774, having been pressed yet again on the subject by his English colleagues, Quier reported that he had taken "pains" to investigate this matter. In so doing, he had discovered that two slave women had, in fact, miscarried shortly after their inoculations. In this final letter, Quier had to admit that inoculation could cause miscarriage.[63]

What is remarkable about this episode is that Quier had not been called to attend, or even been informed about, the abortions—the very aspect of inoculation he had designed his study to investigate. As I discussed in Chapter 3, childbearing, miscarriage, and female maladies among slaves lay generally in the hands of slave midwives. On the Jamaican plantations where Quier served, secrecy surrounding abortion trumped his efforts to test the relationship between inoculation and miscarriage. Social divides (his position as a European doctor to slave women) distanced him from birthing practices to such an extent that he missed data crucial to his study.

Physicians still in the eighteenth century were not on the front lines of fertility management. Concerning slave abortions in Jamaica in 1826, Reverend Henry Beame wrote that "white medical men know little, except from surmise."[64]

Quier's ignorance concerning slave birthing practices is all the more remarkable because he was a man who had put down deep roots in Jamaica, living there continuously from his arrival in 1767 until his death in 1822. As Michael Craton put it, Quier had gone "native." He was assisted by a man of African descent by the name of William Morris, and Quier never married a woman of his own class but cohabited with a number of his slaves (Jenny, Dolly, Susannah Price, and Patience Christian among them) and a free mulatto woman by the name of Catherine McKenzie. It is a mark of his commitment to these women that he recognized the four children he fathered with them and gave them his surname (we don't know if he had other children with slave women). His only son, Joseph Quier (born to Dolly), was manumitted in 1778. After his death, Quier's 250-acre holding, called Shady Grove, along with its seventy some slaves were distributed among his surviving mistresses, children, and grandchildren, with the greater portion going to one of his daughters, Catherine Quier, and granddaughter, Catherine Ann Smith.[65]

It is hard to know if Quier's experiments were unique for their size. Other practitioners experimented on slave populations, but not as extensively. In the process of inoculating some 300 slaves on Antigua, Dr. Fraser, who also reported his findings by letter to Monro, studied whether inoculation "proves mortal" in persons who had already contracted the smallpox naturally. Despite medical opinion to the contrary, Fraser inoculated five slaves already ill with smallpox. When two of the subjects died, he explained this as a result of "want of care."[66] Fraser also experimented with inoculation in forty white people: twenty-one soldiers and nineteen others of different ages and conditions, including "full-grown women," without incident.

Quier's work is interesting for the questions it raised concerning race. Physicians in the seventeenth century rarely phrased their concerns about transferring medical procedures from people to people in terms of "race," as we understand that term today. If they did question whether cures learned from slaves or Native Americans in the Caribbean were valid for Europeans, they were concerned that differences in climate—the West Indies being hot, Europe being cold—might render a drug ineffective. Or physicians, such as William Wagstaffe in London, explained differences in the efficacy of medical treatment by diet: Wagstaffe objected to the introduction of Turkish inoculation practices into England. It would be impossible, he wrote, to "transplant to us [in England] with success, or naturalize to our advantage" a medical procedure deriving from a people (the Turks) who live on a spare diet and in the lowest manner. The English "National Blood," even among the "meanest of our people," was a rich blood produced by the richest diet in the world, which yielded in turn a blood that "abounds with particles more susceptible of inflammation."[67]

Colonial physicians tended to treat black and white bodies as interchangeable in terms of medical experimentation. This was Quier's assumption. Throughout his experiments, he continued to defend the notion that physiological differences between Africans and Europeans resulted primarily from diet, habit, and manner of work. A notion of interchangeability was necessary if testing considered valid for Europeans was to be carried out on African bodies, and vice versa. Colonial doctors like Quier were interested in racial differences primarily from the point of view of caring for and curing the displaced Africans and Europeans under their care: their concern was for how their cures might need to be tailored to people subject to different living conditions and ways of life. The creole Thomson was unusual among eighteenth-century colonial physicians in studying the physical differences of race. Although he often considered Africans and Europeans interchangeable when treating "[European] gentlemen" and "Negro women" with the same drugs, as we have seen above, nevertheless over his years of practice he also devised numerous "experiments regarding the differences of anatomical structure, observable in the European and Negro, but particularly those of the skin."[68]

Abortifacients

European exploration in the West Indies yielded about a dozen known abortifacients. The abortifacients identified by name included the *Poinciana pulcherrima* (reported in Jamaica, Surinam, and Saint Domingue) and the *Caraguata-acanga*, "penguins" (used in Jamaica and Surinam). Edward Bancroft in Guiana identified the "gulley-root" *(Petiveria alliacea)*, ocro *(Hibiscus esculentus)*, and the sensitive plant *(Mimosa pudica)*. As noted in Chapter 3, Descourtilz described four abortifacients in addition to the *Poinciana:* the *Aristolochia bilobata*, the *Trichilie à trois folioles*, the *Veronica frutescens*, and the *Eryngium foetidum*. A Dr. Michael Clare who practiced in Jamaica told the select committee of the House of

Lords (1818) that midwives administered wild cassava to induce abortion. Other physicians mentioned coffee (of North African origins but a West Indian plantation crop), jalap (native to the area), and quinquina (Peruvian bark—also native to the Americas). Were these West Indian abortifacients tested by European physicians? Did testing with these drugs conform to the formal and informal trials physicians were developing to insure the safety of sometimes little-known and dangerous medicaments? Did they eventually become drugs of import in Europe?[69]

When a drug successfully entered mainstream European medicine it appeared in various of the major *Pharmacopoeia*. In the sixteenth century, European city councils working together with royal colleges of physicians began taking steps to regulate drugs by publishing lists of officially recognized medicaments (in London, 1618; Amsterdam, 1636; Paris, 1638; and Brandenburg, 1698). The purpose of these *Pharmacopoeia*, or *Dispensatories*, was to codify acceptable drugs and their ingredients in order to prevent "all deceits, differences, and incertainties in the making and compounding of medicines." According to John Chandler, an eighteenth-century London apothecary, abuses abounded in the use of compounds, such as purgatives and mercurials, and also simples, such as opium, rhubarb, jalap, ipecacuanha, and the Peruvian bark—all imports, and three of the five originating in the Americas. At the very best, he lamented, raw materials of good quality coming into England for preparation were mixed with faulty wares so that great skill and care were required to separate the one from the other. More often, however, he continued, "the whole thing is a cheat": bark of cherry is dipped in a tincture of aloes to make it bitter and then sold as the costly Peruvian bark; inert residues are powdered and sold as the genuine powder of jalap; sal-ammoniac is powdered, colored, and flavored to pass as the salt of amber; cassia of Malabar is substituted for true Ceylon cinnamon; cheap Jamaica pepper is disguised as clove, mace, and nutmeg; and the compound of black cherry water is often prepared with laurel leaves, which are poisonous.[70]

From our point of view these *Pharmacopoeia* often included ineffectual and revolting substances, such as urine or "mummy of man." Nonetheless, they provide a good measure of which drugs were tested and whether a particular drug passed into medical usage in Europe. Many New World drugs, including Peruvian bark, jalap, ipecacuanha, guaiacum, sarsaparilla, cacao, tobacco, and even sugar, all underwent extensive testing and eventually appeared in a number of these reference books. Was this the case with Merian's peacock flower and other West Indian abortifacients?

Use of *Poinciana* in the Caribbean was reported continuously from the late seventeenth century to the early nineteenth century. The Reverend Griffith Hughes wrote in 1750 that in Barbados the root of the "Flower-Fence or Spanish Carnation," as he called it, is burned to ashes, made into lye water, and "looked upon to be good to bring down the Catamenia." Its flowers, when steeped in breast milk, were good for quieting very young children. Edward Long in Jamaica noted that the flowers made a delicate syrup used as a purgative and the roots yielded a lovely scarlet dye. Much cherished as a febrifuge, Descourtilz in Saint Domingue recommended that a pound of its flowers be on hand for this purpose in each habitation.[71]

Merian's peacock flower also enjoyed some popularity in Europe for its effectiveness against fevers, especially the quartan fevers, and was introduced for this use into French military hospitals. Jean Chevalier, who began his career in the West Indies, provided a long description of the plant in 1752 and discussed its extensive use in Martinique. Chevalier used it twice himself, once in an infusion in wine and another time as an infusion in boiling water. He also noted that, when it was taken as a tea with a bit of sugar, it cured lung ulcers, fevers, and flu of all sorts. He also recommended it for use against smallpox. Chevalier reported that he sent *Poinciana* yearly to a Doctor Alais in La Rochelle who used it effectively to cure tuberculosis. Chevalier found *Poinciana* more effective than *quinquina* (quinine) for fever, and much less bitter.[72]

Pharmacological testing often began on the ground in the West Indies, as we have seen elsewhere in Thomson's experiments with Capsicum peppers and prickly yellow wood, for example. In a sense, knowledge of the *Poinciana*'s use as an abortifacient did not make this first cut. We see this in the work of William Wright. A medical doctor and eventually surgeon general of Jamaica, Wright arrived in Kingston in 1764. He serviced the Hampden Estate, 150 miles outside the city, and its 1,200 slaves, along with the white population living within his twelve-mile circuit. By 1767, he had himself acquired property, seven horses, and fifteen slaves. Never married, Wright had sufficient contact with females in the island to know the uses of the *Poinciana*. The leaves, he recorded, which have a highly "disagreeable smell," are said to be "emmenagogue" and cathartic. "Some people," he continued, "make use of them as such, but they are not admitted in the practice of the physician." He did not explain who used the *Poinciana* leaves, or why medical doctors frowned upon them. In Wright's work, one can begin to trace the rupture in the chain of knowledge between the West Indies and Europe. Early on in his years in Jamaica, Wright had been asked by John Hope and a Professor Ramsey of the University of Edinburgh to collect for them. Although he recorded the *Poinciana pulcherrima*'s many uses in his five-volume herbaria, he did not incorporate them into his report on the officinal plants grown in Jamaica published in the *London Medical Journal* in 1787. Nor, after his return to Edinburgh, did he include the *Poinciana* in his reworking of the Royal College of Physicians' *Pharmacopoeia*.[73]

Nor did Sloane, who returned to England to become a renowned practicing physician, introduce the *Poinciana* into his practice, even when the life of the mother might depend on the effective use of an abortifacient. (As noted in Chapter 3, in such cases he preferred to use "the hand.") Sloane did popularize several colonial products in England, including a Jamaican bark and cacao.[74] Although he knew the "flour-fence" (his name for the *Poinciana*) well, he did not collect it in large quantities or introduce it as an abortifacient into his London practice. As his *Voyage* revealed, he was hard set against abortifacients.

It was not only knowledge of the abortive qualities of the *Poinciana* that failed to transfer to Europe: not one West Indian abortifacient was listed in Europe's *Pharmacopoeia* in the eighteenth century. Evidence that knowledge arriving in Europe from around the world was actively suppressed is rare: the Académie Royale des Sciences did specifically move to suppress a report of an abortifacient in their meeting of March 16, 1763. The report, sent by a M. De la Ruë from the Island of Bourbon, indicated that people there *(les gens du pays)* used a poultice made from a plant known as *la patate à deux rangs* (the potato with two roots) to abort dead fetuses. De la Ruë reported that he had experimented with the plant in a European woman *(une Dame)*, a "Negresse," and also with a nanny goat, and found use of the poultice superior to the painful and dangerous surgical removal of the fetus. The report was read to the full Academy and passed to the Comité de Librairie, where it was marked *"supprimé attendu le danger de la publication."* De la Ruë had sought to make his discovery more widely known in order to benefit "the public"; instead, it was suppressed.[75]

Abortifacients, however, were known and used in Europe. Therapeutic abortions were often required, and *Pharmacopoeia* listed Europe's top herbal abortives: savin, pennyroyal, and rue, drugs known for this use since ancient times. In the pharmaceutical race to test and evaluate potentially useful drugs in the eighteenth century, was this class of drugs tested?

Certainly, there was no lack of females to experiment on. The advent of lying-in hospitals across Europe at mid-century allowed the same opportunities for experimentation in females that military hospitals had made available for males.[76]

Before turning to testing with abortifacients, let us look at emmenagogues, drugs closely associated with abortives. Although abortifacients did not rank as a class of drugs in European *Materia medica,* emmenagogues did. Beginning in the late seventeenth century and throughout the eighteenth century, physicians experimented extensively with emmenagogues or menstrual regulators, medicines considered important for women and widely used across Europe and among European populations in its colonies.

The use of emmenagogues in this period is astonishing; every woman, it seems, felt the need to regulate her menses. Studying the efficacy of these medicaments, a physician proclaimed, "their number is almost infinite, new ones are discovered everyday." The importance of menstruation in this period may account for what seems an overindulgence in these drugs. As medical historian Gianna Pomata has shown, menstruation was considered a necessary cleansing process, so much so that artificial menstruation or blood loss was induced in men in the form of blood-letting. A number of methods for "bringing down the flowers" or inducing menstruation in early modern Europe, such as blistering the inside of the thighs or plastering certain resins on a woman's navel, were highly suspect. Others, no doubt, were beneficial.[77]

As understood in this period, the term "emmenagogue" came from the Greek *"emmena,"* meaning menses, and *"ago"* or "drawing forth." In his lengthy discussion of vegetable emmenagogues, Antoine de Jussieu at the Parisian Jardin du Roi noted that women often miss their monthly purgation and that this condition demands prompt attention because it causes a multitude of afflictions, including jaundice, migraines, "vapors," convulsions, vertigo, apoplexy, madness, a globe creeping upward in the throat, spasms of the uterus, and involuntary laughing or crying. He characterized the condition underlying a suppression of the menses as a "loose" and "soft" constitution that causes a woman's respiration and circulation to slow and weaken. Emmenagogues worked, he argued, by thinning the blood, giving it greater force, and dislodging the obstruction: They "elevate the pulse, increase body temperature, and bring a red glow to the face." John Freind, who researched emmenagogues extensively in England, concurred that "there are no emmenagogues, but what, by some means or other, increase the momentum of the blood . . . the momentum of the blood being increased, the menses break forth."[78]

John Riddle has argued that an emmenagogue in early modern Europe was simply an abortifacient by another name—that, as abortion became increasingly criminalized, emmenagogues were often prescribed where abortion was intended. He has urged that the phrases to "promote the menses," "bring down the flowers," "purge the courses," "restore menses obstructed," or "bring on the menses" were euphemisms for abortion. As discussed in Chapter 3, there was in the early modern period little distinction among bringing on the menses, a miscarriage, and what we today would call an early-stage abortion, since a woman was not considered "with child" until the baby quickened in the fourth or fifth month of pregnancy. Riddle has argued that this ambiguity was intentional. Edward Shorter shares Riddle's view that menstrual regulators can be seen as abortifacients. Pharmacologically there was often little distinction between an emmenagogue and an abortifacient, the latter in many cases being merely a stronger dose of the former.[79]

There is much to recommend these arguments. Physicians at the time, however, made many careful distinctions, and those noted in the entry on emmenagogues in Diderot and d'Alembert's *Encyclopédie* are typical. Remedies that empty the uterus, the author stressed, are of three types: *emmenagogues* cause evacuations of the menstrual flux, *ecboliques* call forth the fetus (either to aid in childbirth or to help remove a dead or live fetus), and *aristolochiques* help dislodge the afterbirth. Zedler's German lexicon made similar distinctions and remarked that emmenagoga should not be used when a woman is pregnant or sick. And Carl Linnaeus, both a professor of botany and a practicing physician, distinguished between emmenagogues and abortives in his explanation of terms: an emmenagogue "expels the menses"; an abortive "expels the fetus" (he did not, however, distinguish between expelling a dead or live fetus).[80]

Medical doctors like Sloane were aware of the possible "misuse" of emmenagogues, as I noted in Chapter 3. Pierre Dionis, a French royal surgeon, cautioned midwives not to give a woman emmenagogues until the midwife was certain that the woman was not pregnant.[81] Thus while physicians promoted the use of emmenagogues, they did not intend them to be used as abortifacients.

Emmenagogues were something that naturalists collected and brought home for use by Europe's women. Francisco Hernández gathered and Nicolás Monardes discussed New World treatments for female ailments as early as the sixteenth century. Sloane listed five menstrual stimulators that he had employed in his practice in Jamaica. In his 1826 *Nouveaux élémens de thérapeutique et de matière médicale,* Jean-Louis-Marie Alibert mentioned that a particular *Aristoloche* (not the *bilobata* discussed by Descourtilz, but the *odorante*), used as an emmenagogue "by the Peruvians from time immemorial," had recently been imported into Europe. Joseph de Jussieu included a section on *"histeriques"* in his survey of the medical plants of Saint Domingue. One of these, *Assafoetida,* originally brought into Europe from Persia but growing also in the Caribbean, was subsequently cultivated in Europe. William Woodville of the Royal College of Physicians in London provided an extensive report on this plant, noting that John Hope in 1784 had been the first to cultivate it in Britain ("perhaps even in all of Europe"). *Assafoetida* was a commercial crop in Persia, cultivated and harvested by peasants; it was the gum, extracted from the root in a tedious six-week process, that Europeans prized. In addition to being used as an emmenagogue, the *Assafoetida,* taken either as an enema or an oral tincture, was employed to counteract spasms, dyspepsia, flatulent colic, and nervous disorders. It was also said to be popular in "sauces." *Assafoetida* entered both the London and Edinburgh *Pharmacopoeia,* even though it was known to the British in India as an abortifacient.[82]

Many colonial drugs, such as coffee (both an economic and medical plant), jalap, and *Cinchona* were found also to induce the menses. William Buchan, member of the Royal College of Physicians in Edinburgh, indicated that the Peruvian bark was widely used as a menstrual regulator in his *Domestic Medicine; or, the Family Physician:* "When the obstructions proceed from a weak relaxed state of the solids, such medicines as tend to promote digestion, to brace the solids, and assist the body in preparing good blood, ought to be used." Principal among these medicines were iron, "the Peruvian bark," and other bitter and astringent medicines.[83]

Physicians in Europe also tested emmenagogues. Although unable to engage in self-experimentation, male physicians did try to test these drugs according to standard practices of the time. John Freind in London experimented extensively with menstrual regulators and published his *Emmenologia* in 1720. For Freind, many things could influence the natural working of the menstrual cycle. The menses could be delayed, for example, by "immoderate cold, sorrow, a sudden fright, too great an evacuation, incrassating diet, crudity of humors, or astringent medicines, all things that diminish the momentum of the blood." Conversely, the menses could be "brought down" in an untimely fashion by "fever, smallpox, coition, drinking too much, violent motion, vomiting, sneezing, anger, the hysterick passion, passion of the mind, and plants called emmenagogues"—all of which increased the motion of the blood or stimulated the vessels.[84]

In his "laboratory," Freind identified different classes of emmenagogues. First were the "bitters," with hot and odoriferous qualities, such as opium, gentian, myrrh, arum, wormwood, savin, rue, and pennyroyal. To this he said should be added the Peruvian bark, "which tho' as yet, it obtains no place among the Emmenagogues, ought however to be ranked with them, upon the account of its remarkable effect in attenuating the blood." Second were the "aromatic volatile salts," notably saffron and cinnamon. Third were the "hot and cardiac substances" that quicken the pulse, such as steel, hydrarg, and so forth. To demonstrate the effectiveness of these substances he mixed various simples—savin, rue, pennyroyal, sage, lavender, wormwood, jalap, Peruvian bark—with the blood, "fresh drawn" out of an artery of a dog to see if the fluidity of the blood were increased. A test of savin, on February 12, 1702, revealed that the blood turned intense red and thinned dramatically. These same substances tested with the "serum of the human blood" all increased the "attenuation" or fluidity of the serum, indicating their potential value as emmenagogues.[85]

Finally, Freind tested "the attenuating quality of emmenagogues" on animals, sacrificing no fewer than fifty dogs in this particular set of experiments. By means of a syringe, he injected decoctions of savin and Peruvian bark, water of cinnamon, and syrup of violets directly into the jugular veins of these animals, "so that," he wrote, "if any are so unbelieving, as not to be satisfied with the reason of our former experiments, let them now *behold* and be convinced." It took only a short time (four to fifteen minutes) for the injected animals to die. Freind observed that the limbs of the animal did not stiffen until a long time after death. Upon opening the dog's *vena cava* and *aorta descendens*, very thin and voluminous blood poured forth. In this way, Freind believed that he had proven that the effective quality of emmenagogues lay in their power to increase the fluidity and thus the motion of the blood.[86]

Freind also provided six lengthy case histories of the use of emmenagogues drawn from his medical practice. The first from October 16, 1700 involved an eighteen-year-old female who had never had a menstrual period. She sought help because she suffered from pain in the loins, knees, and ankles; nausea; "grippings" of the stomach; and palpitation of the heart. To induce menstruation, Freind sought to furnish the blood with "a momentum strong enough to break through the uterine vessels." This he did with a standard compounded emmenagogue, in this case a purging agent, which gave her two stools and somewhat eased her pains. To increase the momentum of the blood further, he prescribed a second emmenagogue of such complex composition as to rival the highly regarded theriac, a popular antidote for poison compounded from over sixty-five separate substances. Because the young woman was already weak, he dared not bleed her. By October 28th Freind's patient had an increased pulse, less pain in her stomach, and more strength. On the 30th of October her menses—"of a laudable colour"—came down. The pain in her loins and ankles vanished, her flux continued for eight days. The medicines were repeated after two weeks, her menses again flowed regularly, and the woman entirely recovered her health.[87] Freind's "emmenagogue" included savin and aristolochia, and although he does not comment on the possibility, he might have induced an abortion along with her menses.

Of the five other cases he treated—identified only as "a woman aged 30," "a washer woman, aged 24," and so on—only one was possibly pregnant. This woman, aged twenty-five and married, had had a decreased menses for almost a year. The neighboring "old women" and the woman herself believed that she was "with child." For his part, Freind was inclined to think otherwise and began treating her in October 1702. By April of the next year, after repeated use of an emmenagogic infusion, her menses came down and her health was restored.[88]

Physicians conducted numerous tests with savin, one of Europe's top abortifacients, but only as an emmenagogue. Edinburgh experimentalist William Cullen, the Boerhaave of his age, began his testing with an already low opinion of emmenagogues. In his massive 1789 *Treatise of the Materia Medica,* he accused the ancients, who identified many drugs in this class, of not writing from firsthand experience. He concluded that there was no medicine that had "any specific power in stimulating the vessels of the uterus." Under his section on "antispasmodics" he reported that savin *(sabina)* showed "a more powerful determination to the uterus than any other plant I have employed." He did not find it reliable, however, and this he attributed to its great "acridity and heat," which prevented him from employing it in high enough doses.[89] He was similarly disappointed with the effects of rue, another common emmenagogue.

Cullen's colleague Francis Home, fellow of the Royal College of Physicians of Edinburgh and professor of materia medica at the University, also tested savin as an emmenagogue. In his 1782 *Clinical Experiments, Histories, and Dissections,* he tells that he selected savin for his patients despite its reputation: "The Sabine [savin] is infamous, for its strong effects on the uterus. It is often used to procure abortion; and is said to endanger the mother by the violent hemmorrhagy which it occasions. In many countries it is not allowed to be sold, but by order of the physician." These bad effects, he noted, could be avoided by administering a small dose. While many authors mention a dram (or 3.9 grams) as the proper dose, he wrote, "I gave generally half that quantity, and have found it useful and safe." Home's first experiment concerned a certain woman by the name of Jean Mason, whose amenorrhea had failed to respond to an earlier treatment consisting of a tincture of hellebore *(Helleborus niger)* and compression of the crural artery (then also a common way to induce abortion). Home prescribed for the woman a half dram of the powder of savin twice each day, and "in four days the menses appeared, and continued for two days."

He experienced similar success with Janet Dallas, age twenty-eight, who had failed to respond to a ten-day course of the elixir of sacrum or pills made of "filings of steel and extract of gentian." Two scruple of savin administered twice a day, however, produced her menses on the third day of treatment. All in all Home judged that "from these trials the sabine appears to be a powerful remedy, as it succeeded in three of five cases, or rather of four." In one case, he noted, the malady resisted every application.[90] By the 1780s, physicians had worked out the dosage for savin employed as a menstrual regulator at between a half and whole dram given twice a day. More than that was considered dangerous.[91]

In contrast to the extensive medical testing with emmenagogues in this period, testing with abortifacients was minimal. There is overwhelming evidence that abortion techniques existed in Europe from ancient times, and that they were often used in the seventeenth and eighteenth centuries. The rise of "scientific" medicine and systematic experimentation with medical techniques in the late eighteenth century did not, however, lead to the development and testing of abortifacients. Abortifacients (except when the same plant was also used as an emmenagogue) did not enter mainstream European drug testing and did not become a part of academic medicine or pharmacology as these fields developed in the eighteenth and nineteenth centuries. As a result, drugs that were potentially dangerous remained so.

Abortifacients had long been considered dangerous, and this was a message expressed with renewed urgency after the 1750s. The author of the article on savin *(savine)* in Diderot and d'Alembert's *Encyclopédie* found its effectiveness woefully exaggerated. "This plant . . . regulates the menses and chases the fetus from the uterus . . . an excess, however, does not procure abortion dependably and swiftly as it is the custom to believe" but often causes "violent hemorrhages that kill both the mother and child." The author continued that it would be desirable to destroy the "disastrous opinions that run rampant through the public concerning savin." Physicians from the early part of the nineteenth century taught that there was "no abortive simple, no medication that leads to abortion and only abortion in a manner direct and specific." In his 1831 *Manual of Medical Jurisprudence,* Michael Ryan concluded similarly that there was "no medicine or abortive means, which always produce abortion, and nothing but abortion: Every woman who attempts to promote abortion, does it at the hazard of her life. There is no drug which will produce miscarriage in women who are not predisposed to it, without acting violently on their system, and probably endangering their lives."[92]

At the same time savin and other of Europe's stock abortifacients, such as pennyroyal and rue, were increasingly vilified. From the seventeenth until the last quarter of the eighteenth century, savin, rue, and pennyroyal had enjoyed the reputation of being effective abortifacients and menstrual regulators, and were used to provoke the menses, induce abortion, expel a dead fetus, and expel the afterbirth. Doctors in Halle agreed in 1738 that a patient's ingestion of savin would provoke an abortion. By the turn of the nineteenth century, however, physicians were evaluating these drugs negatively, suggesting that they were ineffective, their abortive qualities largely imagined. Cullen sounded the general theme, writing that the "abortive qualities frequently ascribed to medicines by the ancients, seem to me, and perhaps to most physicians of these days, to be imaginary, and accordingly such medicines are now hardly ever employed." Many physicians, including William Thomas Brande, joined in this refrain, saying of pennyroyal that "the old physicians had a high opinion of [its] virtues in hysteria and uterine obstructions . . . it is now, however, rarely prescribed." In France, Alibert emphatically stated that savin simply does not work. Even though "one says" that women use savin to procure abortions, "happily for humanity" that plant often fails to produce an abortion. Its powers, he concluded, had been greatly exaggerated. William Smellie, the English man-midwife and encyclopedist, claimed that the seeming effectiveness of these drugs could be explained by the strength of the female imagination: if a physician gave a woman any type of cordial or infusion in the course of a difficult labor, her imagination sped her along to a timely birth—even if the drug itself did nothing to encourage labor.[93]

That physicians induced abortions in the course of their practice is clear. Perhaps they recorded their findings for their own reference or even communicated them to students, but they did not publish these materials. In a sense physicians were intruding in the business of birthing (which passed from the hands of midwives to obstetricians over the course of the seventeenth and eighteenth centuries) and in the business of aborting unwanted pregnancy. The well-known Jean Astruc, who noted that he was "on occasion" called to help a desperate woman, discussed the means that "wicked" women employ to "lose their fruit." "One says," he wrote, "that there are many methods, but I have not been curious to know them, and because of that I am happy." For more than a century now, he continued, physicians have said nothing about aborting an unwanted pregnancy in their books and lectures on birthing. Although he knew the means by which to induce abortion, he refused to instruct young physicians in these matters that, in his estimation, almost always proved fatal. The infamous anti-abortionist of the nineteenth century, Ambroise Tardieu, concurred that "there is no physician who does not know the methods of abortion." They do not wish, however, to expose these methods for fear that the "ill intentioned" will commit new crimes.[94]

In some cases, authorities pleaded with doctors *not* to research what they considered dangerous drugs. In his 1784 *System einer vollständigen medicinischen Polizey,* Johann Peter Frank painted a picture of abortifacients as grown freely in farmers' gardens and sold freely by apothecaries, surgeons *(Barbieren),* and midwives; that is, by anyone except licensed physicians. States wishing to do away with abortion increasingly used medical doctors as vehicles for controlling these substances. Frank suggested that physicians were in agreement *(einstimmigen)* that abortifacients are dangerous and should not be taken internally (though he provided no evidence for his statement). Because of their great danger, he exhorted doctors "not to tolerate" *(sich verbitten)* investigation of such medicines. In some cases, physicians themselves suppressed knowledge of abortifacients. Gabriel-François Venel, professor of medicine in Montpellier, wrote, "the abuse of this plant [savin] does not permit us to write more about it." In some cases, city fathers implemented new laws to suppress the use of abortifacients. The apothecary Philippe Vicat, translating

Haller's *Materia medica,* reported that a new law forbade apothecaries from selling savin to the general public because "persons without integrity and the destitute." [*gens sans probité & des malheureuses*] used it to abort, and women often died from it.[95]

The conflict between women seeking help to end unwanted pregnancies and physicians who abhorred abortion continued to sharpen over the course of the eighteenth century. Physicians sometimes accidentally induced abortion by some medicines prescribed for another condition. "These women deny that they are pregnant," one disgruntled physician wrote, "even after the doctor has received the aborted fetus into his hands." This same physician reported that a sixteen-year-old woman visited a doctor to confirm that she was pregnant. She asked him for an abortive, because she had had difficulty giving birth some years earlier and feared for her life. The physician, however, refused to help her. The girl found a woman who told her to take three fresh roots of rue, each the size of a finger, to slice and boil them in a pound and a half of water. The liquid was to be divided into three glasses and taken in the evening all at one time. She became sick to her stomach and "so generally and profoundly ill that she believed she would die." The next night, forty-eight hours after taking the rue, she aborted. Visiting her physician again some days later, she reported that apart from some fatigue, she felt fine.[96]

Developing and testing abortifacients in order to provide women with safe methods for abortion was simply not a priority among physicians in this period. The numerous theses and books produced on "abortion" in the early nineteenth century were devoted to preventing miscarriage, not safely inducing abortion. John Burns, a member of the faculty of physicians and surgeons in Glasgow, for example, published his lectures on midwifery in 1806, wherein he devoted considerable text to the prevention of spontaneous abortion caused by advanced age; death of the fetus; smallpox or other disease; violent exercise, such as dancing, vigorous walking, excessive laughing or singing; the "fatiguing dissipations of fashionable life (tight clothing and immoderate foods)"; strong passion of the mind, such as fear, joy, or other emotions; and any sudden shock, including having a tooth pulled, exposing the neck and arms to cold, or immersing the buttocks or feet in cold water.[97] Burns also acknowledged that emmenagogues, such as savin, might produce abortion, and counseled their use only when necessary to save the life of the mother.

Along with developing methods in this period to prevent miscarriage, physicians also honed their skills to recognize the marks of provoked abortion in (female adult) cadavers. This task was so difficult that Tardieu threw up his hands in despair—if the woman did not die when inducing abortion, physicians, and hence the authorities, had no way of knowing about it and "the crime of abortion" went unpunished.[98]

In sum, late eighteenth-century experimental physicians stood at a fork in the road with respect to abortifacients. They could choose the road toward development and testing of safe and effective abortive techniques, or they could choose the road toward suppression of these knowledges and practices. Ambroise Tardieu spoke for many when he proclaimed that the medical-legal community was in the best position to suppress abortion, especially after the passage of the French code in 1810 outlawing "criminally induced" abortion (*l'avortement criminel provoqué*). One way to do this was to exclude abortifacients from pharmaceutical testing. Johann Andreae Murray's six-volume *Apparatus medicaminum*, published in Göttingen in 1793, provides a good comparison of the quantity of research done on various drugs. Murray summarized research on savin in seven pages and on rue in six pages. By contrast, his annotated bibliography on quassia (arguably a relatively insignificant drug) comprised forty-two pages, on guaiacum (a drug thought to cure syphilis) thirty-three pages, on ipecacuanha (important militarily to act against dysentery) thirty-two pages, and on *Cinchona* (important in efforts to colonize tropical areas) 108 pages. The same pattern of neglect of anti-fertility drugs can be found in Brande's *Dictionary of Materia Medica and Practical Pharmacy* (1839), in which he pointed out that the essential oil of savin had never been examined chemically—which by 1839 demonstrated a serious lack of research. That savin was "little studied" as an abortifacient was also noted by Théodore Hélie in 1836.[99]

The suppression of abortion came to fruition in the nineteenth century, when European states passed centralized statutory laws making abortion criminal. The 1794 Prussian Legal Code (sections 985 and 986) negated a woman's traditional prerogative to determine for herself when she was pregnant (at quickening). The Napoleonic Code (article 317) of 1810 and England's Lord Ellenborough's Act of 1803, however, continued to adhere to the unquick/quick distinction, restricting willful "miscarriage" to cases in which a woman was "quick with child." In England, inducing "miscarriage" in a woman "not being quick with child" was a felony punishable by fine, imprisonment, whipping, or transportation to the colonies for up to fourteen years. Should the woman be proved to have quickened, abortion was punishable with death. By 1837, however, English law made no distinction between a quick or nonquick fetus. Both were now essentially murder (punishable by transportation or imprisonment, but not death). In the United States, no state had a statutory law against abortion prior to 1821; by 1850 seventeen states had criminal abortion laws. Abortion laws were passed in Austria in 1852, in Britain in 1837, Denmark in 1866, Belgium in 1867, Spain in 1870, Zürich in 1871, Mexico in 1871, the Netherlands in 1881, Norway in 1885, and Italy in 1889. The majority of these outlawed abortion (whether of a nonquick or quick fetus) unless required to save the life of the mother.[100]

At the same time that laws against abortion were tightened, abortifacients were outlawed. In Germany, public hygienists (the *medicinische Polizei*) discussed how best to decrease the number of abortions and thus favorably increase population. Methods proposed included: building state-funded foundling hospitals and orphanages on the model of the French; restricting public access to abortifacients *(Fruchtabtreibungmittel entziehen)* by allowing apothecaries and druggists to sell abortifacients only by prescription; policing "old women" *(alten Weiber)* and "matchmakers" *(Kupplerinnen)*, since they were generally thought to be the ones who supplied such remedies; removing savin trees and similar plants from public gardens; allowing surgeons and bathers *(Bader)* to bleed single women only as prescribed by physicians; and outlawing all medicines sold by itinerant healers and quacks.[101]

In France, similar measures were prescribed to "prevent the knowledge of the proper means of procuring an abortion from spreading among the people *(le peuple)*." These included classifying any discussion of abortifacients as a "public nuisance." For physicians, this meant not publishing details of abortives and their methods of use. Although the author of the lengthy article on *avortement* in the 1812 *Dictionaire des sciences médicales* noted that the number of "vegetable substances more or less active" that induce abortion was infinite, he refrained from giving any details of their names or usages; "my pen," he wrote, "refuses to record any more details." Further, he recommended that popular books offering recipes for abortives be suppressed and that "vegetables known to the public as abortives" be carefully weeded out of all public and private gardens, and not be allowed to be sold in public markets. Their seeds should not be sold by florists, nurseries, or seed merchants.[102]

In England in 1861, the Pharmacy Act finally outlawed savin, long present as an official drug in Europe's leading *Pharmacopoeia,* by classifying it as a poison. The act read: "Every woman being with child who, with intent to procure her own miscarriage, shall unlawfully administer to herself any poison or other noxious thing, or shall unlawfully use any instrument or other means whatsoever with the like intent, and whosoever, with intent to procure the miscarriage of any woman, whether she be or be not with child, shall unlawfully administer to her, or cause to be taken by her any poison or other noxious thing, or shall unlawfully use any instrument or other means whatsoever with like intent, shall be guilty of felony . . . to be kept in penal servitude for life."[103] Savin (along with ergot of rye, belladonna, and some fifteen other substances) was included among those drugs classified as poisons. This law was coordinated with the Offences Against the Person Act of 1861 (section 59) that deemed it a misdemeanor punishable with imprisonment for three years to supply instruments, poison, or "noxious things" to cause an abortion.

Laws against abortion in the nineteenth century hardened at the same time that abortifacients were being discredited. The French physician E.-N. Cotte noted physicians' increasing belief that specific remedies could not actually cause abortion. Cotte argued that the "prejudice" that specific herbs cause abortion had for many centuries made doctors afraid to prescribe certain medicines for a pregnant woman. These remedies, he asserted, "no longer have any credit among modern doctors." As these drugs were being discredited in medical circles, the real drugs were being pulled from shop shelves. Edward Shorter cited a study done in 1923 of thirty-eight samples of commercial savin from five different countries: the "savin" sold in England, France, and Spain was found to be counterfeit. Only the Swiss and German savin was genuine. A study done in 1989 also found that commercial samples of the essential oils of savin were falsified and contained no active ingredient. When savin did not work, it was often because the oil was old, not properly prepared, or counterfeit, so that unsuspecting women took it but with no effect.[104]

As midwives lost ground to newly minted obstetricians (male surgeons) over the course of the eighteenth century, herbal abortifacients gave way to surgical instruments designed to induce abortion. Although knitting needles and other sharp instruments or "the hand" had long been employed, herbal abortifacients predominated in the eighteenth century. In the nineteenth century, as Tanfer Emin has pointed out, surgeons rushed to patent cervical dilators, curettes (specially curved knives), new-styled forceps (for pulling the fetus through the birth canal), and crochets (hooks used for craniotomy), as they had increasing need of these technologies. As midwives were run out of the profitable end of their profession (there was always work for them among the poor), abortifacients gradually disappeared from mainstream medicine.[105]

Abortifacients, then, were not part of the pharmacopoeia scrutinized by experimental physicians and pharmacists in the eighteenth century. As research on drug mechanisms, effects, and reactions increased, certain undesirable drugs were pushed aside. Abortifacients did not enter mainstream European academic drug testing: They were considered dangerous and—in the absence of research—destined to remain so. Criminalizing abortion made taboo medical research into their usage.

Although many drugs entered Europe from abroad, abortifacients did not. The *Poinciana*—Merian's peacock flower—was never tested as an abortifacient in Europe and it never found its way into any of Europe's major *Pharmacopoeia*.[106] Scientists since the time of Francis Bacon have often been portrayed as standing on the shoulders of giants, each building new knowledge firmly upon old foundations. With the case of exotic abortifacients, however, the foundations of traditional knowledges were shattered and left to fall into ruin.

5

Linguistic Imperialism

> The names bestowed on plants by the ancient Greeks and Romans I commend, but I shudder at the sight of most of those given by modern authorities: for these are for the most part a mere chaos of confusion, whose mother is barbarity, whose father dogmatism, and whose nurse prejudice.
>
> CARL LINNAEUS, 1737

Voyaging outside Europe, naturalists, planters, missionaries, and traders encountered all manner of new and strange fruits, vegetables, medicines, spices, and other economically valuable plants. Known plant species burgeoned from the 6,000 described by Caspar Bauhin in 1623 to the 50,000 recorded by Georges Cuvier in 1800. Botanists, doctors, pharmacists, gardeners, and collectors all had an interest in establishing systems of classification capable of managing this fabulous floral profusion.

For Carl Linnaeus, the celebrated father of modern systematics, botany had two foundations: classification and nomenclature. Nomenclature, he taught, is a convention agreed upon by a community of practitioners: like coins and currency, botanical names are used only by "agreement of the commonwealth." Never a modest man, Linnaeus fancied himself a lawgiver bringing order to the botanical commonwealth threatened by the "invasion" of "vast hordes" of foreign plants and their "barbarous" names.[1]

Here we shall explore the rise of botanical nomenclature through the linguistic history of plant names, stepping only lightly on issues concerning the well-trodden terrain of taxonomy. Scientific nomenclature can be an exacting task: it is crucial that a name refer to one and only one plant if it is to be discussed over large units of space and time. Stability in nomenclature today is generally fixed by historical priority; he (or she) who first publishes a name—accompanied by an exact description of the plant—is considered to have named that plant for all time. A botanical description (whose function it is to distinguish a particular plant from all other plants) is often guaranteed by a recognized herbarium specimen that "typifies" the plant. Once recognized, a generic plant name can be changed only if the plant is reclassified, that is, moved from one genus to another.

Nomenclature interests us here in its technical aspects, but also for what it can reveal about the cultural history of plants: how plants and the knowledge of them traveled across early modern botanical networks, how European cultures understood the biodistribution of plants, and how European botanists evaluated knowledge systems of other peoples.

Historians have tended to celebrate the rise of Linnaean systematics as the birth of scientific botany. Between Alphonse de Candolle's *Lois de la nomenclature botanique* (1867) and the first *International Code of Botanical Nomenclature* established by the 1905 International Botanical Congress held in Vienna, botanists fought bitterly over whether 1737 or 1753 (both dates referring to works by Linnaeus) was to serve as the recognized "starting point" of botanical nomenclature. This story is well known and will not be recounted here.[2] One could, however, see the rise of Linnaean systematics also as a form of what some botanists have called "linguistic imperialism," a politics of naming that accompanied and promoted European global expansion and colonization.

Naming, the way cultures come to refer to objects whether animate or inanimate, is a deeply social process. It is also highly political, and botanical nomenclature should be considered in a larger context of the history of naming. When, for example, Caribbean or African countries finally shook off the yoke of European rule, many chose names for their new republics to highlight what they considered indigenous cultural traditions. Thus when French rule collapsed in Saint Domingue in 1804, making it the second such country in the Americas to win its independence from Europe (after the United States), the former French part of Hispaniola selected the Arawak-derived name "Haiti," despite the fact that few indigenous people survived there.

The process of the loss and rebuilding of cultural identities through naming can also be seen in social practices surrounding names given West Indian slaves during the colonial period. When slaves were first transported from Africa to the Caribbean, many retained their African names and were entered on plantation lists as Quashie (Sunday), Phibah (Friday), Mimba, Quamino, and so forth. By the 1730s, however, the proportion of African names had diminished. Slaves in British territories often had several names, one given them by their master or overseer (at the time of purchase or at birth) and others given by their parents or neighbors in their community (which masters might be aware of but rarely used). The masters' names for slaves were the ones recorded in plantation records and were generally limited to first names (Dolly, John, Samuel, Betsy, or Jenny, for example), or nicknames derived from some perceived characteristic of the slave (Run-Away Mary or Big Tom), or whimsical names (such as Time, Fate, Badluck, or Strumpet). Masters often assigned names arbitrarily, disregarding a slave's family ties or geographical origins. By contrast, masters had at least two names and often three or more, acknowledging a set of family connections important for social status and property transfers.[3]

With emancipation in the 1830s, former slaves in British territories began taking legal surnames for the first time. In choosing these names, they rarely drew from African or slave languages but tended to model their names on those of their former owners, sometimes their white fathers or persons of European heritage whom they admired. In French territories, however, free people of color were for many years barred from adopting European names, and after 1773 were actually *required* to take names of obvious African origin to emphasize their subordinate status.[4]

Names offer a sense of identity, cultural location, and history; this is true, of course, for people, but it is also true, to a certain extent, for plants. This chapter explores the extent to which plants were uprooted from their native cultures and acclimatized to colonial rule by being given European names. This history is epitomized, as we shall see, in the onomastic history of the flower called the *flos pavonis* or peacock flower by Europeans in Surinam, the *tsjétti mandáru* on the Malabar coast in India where it also flourished, and the *monarakudimbiia* in seventeenth-century Ceylon (today Sri Lanka). In the course of the eighteenth century, the variety of (published) names for this flower—many of them East Indian and emphasizing the plant's beauty—was reduced to a single scientific (Linnaean) term still used internationally, *Poinciana pulcherrima,* a name commemorating General Philippe de Lonvilliers, chevalier de Poincy, a seventeenth-century governor of the French Antilles.[5]

Linnaeus' system encountered a great deal of opposition in the eighteenth century, precisely surrounding issues concerning naming. We shall look at the efforts made by his contemporaries to develop botanical nomenclature that incorporated names from the cultures within which the plants grew. I will argue that modern botanical nomenclature might have developed substantially differently if the system of Linnaeus' ardent opponent, Michel Adanson of France, had been chosen as the starting point of modern systematics. Adanson, along with others, attempted to conceptualize plants globally and often chose to retain plant names indigenous to the areas where they are (or were) found. Linnaeus was not impressed, complaining vis-à-vis Adanson's nomenclature: "All my generic Latin names have been deleted and instead come Malabar, Mexican, Brazilian, etc., names which can scarcely be pronounced by our tongues."[6] Even within Europe there were alternative practices and viewpoints, and the one that won triumphed more for convenience than from a primary kind of iron necessity.

Although I will (for lack of materials) focus on European botanical nomenclature, it is important to keep in mind that Amerindians and African slaves in the Caribbean actively engaged in naming practices of their own. Alexander von Humboldt reported that the native peoples of the Upper Orinoco—the Marepizanoes, Amuizanoes, and Manitivitanoes—had their own elaborate names for every nation of Europe: Spaniards were the *Pongheme* or "clothed men"; the Dutch were *Paranaquiri* or "inhabitants of the sea"; and the Portuguese were *Iaranavi* or "sons of musicians." Pierre Barrère, sent to Cayenne by the French Crown in 1722, noted that the "Negresses" who educated children (white as well as black), had introduced many words from their own country into the creole language that dominated that island. Some of these words appear in European documents, but neither the Caribs, Arawaks, nor "Negresses" of Cayenne recorded botanical names in any way that has survived; those names have been lost to history.[7]

Empire and Naming the Kingdoms of Nature

What's in a name? In Spain's Peruvian territories of the seventeenth century, the University of Lima turned down a proposed new chair of medicine devoted to botanical studies on the grounds that physicians should instead study Quechua. In that ancient Native American language, plants were said to have been named for their medical virtues. Physicians, it was suggested, could learn the uses of plants more quickly by studying this Incan language than by investigating the plants themselves. Martín Sessé, director of the Spanish Royal Botanic Expedition to New Spain (1787–1802), made similar claims for Nahuatl, the language of the Aztecs and other Native Americans living in Mexico and Central America. It was precisely this type of information—medicinal usages, biogeographical distribution, and cultural valence—that was to be stripped from plants in Linnaean binomial nomenclature as it has come down to us.[8]

Michel Foucault has defined the eighteenth century as the "Classical Age," the age that fashioned new conceptual grids to discipline the unwieldy stuff of nature. Within these grids, names became technical reference tools—said to be simple tags or neutral designators—no longer burdened by Baroque notions of resemblance. A name, in other words, was to have no essential connection to the plant, but was to be something agreed upon by convention. The illustrious B. D. Jackson, secretary of the Linnean Society and keeper of the Kew Index in the 1880s, wrote that a plant's name was merely a "symbol," and that if it and the plant to which it belonged were firmly united leaving nothing to doubt, it mattered little what the name might be. Nomenclaturists today often recognize this and emphasize the playfulness of names: a fossil snake might be called Monty Python, or a name, such as *Simiolus enjiessi,* might encode reference even to a granting agency (*Simiolus enjiessi* when properly pronounced contains a hidden reference to "NGS" in honor of the National Geographic Society). And it is certainly the case that innumerable natural objects have been named for spouses, lovers, and so on.[9]

Though names today may be abstract and arbitrary in this way, naming practices are not. They are historically and culturally specific, growing out of particular contexts, conflicts, and circumstances, and it is the job of the historian to ask why a particular naming system and not another came to be. My argument is that what developed in the eighteenth century was a culturally specific and highly unusual practice of naming plants from around the world after prominent Europeans, especially botanists. Linnaeus argued long and hard for this relatively new practice—a practice validated when his work was made the starting point of modern botany at the beginning of the twentieth century, a period that marked the apex of European imperial power. Naming practices devised in the eighteenth century assisted in the consolidation of Western hegemony and, I will argue, also embedded into botanical nomenclature a particular historiography, namely, a history celebrating the deeds of great European men.

The grand achievement of the early modern period was the invention of *binomial nomenclature,* emerging in nascent form with the Swiss Bauhin in the seventeenth century and systematically developed to a fine art by Linnaeus in the eighteenth century. Binomial nomenclature refers to that system of naming whereby a species of plant is designated by a two-word name, consisting of a generic name followed by a one-word specific epithet, as in *Homo sapiens, Notropis cornutus,* or *Poinciana pulcherrima.* A plant is considered fully named when it is furnished with a generic and a specific name.[10]

There is no doubt that in Linnaeus' day some kind of taxonomic reform (and standardization) was needed. Seventeenth-century botanists raised a cacophony of botanical names. Until the end of the fifteenth century, *Materia medica* were commonly Arabic versions of Greek texts in Latin translations that provided five types of names for any given substance: Latin, vernacular, Arabic, apothecary, and "polynomials" or descriptive phrases. An entry in John Gerard's 1633 *Herbal* typically listed names for "Sow-Bread": in Latin, *Tuber terrae* and *Terrae rapum;* in shops, *Cyclamen, Panis porcinus,* and *Arthanita;* in Italian, *Pan Porcino;* in Spanish, *Mazan de Puerco;* in High Dutch, *Schweinbrot;* in Low Dutch, *Uetckinsbroot;* in French, *Pain de Porceau;* and in English, "Sow-Bread." These names were often—but not always—based in literal translations from one language into the next. To handle this burgeoning Babel, botanists published dictionaries and vocabularies listing synonyms across numerous languages. Christian Mentzelius' *Index nominum plantarum multilinguis,* for example, provided 350 pages of botanical synonyms calibrating names across one hundred and eight languages—including Latin, Greek, German, "Scottish," "Bangalenese," Chinese, Mexican, and "Zeilanense."[11]

Naturalists in the sixteenth and seventeenth centuries expanded these unwieldy practices as they came into contact with new plants abroad. Francisco Hernández, collecting in New Spain in the 1570s, diligently recorded Nahua names for many of the plants he encountered. The military man Charles de Rochefort, working in the West Indies in the 1650s, employed the European method of collecting synonyms for particular plants across diverse Indian languages. Writing of the *manyoc* of the Caribs, for example, he gave equivalent names in Toupinambous *(manyot)* and other Amerindian tongues (for example, *mandioque*). Charles Plumier in his 1693 description of plant life in the Americas also gathered and recorded Taino and Carib names. Hendrik Adriaan van Reede tot Drakenstein, while exploring the coast of Malabar, provided "Brahmanese" and Malayalam names for the plants he encountered. Pierre Barrère in Cayenne offered names for plants in Latin, French, and "Indian." Jean-Baptiste-René Pouppé-Desportes, working in Saint Domingue, supplied names in Latin, French, and Carib.[12]

While many naturalists happily incorporated names from other continents and cultures into the European botanical corpus in order to accommodate a burgeoning flora, others were less sanguine about the practice. Linnaeus emphasized in his 1737 *Critica botanica* the urgency of developing a strict and standardized "science of names," by which he meant a set of rules regulating how names should be created and maintained. He judged the reigning practices a "Babel" of tongues and with characteristic flourish warned, "I foresee barbarism knocking at our gates." His *Critica botanica*, a two-hundred-page work setting out rules for standardizing botanical nomenclature, might be viewed as a first code of botanical nomenclature. But Linnaeus' extensive prescriptions also banished many things: European languages except Greek or Latin; religious names (though he allowed names derived from European mythology); foreign names (meaning foreign to European sensibilities); names invoking the uses of plants; names ending in *oide;* names compounded of two entire Latin words; and so forth. He was especially emphatic that "generic names that are not derived from Greek or Latin roots must be rejected." Expressly targeting van Reede's *Hortus Indicus Malabaricus,* Linnaeus declared all foreign names and terms "barbarous" (though for some reason he preferred these barbarous names to what he considered the "absence of names" in Maria Sibylla Merian's account of the plants of Surinam, another text he mentioned). Linnaeus retained "barbarous names" only when he could devise a Latin or Greek derivation, even one having nothing to do with the plant or its provenance. *Datura* (a genus in the potato family) he allowed, for example, for its association with *dare* from the Latin meaning "to give, because it is 'given' to those whose sexual powers are weak or enfeebled."[13]

In setting conventions, Linnaeus made clear that Latin was to become the standard language of botany. That is to say, that all names and descriptions or diagnoses were to be published in Latin. "Long ago the learned men of Europe met and chose the Latin language as the common language of learning," Linnaeus wrote. "I do not object to any nation retaining its own vernacular names for plants," he continued. "[W]hat I do earnestly desire is that all learned Botanists should agree on the Latin names." In fact, Linnaeus' preference for Latin may have derived from the fact that he himself wrote no other European language but Swedish, a language few Europeans could read. Latin of course was the *lingua franca* of academic exchange; botanist William Stearn has suggested, however, that Latin was chosen for international communication between scholars precisely because few women read it. Stearn has also suggested that because Latin was the language of educated men, its "neutrality" facilitated communication worldwide.[14]

In respect to other cultures, of course, Latin was not value-neutral. Linnaeus' devotion to Latin displaced other languages; he explicitly chose as the "Fathers of Botany" in this regard the ancient Greeks and Romans, not the "Asiatics or Arabians" whose knowledge of plants even Linnaeus would have recognized as ancient and extensive but whose languages he considered "barbarous." Botanical Latin did not come ready-made in the early modern period; it was made and remade—new terms were introduced, others stabilized—to suit botanists' purposes. Botanical Latin has best been described by Stearn as a modern Roman language of special technical application, derived from Renaissance Latin with much plundering of ancient Greek, which has evolved, mainly since 1700.[15] As we shall see, local and global politics came to be embedded in this scientific language as it developed in this great age of scientific voyaging.

In creating Latin names for plants, Linnaeus perfunctorily gave a nod to the Baroque notion that the best generic name was one to which "the plant offered its hand"—that is, a name that highlighted a plant's essential character or appearance. Thus *Helianthus* (Flower of the Sun) designated a plant "whose great golden blossoms send out rays in every direction from the circular disk," or *Hippocrepis* (Horseshoe) denoted "the marvelous resemblance of the fruit of this plant to an iron horseshoe." At the level of class and order, naturalists, including Linnaeus, often derived names from what they defined as some essential character for that class or order. Thus Linnaeus named that class of animals characterized by having mammae, *Mammalia*. In botany, Linnaeus named his classes and orders for the number of male sexual parts (stamen) and female sexual parts (pistils), which he considered essential characters. Merian's *flos pavonis,* for example, figured in Linnaeus' class *Decandria* (having ten "husbands" or male parts) and the order *monogynia* (having one "wife" or female part).[16]

But Linnaeus understood that all too often what is considered an essential character in a plant may lie merely "in the eye of the beholder." (Modern nomenclaturists avoid coining names that attempt to capture essential attributes of taxa because their choice may prove incorrect in the future; ambiguity, in this case, precludes embarrassment.) What Linnaeus proposed was a naming system abstract in relation to the properties of plants but concrete in relation to the history of botany in Europe: "as a religious duty" he intended "to engrave the names of men on plants, and so secure for them immortal renown." He devoted an uncharacteristic nineteen pages to this point; most entries in his *Botanica critica* are one to three pages long. This practice of naming plants after men was an ancient one, he argued, one practiced by Hippocrates, Theophrastus, Dioscorides, and Pliny, and newly revived by his immediate predecessors Charles Plumier (1646–1704) and Joseph Pitton de Tournefort (1656–1708). Linnaeus noted that he preferred the names Plumier devised to commemorate heroic botanists to those "barbarous" coinages he (Plumier) formulated from Amerindian languages.[17]

Men immortalized in the Linnaean system included Tournefort *(Tour-*

nefortia), van Reede *(Rheedia)*, the Commelins *(Commelina)*, Sloane *(Sloanea)*, and André Thouin, gardener at the Jardin du Roi *(Thouinia)*. In his *Critica botanica* Linnaeus counted 144 genera named after prominent botanists, fifty of which were coined by Plumier, five by Tournefort, and eighty-five by himself. Few women's names appear in this 1737 list; surprisingly absent is the name of Maria Sibylla Merian, whose work Linnaeus often did in fact cite. Only when softened by old age did Linnaeus commemorate a few women (see below). *Meriania* was introduced in the 1790s by Olof Swartz, a Swedish botanist who worked extensively in Surinam. Eventually six species of plants, nine butterflies, and two beetles were named for her.[18]

Linnaeus often gave fanciful reasons for naming a particular genus after a particular botanist. *Bauhinia,* he wrote, has two lobed leaves growing from the same base and appropriately bears the name of the noble pair of brothers Bauhin, Jean and Caspar. *Hermannia,* which produces flowers that are very unlike any others and belongs to Africa, is appropriately named for Paul Hermann, a botanist who introduced Europe to African flora. *Hernandia,* an American tree with handsome leaves, was named for a botanist (Hernández) who was highly paid to investigate the natural history of America. *Magnolia* was "a tree with very handsome leaves and flowers, recalling that splendid botanist [Pierre] Magnol." With not a little feigned modesty Linnaeus wrote of the *Linnea* (a small flowering plant): it was named by the celebrated Johan Frederik Gronovius and is a plant of Lapland, "lowly, insignificant, disregarded, flowering but for a brief space—from Linnaeus who resembles it."[19]

Linnaeus himself expected resistance to his practice of naming plants after botanists: "if any of my aphorisms should provoke opposition, it will assuredly be this one." He justified his naming procedures in four ways. First, the ceremony involved in bestowing a man's name on a plant "aroused the ambition of living [botanists] and applied a spur where it is suitable." Second, such practices were sanctioned by other sciences: physicians, anatomists, pharmacists, chemists, and surgeons customarily attached their names to their discoveries (he mentions Harveian circulation, Nuck's canal, and Wirsung's duct). Third, this procedure conformed to the customs of voyagers, who often lent their names to lands they discovered. "How many islands," Linnaeus pondered, "have not obtained their names from their first European visitors? Indeed a quarter of the globe has received from that insignificant specimen of humanity Amerigo [Vespucci] a name which no one would refuse to give it." Who, he continued, would deny a botanical discoverer his discovery? Finally, and most significantly, this practice of naming plants after great botanists seamlessly folded a history of botany into botanical nomenclature itself: "it is necessary for every Botanist to treasure the history of the science which he is passing on, and at the same time to be familiar with all botanical writers and their names." Nomenclature was mnemonic and honorific. It was also simply "economical" for the names of plants and of great botanists to be one and the same. Linnaeus therefore admonished botanists to lend their names to genera with care. One should, he noted, lend one's name only to natural genera; an artificial and provisional genus would soon pass away and with them one's name.[20]

Linnaeus could have chosen many things to highlight in naming practices—for example, the biogeographical distribution or the cultural uses of plants. In fact, he chose to celebrate botanists known to him—a practice that reinforced the notion that science is created by great individuals, and in this case European men. In so doing, he inscribed a particular vision of the history of botany into the very names by which we know the world. Linnaeus' naming system retold the story of elite European botany—to the exclusion of other histories.

It is important to recall that Linnaeus' naming practices developed at a time when naturalists were newly regulating who could and could not do science. It was a time when the informal exclusion of women was formalized, for example. It was also a time when European science was establishing its power vis-à-vis other knowledge traditions. As part of this, Linnaeus closely guarded the power to name. Accordingly, "no one ought to name a plant unless he is a botanist" (*de facto* not a woman or a person from another culture). Linnaeus furthermore admonished that only "he who establishes a new genus should give it a name," strengthening priority of discovery as a chief scientific value. Linnaeus' system also served to solidify growing professional divides. "I see no reason why I should accept 'officinal' names, unless one should wish to place the authority of pharmacists unnecessarily high." Finally, the aging Linnaeus required that only "mature" botanists—not rash young men or "newly hatched botanists"—name the various parts of nature's body.[21]

Linnaeus argued that the botanists commemorated in his system (Sloane or Tournefort, for example) deserved such high honor because many were martyrs to science, having suffered "wearisome and painful hardships" in the service of botany. First of the beleaguered "officers in flora's army" was himself: "In my youth I entered the deserts of Lapland . . . I lived on only water and meat, without bread and salt . . . I risked my life on Mount Skula, in Finmark, on icy mountains and in shipwreck."[22] Linnaeus also promoted generic names celebrating European kings and patrons who had contributed to the cost of oceanic voyages, botanical gardens, extensive libraries, academic professorships in botany, and textual illustrations.

The point again was glory, or immortality. Anyone whose name was "gloriously" immortalized by science had, Linnaeus maintained, "obtained the highest honor that mortal man can desire." This little bit of glory compensated the botanist for pursuing a passion that otherwise yielded little worldly gain. This higher prize, Linnaeus admonished, was to be guarded jealously; it was too "priceless, brilliant and valuable" to be wasted on "the uneducated, florists, monks, relations, friends and the like."[23] More stringent on this matter than most taxonomists, Linnaeus preferred also to exclude the names of men, saints, and public figures not directly connected to botany.

I myself lavish attention on Linnaeus because his naming system was consecrated as the starting point of modern botany in 1905 at the International Botanical Congress, in article 19 of the *Règles de la nomenclature botanique*, and has remained the standard. Since the 1860s, botanists have met approximately every five years (except when interrupted by war) in order to standardize the rules of nomenclature. Charles Darwin deemed such activities so important that he bequeathed substantial funds from his own estate to establish a global list of plants, their authors, and geographic locations. Darwin's gift launched Sir Joseph Hooker's *Index Kewensis* that by the 1880s included a set of index cards and boxes weighing more than one ton.[24]

An overriding concern of these meetings has been to provide stability in botanical nomenclature; that is to say, for botanists to agree upon principles, rules, and laws to ensure that each plant is known by one and only one name internationally. Between 1867 and 1905 many solutions for this desired stability in naming practices were devised. In 1905, botanists settled on several principles, including the setting of priority of publication as the fundamental principle for establishing names, and recognizing Linnaeus' *Species plantarum* (1753) and its 6,000 species names as the starting point of botanical nomenclature for all groups of vascular plants. All works published before 1753 (by Plumier, van Reede, or Merian, for example) were declared invalid for purposes of naming plants; these naturalists' naming practices, strongly based in local cultures both domestic and foreign (see below) were replaced by Linnaeus' European-centered system.

Linnaeus' *Species plantarum* was chosen not because Linnaeus had devised the system of binomials or named *de novo* the species it includes. Stearn has cogently argued that Linnaeus' accomplishment was his taking of his predecessors' names and methods and applying them consistently, methodically, and on a large scale to the whole of the then-known floral world. Linnaeus served, in a sense, as a one-man tribunal, sorting and compiling what he deemed appropriate names from earlier authors. In the process, however, he produced the largest one-time change in the history of botanical naming, replacing old and coining new names as he saw fit, sometimes with caprice. The twentieth-century botanist S. Savage notes that in some cases Linnaeus' abrupt interventions were unfortunate; his classical names led some botanists mistakenly to assume that indigenous American plants, for example, grew in ancient Greece.[25]

Eighteenth-century naturalists were keenly aware of the glory attached to having a plant bear his or her name. In preparing for his voyage to the West Indies, Sir Hans Sloane wrote: "some men seem to have a great desire to be the first authors of discovering such or such plants, and to have them carry their names in the first place, but I endeavour'd rather to find if anything I had observed was taken notice of by other persons." Returning to England, he published his 1695 catalogue of Jamaican plants in order "to do right to the first authors and the publick."[26] Botanists in this era often jockeyed to be first to publish a description of a plant and an appropriate plant name. They fought over names, and even on occasion insulted their enemies through the names they chose for particular plants. Linnaeus, for example, called an odious weed *Siegesbeckia* after Johann Siegesbeck, an outspoken critic of his sexual system.[27]

As noted above, modern botanists, paleoanthropologists, and zoologists sometimes portray naming practices as politically indifferent or trivial. The writers of the 1981 *International Code of Botanical Nomenclature,* for example, claimed that "the purpose of giving a name to a taxonomic group is not to indicate its characters or history, but to supply a means of referring to it and to indicate its taxonomic rank."[28] But even in the twenty-first century, certain political considerations creep into naming practices. A paleontologist once told me that he had intended to name a newly-found fossil after his wife, but was persuaded by his colleagues to utilize instead an African name (honoring the place where the fossil was found) in order to ease relations for Americans and Europeans working on that continent.

The 1905 code's insistence on priority also focused attention on authorship (meaning publication), reinforcing the preeminence of well-educated European males over collectors, gardeners, informants, and others who quietly served the cause of botany. The 1905 Vienna Congress also reinforced Linnaeus' choice of Latin as the appropriate language for botany, despite protests from the American delegation that this choice was "arbitrary and offensive."[29] Interestingly, these international meetings were transacted in French (not Latin)—even at the 1924 Imperial Botanical Conference, held in London. Not until 1935 did English became the *lingua franca* of these meetings.

Not to leave any stone unturned, nomenclaturists in the 1960s established rules for naming extraterrestrial taxa. Naturalists fought bitterly over whether 1961 or 1962 should be recognized as the starting point for universally valid names for extraterrestrial fossils, and whether Latin should be required for their diagnoses.[30] The rule has not of course been tested yet in the field.

Naming Conundrums

In several chapters in this book, we have looked at a particular flower, Merian's *flos pavonis*, and the cultural politics surrounding its use as an abortifacient. The history of this same flower epitomizes the shift away from multicultural naming practices that flourished in the late seventeenth century to the onomastic imperialism developing in the eighteenth century. Although Merian was very much aware of the tumult over abortion surrounding her *flos pavonis*, she was largely oblivious to the complex politics involved in the naming and renaming of this plant.

Despite her rarity as a woman naturalist, Merian's collecting practices were similar to those of her male colleagues. Like Sloane, her contemporary, she was keen to collect "the best information" concerning the exotic plants and insects she encountered from the local inhabitants.[31] Like the astronomer Peter Kolb, who wrote an early ethnology of the Africans at the Cape of Good Hope, Merian developed deep friendships with several Amerindians and displaced Africans in Surinam who served as her guides to desirable specimens and provided access to dangerous, often impassible regions. Merian also followed the practice common up to that time of retaining native names and recording much else that native peoples told her about the plants and animals she studied. In the introduction to her *Metamorphosis insectorum Surinamensium* she made a point of noting that "the names of the plants I have kept as they were given to me by the natives and Indians in America."[32]

Reliance on local peoples and their knowledge made sense as more European naturalists ventured farther into Africa, India, China, Japan, and the Americas. It is therefore curious that in this environment, where voyagers often directly transcribed plants' indigenous names, Merian rendered this plant in Latin as the *flos pavonis*.[33] Given that Merian often recorded the personal experiences of her informants in vivid detail, why did she not report a local Arawakan or transplanted Angolan or Guinean name for the plant?

The gap between Merian's professed purposes and her naming practices shows that the history of the *flos pavonis* is not a simple one. We do not know why Merian did not record an indigenous American name for the plant, but it may have been because she did not think it native to the West Indies. Where this plant originated and how its seeds actually traveled, whether on board merchant or slave ships, or adrift at sea, is still uncertain (see Conclusion). A plant's many historical names, however, can sometimes provide clues to its origins and geo-distribution. Merian probably chose the name *flos pavonis* because she had seen this tropical tree in Amsterdam's ostentatious (by standards of the time) botanical garden, the Hortus Medicus. Most of the Latinized East Indian names for this brilliantly flowering plant associated it with the peacock: Jakob Breyne, a Danzig merchant and sometime botanist, reported that in Ambon, an island of Indonesia, the luxuriant tree was called *crista pavonis* for its "distinguished stamen . . . that bursts forth to form the proud crest of the peacock." This flaming red, yellow, and orange flower was also called *flore pavonino* (peacock flower) and *flos Indicus pavoninus*. The Dutch living in the East Indies called the plant "peacock tails" *(paauwen staarten)* and the Portuguese labeled it the *"foula de pavan."* Less poetically, the plant was sometimes known by the Latin *frutex pavoninus,* or "peacock bush."[34]

Merian may simply have followed colonialist practices of applying the name by which she already knew a plant to that or a similar one found in the New World. In this case, the East Indian names associating the plant with the peacock would have moved to the West Indies after transit through Amsterdam. This was not an uncommon practice: the Dutch, who held both Malabar and Surinam, recognized in Surinam a plant that had been of use to them in Malabar for "diminishing the humors"—that is, promoting sweat and urination. In Surinam, they called this same plant "the leaf of Malabar." These practices led Georges-Louis Leclerc, comte de Buffon to complain about the confusions created when llamas, for example, were called "camels of Peru."[35]

But Merian's failure to record an indigenous American or displaced African name for the plant may indicate that it did not have such a name, or at least not one commonly known to Dutch planters in Surinam. Sloane found the peacock flower or a plant resembling it growing in Barbados, and labeled it *Tlacoxiloxochitl*, an Aztec name from Mexico. Today, the *flos pavonis* is also recognized under a Spanish rendering of a Nahua name, *Tabachin* or *Tabaquin*. Though more local, none of these names derives directly from Arawak, Taino, or Carib languages. The French botanist Pouppé-Desportes, who assiduously recorded "Caraïbes" plant names current in Saint Domingue, gave no Carib synonym, noting instead only the French and Latin names for this plant. Given the lack of sources, we cannot know if Merian's native informants had a name for this plant, or whether the enslaved populations from Western Africa learned the name *flos pavonis* from Dutch settlers. We do not know whether slave women, whom Merian observed using the plant, brought an African name for it with them when they were taken from Africa.[36]

If the peacock flower did not have a local West Indian name, it did enjoy indigenous names in Malabar and "Zeylon" (Sri Lanka). Upon her return to Amsterdam, Merian, whose knowledge of Latin was weak, employed Caspar Commelin, a friend and the director of the Hortus Medicus in Amsterdam, to add bibliographical references to the text of her *Metamorphosis* in order to place the Surinamese plants and insects she so elaborately illustrated into the world of European classical learning. What Commelin added to her paragraphs discussing the *flos pavonis* was the term *tsjétti mandáru*, a Latinization of the Malayalam name. Commelin drew his information from van Reede's *Hortus Indicus Malabaricus*. In addition to the Malayalam term *tsjétti mandáru* cited by Commelin, van Reede and his team presented names for the plants of Malabar in all the languages used there: Arabic, Portuguese, Dutch, and "Brahmanese" or Konkani (transcribed as *tsiettia*). Paul Hermann, who in his youth served as a medical officer in Ceylon for the Dutch East India Company, also reported its colorful "Zeylonese" (Sinhalese) name: *monarakudimbiia*.[37] Later in the eighteenth century, Michel Adanson called Merian's *flos pavonis*, which he had found in Africa, by the indigenous name *Kamechia* (in volume II, he rationalized the spelling to *Campecia*).

But a new fate awaited Merian's *flos pavonis*. In 1694, this flamboyant flower was included within Tournefort's abstract typology—the classification widely regarded today as an important forerunner of Linnaean systematics. Tournefort placed the plant in his Class 21, Section 5, encompassing "trees and shrubs with red flowers and seed pods." As was typical of many of the new schema being introduced at this time, Tournefort's classification focused on the physical characteristics of the plant, in this case the corolla and the fruit. The plant's East and West Indian connections, which had played a significant role in earlier European accounts, were not discussed.

In the process of anchoring Merian's *flos pavonis* (van Reede's *tsjétti*

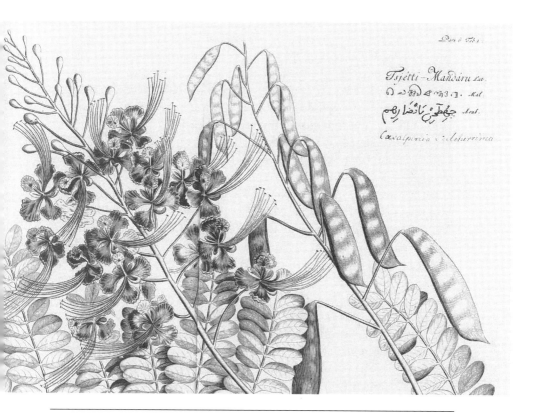

Figure 5.1. Hendrik Adriaan van Reede's *tsjétti mandáru*, the Malayalam name for Merian's *flos pavonis*. Van Reede offered names in Latin, Malayalam, and Arabic. *Caesalpinia pulcherrima*, one of the two approved names for the *Poinciana*, is penciled in Harvard University's copy of this work.

mandáru and Hermann's *monarakudimbiia*) in the European world, Tournefort devised a wholly new name, *Poinciana pulcherrima*, celebrating his countryman, de Poincy, governor of the French Antilles, who used the plant to treat fevers. Tournefort's name celebrated French colonial rule in the Caribbean rather than the plant's own virtues, its East and West Indian heritage, the peoples who used it, those who "discovered" it, or those who supplied Europeans with information about it. Following Tournefort's lead, Linnaeus approved this name and added only that the plant grows in the Indies (apparently both East and West) and under the sign of Saturn. In 1791 Swartz reassigned the plant to the genus *Caesalpinia* (named for the sixteenth-century Italian botanist Andrea Cesalpino) in efforts to do away with the genus *Poinciana* because it included only this one plant. Linnaeus' name, however, remains the basionym (base name) since the species concept has not changed. Botanists today use both *Poinciana pulcherrima* and *Caesalpinia pulcherrima;* among these two, the French still prefer the former *(Poinciana)*, honoring their countryman.[38]

Merian and van Reede's purposes, we should keep in mind, were to collect for the sake of medical and economic utility, not to classify for the sake of establishing a taxonomic system. Merian expressly refused to classify her plants. Discussing her *Metamorphosis,* she wrote, "I could have given a fuller account, but because the views of the learned are so at odds with one another and the world so sensitive, I have recorded only my observations."[39] Merian was not a botanist; she was interested in plants primarily as the habitat and food for the caterpillars that were the focus of her study.

The story of Merian's *flos pavonis* captures well naming practices in this age of colonial expansion. In his study of plants in Spanish realms, historian Jean-Pierre Clément found that of the 175 plants named in the eighteenth century, 111 were named for scientific persons (sixty-five for botanists and naturalists, twenty-five for illustrators, sixteen for physicians, five for astronomers or other scientists), twenty were named for writers, and thirty-eight for influential persons (four royalty, eleven government ministers, seven viceroys, six prelates, and ten others).[40] Such naming was not taken lightly. The Spanish naturalist Gomez Ortega—then director of the Botanical Garden in Madrid—supervised meetings whose sole purpose was to select names for new genera. In their 1794 *Florae Peruvianae et Chilensis,* documenting Spanish expeditions to those countries, Hipólito Ruiz and José Pavón included a note in the description of each genus explaining the individual honored by the name. Historian Mauricio Olarte has documented how the names in this work encapsulate the political history in Spain: prior to 1810 New World plants were named to commemorate the deeds of Spanish botanists, after 1810 and New World movements for independence, new genera were defiantly named by creole botanists for creole botanists.[41]

Even into the nineteenth century, immortalization in botanical nomenclature was much sought after by prominent families. Katharine Saunders, a noted botanical illustrator living in South Africa, wished for various members of her family to be immortalized by having the new species she discovered named after them. Joseph Hooker at Kew Gardens handed the matter over to Professor Oliver with the comment: "Her husband is a member of the legislative council so we must bear with her. Her passion is to have something named after her. Can this be managed?" In the end, plants were named for her, her son, and her daughter-in-law.[42]

Exceptions: *Quassia* and *Cinchona*

To what extent did the names of plants used by African slaves in the West Indies, Arawaks, Caribs, and others who were not European botanists enter into European taxonomic schema? As we have seen in Chapter 1, botanists were not lone wanderers but, like modern-day directors of laboratories, heads of large expeditions, gardens, and herbaria. For botanists to be successful abroad, they needed financiers, ship captains, assistants, illustrators, local guides, and carriers. For success in Europe, they needed gardeners, correspondents, keepers of herbaria, and suppliers of specimens. When it came to naming plants, however, these support people were often overlooked in favor of more prestigious European-educated males.[43]

There were exceptions. Several celebrated plants were named for persons who did not fit Linnaeus' general profile. Take, for example, the tree supplying a much admired stomach tonic that came into Europe bearing the exotic name *Quassia amara*. The plant was named for a freed Surinamese slave, "Graman" (Greatman) Quassi. According to Linnaeus' own criteria, Quassi should not have qualified as a recipient of botanical fame. He was a healer, a medical man, and not strictly speaking a botanist. Linnaeus, however, like other botanists at the time, celebrated the former slave as a heroic individual, immortalized, in this case, for having been the first to discover the plant's medicinal use. Toward the end of the eighteenth century, with the rise of the *Amis des noirs* in France and the movement across Europe to end the slave trade, naming this plant for a former

Figure 5.2. Quassi shown in European dress with the gold medal presented to him by the Prince of Orange. Like other learned Africans celebrated in the age of Enlightenment, Quassi accommodated to the European world in dress, manners, and mores. By William Blake.

slave became a cause célèbre; many reports fêted Quassi as "absolutely the first discoverer" of the Quassia tonic.

If we piece together various eighteenth-century accounts of how the tonic was discovered, however, we find that Quassi did not, in fact, discover the cure for which he was celebrated. At best, Quassi served as a middle man, the individual who brought a widely used remedy to the attention of learned Europeans. William Lewis, a medical doctor and fellow of the Royal Society in London, provided some clues into the development of this drug. According to Lewis, the Amerindians of Surinam first recognized *Quassia* as an excellent tonic. This knowledge passed somehow (we are not told how) to Quassi, then a slave, who developed the root into a secret remedy against the "fatal fevers" of that country. Lieutenant John Stedman, who lived for many years in Surinam, dated this to about 1730. Quassi's secret was purchased for a considerable sum by Daniel Rolander, one of Linnaeus' students, who then took it back with him to Europe in 1756. A specimen of the tree from which the remedy comes was presented by Carl Gustav Dahlberg, a Swedish plantation owner in Surinam, to Linnaeus in 1761. Linnaeus immediately published a dissertation that named, described, and provided an illustration of the plant, thus establishing it within European botany. Interestingly, Dahlberg was appalled when Linnaeus commemorated Quassi in the plant's name; he had hoped for this honor himself.[44] No one at the time considered naming the plant for the Amerindians from whom the cure seems to have originated.

Quassia became a popular "bitter," praised for its efficacy in suppressing vomiting and removing fever, both in the Caribbean and in Europe. Experiments by European physicians showed it to be as potent as the Peruvian bark without any of the bark's main side effects (notably diarrhea). Deemed safe and effective, *Quassia*—used in infusion, extract, or pills—found its place in the various European *Pharmacopoeia*. Only rarely did physicians, such as Edinburgh's William Cullen, find the enthusiasm for the bitter overdrawn and characteristic of what he judged to be an unhealthy vogue for exotic medicines.[45]

Physicians on the ground in Surinam did not appreciate Quassi's being given credit for having discovered the drug. The experimentalist Philippe Fermin remarked in 1769 that "this wood has been known for forty years to nearly all the inhabitants of Surinam." Stedman, the military man, however, painted a picture of the "Graman Quacy" as "the most extraordinary black man in Surinam, or perhaps in the world." By his "industry," "artifice," and "ingenuity," Stedman wrote, "he obtained his freedom from slavery" and, from his healing arts, also "a very competent subsistence." He made his considerable fortune by cultivating a reputation as a *looco-man*, or sorcerer, among the vulgar slaves and by selling his *obias*, or amulets (made, Stedman charged, with "trash"—fishbones, eggshells, etc.—that cost him nothing) to the free African soldiers in Surinam who, under the power of this superstition, fought "like bulldogs" for the Dutch. This "filled his pockets with no inconsiderable profits." On top of all this, Stedman continued, Quassi had the good fortune of discovering the bitter that made him famous. But all was lost because, Stedman judged, Quassi was "an indolent dissipating blockhead" who fell into "loathsome disorders" and finally contracted leprosy, for which there was no cure.[46]

The Jewish physician in Surinam, David de Isaac Cohen Nassy, also wrote that Quassi had the confidence of the whites, who consulted him on many things, including finding and prosecuting plantation slaves who poisoned their masters.[47] One might conjecture that it was for his service in supplying troops to the Dutch colonists and quelling rebellions among his own people as much as for his priority in discovery that Quassi was eventually rewarded with immortality in the European system of botanical nomenclature.

The Peruvian bark *(Cinchona officinalis)* serves as another example of a plant—in this case a plant of strategic importance to Europeans in tropical areas—named for a botanical outsider. *Cinchona*, the source of quinine, was named by Linnaeus after Francisca Fernandez de Ribera, wife of the fourth Count of Chinchón, the Spanish Viceroy Luis Geronimo Fernandez, for her role in drawing attention to the virtues of this "miraculous cure" sometime between 1632 and 1638. The countess hardly qualified for such an honor in that she was neither a botanist nor a patron of the emerging science. Though Linnaeus was taken to task for misspelling the countess's name when he coined the term in 1742, it was widely held that he rightly immortalized this woman's heroic risking of her life by trying the unknown anti-malarial.[48] Much like Lady Mary Wortley Montagu, the countess helped to popularize an exotic medicine in Europe.

Linnaeus of course had many names to choose from when coining the term *Cinchona*. Before 1742, the bark was most often known by the Quechua name: *Quinquina*. Charles-Marie de La Condamine, from whom Linnaeus received an illustration of the plant, reported that *Quinquina* derived from the "ancient of language of Peru," but that no one, either in Lima or elsewhere, could tell him its literal meaning. Confounded, La Condamine consulted a 1614 "Quichoa" dictionary and found *"Quina ai,"* a term no longer in use since, as La Condamine lamented, by his day the Quechua language was "strongly altered and mixed with Spanish." This term translated into Spanish as designating a kind of Indian cape or shawl; La Condamine suggested that this could also refer by analogy to the bark taken as the "mantle" of a tree. Repetition was common in Quechua, especially in the names of plants, and the Frenchman concluded that *Quinquina* was best rendered "bark of bark." La Condamine took pains to point out that *Quinquina* was also the indigenous name for the Peruvian balsam tree (later *Myroxylon peruferum*), known to Europeans as early as 1565 and often confused (sometimes purposely) with true quinine.[49]

Joseph de Jussieu, who accompanied La Condamine to Peru, reported a different Quechua name for the plant: *Yaracucchu Carachucchu,* with *yara* signifying tree, *cara* bark, and *chucchu* the shiver of the malarial fever. It was also known in Spanish as *cascarilla*. Widely known in English as "the Peruvian bark" or simply "the bark," *Quinquina* was sometimes also called the *Countess's Powder,* after the Countess of Chinchón, or the Jesuits' Powder, after the Jesuits who were instrumental in introducing it into Europe. Finally, it was known as the *pulvis patrum* or the *pulvis cardinalis de Lugo* (the cardinal's bark) after Cardinal de Lugo, who distributed it to the poor at Rome. Quinine's close association with Jesuits aroused distrust among Protestants, some of whom refused to use it.[50] From these many possibilities Linnaeus chose to immortalize the countess in his system.

The story of who discovered quinine and introduced it into Europe has many variants and embellishments. According to La Condamine, who traveled extensively in present-day Ecuador, the Incas knew *Quinquina* long before the Spanish arrived but, given their great hatred of their conquerors, hid this "most precious and useful" cure from them. After guarding their secret for some hundred and forty years, the "naturals," as La Condamine called them, revealed their remedy to the Europeans. Jussieu testified that the Europeans learned of the cure when a Jesuit, laid low by fever when passing through the village of Malacatos, was pitied and treated by an "Indian chief." Another version has it that the discovery was made accidentally when a Spanish soldier, seized with malaria and overcome with thirst, happened to drink from a lake impregnated with quinine from some fallen trees.[51]

According to La Condamine's account, *Quinquina* was widely used by the Indians and also the Spanish at Loja but ignored by the rest of the world. This was to change in 1638 when the Countess of Chinchón was struck by a severe fever in Lima. Her physicians could find no cure. An official in Loja, hearing of her illness, sent some bark to the viceroy in Lima. The man was immediately summoned to the countess's bedside to prepare the remedy and regulate the dosage. La Condamine quietly stated that before the drug was given to the countess "some experiments were made with success on other ill persons." In some versions these persons were "patients of inferior rank"; the fresco painted for Santo Spirito in Rome shows that to guard against poisoning, an Indian messenger was first made to drink the unknown infusion. The countess took the medicine and was cured. After her recovery (per La Condamine's account), the countess immediately requested large supplies of the bark and (again like Lady Montagu in Adrianople) distributed the cure widely among the Spanish settlers in the New World. A fictional version of the story, published by Stéphanie Félicité, comtesse de Genlis, in the early nineteenth century, made Zuma, the beautiful Indian princess and chambermaid to the Countess of Chinchón, the heroine. On pain of her own life, Zuma revealed the cure to the countess, testing the suspect powder herself before giving it to her mistress.[52]

In coining the term *Cinchona*, Linnaeus laid laurels at the feet of a woman of European extraction, celebrating European colonial rule. He also cut the plant's ties to its South American home and the Incans who first came to know it and who chose to inform the Europeans about it. The physician Thomas Skeete, experimenting with the bark in the 1780s, noted that the Quechua term, *quinquina*, was seldom heard anymore.[53]

With both Quassi and the countess, we see Linnaeus celebrating individuals who served as "intellectual conduits" between the New World and the Old. Quassi did not discover the drug for which he is celebrated any more than did the Countess of Chinchón, but both helped to introduce the drug into Europe. Priority and discovery, so much celebrated in scientific nomenclature, was obscured in each of these cases.

Linnaeus went to great lengths to name *Cinchona* after a woman, in this case as in the others discussed below, of high social standing. It is unclear what his motivations were. Elsewhere I have discussed the social conservatism apparent in his desire to see his own daughters grow up to be hearty, strong housekeepers, not "fashionable dolls" or bluestockings. Historian Gianna Pomata, however, might interpret this preference for a woman's name consistent with Linnaeus' ovism (the belief that new humans exist preformed in the ovum) that resulted from a seventeenth-century revaluation of female contributions to generation. In either case,

Linnaeus used the status of wellborn women to enhance the reputation of botany, a common occurrence in early modern science.[54]

Linnaeus had no reason to doubt the veracity of the Chinchón legend. Humboldt, however, in the nineteenth century and historians today have shown the story to be a fabrication. The Count of Chinchón's diary made no mention of his wife's illness or miraculous cure; nor did the writings of other Europeans living in Peru at the time.[55]

I have found only the one male freed slave, Quassi, immortalized in eighteenth-century botanical nomenclature. As far as I am aware, no Amerindians were so honored, although a number of scientific plant names do derive from Native American languages. Several European women, however, number among the "heroic" individuals honored in botanical nomenclature. In his *Critica botanica,* Linnaeus highlights the names of women that he retained from classical Greek literature: *Helenium* for Helen, Queen of Menelaus, King of Sparta; *Artemisia* for Artemisia, Queen of Caria; and *Althea* for Althea, wife of King Aeneas of Chalcedon. In addition to the Countess of Chinchón, Linnaeus also immortalized in his system a second highborn eighteenth-century Englishwoman, Lady Anne Monson, great-granddaughter of Charles II, for whom he named the genus *Monsonia*—a type of South African geranium. It is possible that Lady Anne contracted her second marriage to a colonel of the East Indies Company with the thought of botanizing in exotic areas. Carl Thunberg reported from the Cape of Good Hope in 1774 that "there has arrived from England, a Lady Anne Monson, who had undertaken this long and tedious voyage, not only for the purpose of accompanying her husband . . . but also with a view to indulge her passion for natural history." Thunberg described her as about sixty years of age, an avid collector, especially from the animal kingdom, and versed in languages, including Latin, who had "at her own expense, brought with her a draughtsman, in order to assist her in collecting and delineating scarce specimens of natural history."[56]

In requesting permission to confer this honor upon Lady Anne, Linnaeus wrote a florid letter full of eighteenth-century courtesies that his biographer Wilfrid Blunt has suggested may never have been sent: "I have long been trying to smother a passion which proved unquenchable and which now has burst into flame. This is not the first time I have been fired with love for one of the fair sex, and your husband may well forgive me so long as I do no injury to his honor . . . So far as I am aware, Nature has never produced a woman who is your equal—you are a phoenix among women." He intended to enclose in the letter some seeds—"a few rare and genuine pearls"—of the *Alströmieria*, which he judged no one in England had ever seen. Concerning the *Monsonia*, Linnaeus continued: "But should I be so happy as to find my love for you reciprocated, then I ask but one favor of you: that I may be permitted to join with you in the procreation of just one little daughter to bear witness of our love—a little *Monsonia*, through which your fame would live forever in the Kingdom of Flora."[57]

It should be noted that other women of botanical stature were not so honored by the father of modern botany. As mentioned above, Merian was not recognized for her many contributions by Linnaeus; nor was Mary Somerset, Duchess of Beaufort, who maintained extensive botanical gardens at her estate in Gloucester. *Beaufortia* was first introduced as a generic name by Robert Brown in the nineteenth century. Philibert Commerson named the genus *Baretia* after his voyaging assistant, Jeanne Baret, but botanists did not retain this name. The genus is today the *Turraea*. Commerson dedicated the *Peautia coelestina* to the astronomer Nicole-Reine Lepaute, but this name too was soon discarded.[58]

Thus one freed slave in Surinam and several well-born women were commemorated in Linnaeus' system in the course of the eighteenth and nineteenth centuries. These examples, however, did not challenge the spirit inflaming such choices of names: the African and the women were assimilated to the same type of heroic individualism celebrated in eighteenth-century naming practices.

Despite Linnaeus' stringent rules of nomenclature, he did retain a number of well-established "barbarous" names. These, it should be emphasized, were used only if the word was short, attractive when Latinized, and not difficult for a European speaker. I have given the example of *Datura* above; other non-European names Linnaeus adopted include: *Alchemilla, Areca, Berberis, Coffea, Guaiacum, Tulipa,* and *Yucca*. In spite of Linnaeus' disdain for the "barbarity" of van Reede's names, he retained Malayalam names for twelve of his genera, including *Averrhoa bilimbi, Euphorbia tirucalli,* and *Piper betle*. And although Latin dominated naming practices (the Latin names were used to organize and index plants), names in "Telinga," "Bengalese," "Hindoo," Sanskrit, Arabic, French, Spanish, Dutch, Carib, and many other languages were still listed as synonyms.[59]

It is also important to note that Linnaeus' rules often were not followed. He inveighed against utilizing terms already in use by another science, such as anatomy or zoology, and ridiculed in particular the term *Clitoridis*, but this did not remove *Clitoria* from use. This flower name remained until it simply became too much for early Victorian sensibilities.[60] In the 1830s Michel-Étienne Descourtilz wrote of the impropriety of the name *Clitore sensible* given to the plant "whose flowers so greatly resemble that organ in the female body." He offered a new name, *Nauchea pudica*, commemorating the physician Jacques-Louis Nauche and his professional modesty *(pudor)*.[61]

Alternative Naming Practices

Linnaeus' system encountered many forceful foes in the eighteenth century, precisely surrounding this issue of naming. It was not clear to anyone at the time that the Linnaean system would serve as the foundation for modern systematics. According to contemporaries, the "system-madness" of this age was truly "epidemical." Michel Adanson in France counted sixty-five unique systems of botany in 1763; his counterpart in England, Robert Thornton, enumerated fifty-two different systems in 1799. Whereas Linnaeus' system dominated Sweden and England, its acceptance on the Continent, especially in France and Germany, was never complete. The Linnaean system, one among many, came to be regarded as the "starting point" of modern botany (as discussed above) only in 1905.[62]

Chief among Linnaeus' rivals was Buffon, his contemporary and director of the Parisian Jardin du Roi. Buffon opposed system building in general and ridiculed Linnaeus' system in particular for being too abstract and, most grievously, too artificial. In the first volume of his wildly popular *Histoire naturelle, générale et particuliére* of 1749, Buffon decried the proliferation of systems, each with an accompanying system of names. "To speak truthfully," he continued, "each method is only a dictionary in which one may find names arranged according to an order relative to a certain idea, and, consequently, arranged as arbitrarily as if in alphabetical order." The many methods of naturalists were just so many "systems of artificial signs." Buffon emphasized that no method could ever capture nature itself. "Nature," he wrote, "advances by imperceptible nuances so that it is impossible to describe it with full accuracy by strict classes, genera, and species." Buffon nonetheless acknowledged the usefulness of such methods as heuristic devices: they produce a common language assisting mutual understanding, they shorten the work, assist the memory, serve as an aid for studying, and provide an imaginary goal that sustains naturalists in their true "labor," which is the accurate description of natural objects. "Nothing is so rare," he opined, "as to discover exactitude in descriptions, novelty in details, and subtlety in observations."[63]

In respect to nomenclature, Buffon took the traditional approach of listing all known names for a particular species. He cited names given by the ancients, Aristotle and Pliny; authors of the sixteenth century, Gesner, Aldrovandi, Belon; and finally the moderns, John Ray, Linnaeus, Klein, Brisson. Equally important for Buffon were "common" names, whether in Greek, Latin, Italian, Spanish, Portuguese, English, German, Polish, Danish, Swedish, Dutch, Russian, Turkish, Persian, "Savoyard," Old French, or Grison (Romanche). For New World fauna or flora he provided names in "Indian," "Mexican," and "Brazilian" along with names used by the French living there, which were often simply indigenous American names transcribed in French.[64]

Buffon opposed European practices of assimilating exotic plants or animals to Old World taxonomies. Felines with striped fur found in South America, he cautioned, should not be called tigers. Such practices led to assuming they existed there, when in fact they did not. Names poorly adapted, borrowed, badly applied, or newly invented confused nature's order. In direct opposition to Linnaeus, Buffon advocated using native names. Native names (as opposed to newly coined Latin names) gave a clue to the geographic distribution of species. The Greeks and Romans had no name for the buffalo, for example, because that animal is not found in the Old World. As Buffon pointed out, the exotic name "buffalo" indicated a foreign origin. Another advantage of exotic names was that they signaled relationships among animals found in different locations. Thus one might conjecture that the Cayenne *cariacou* is perhaps the *cuguacu* or *cougouacou-apara* of Brazil because of the similarity of names. Félix Vicq-d'Azyr, Buffon's successor at the Académie Royale des Sciences in Paris, also entered the fray, asking why Linnaeus restricted botanical names to Greek and Latin roots. Would it not be preferable, he queried, to use names for plants given to them by "the naturals of different countries"?[65]

Michel Adanson, born two decades after Linnaeus, was also highly critical of Linnaeus' artificial system, which, according to Adanson, gave "unnatural" attention to the numerical proportions of the parts of fructification with undue attention lavished on the male parts. He was no doubt referring to the fact that Linnaeus used the number, relative proportions, and position of a plant's stamen (or male parts) to determine the taxonomic *class* to which it was assigned, while the number and relative position of a plant's pistils (the female parts) determined its *order*. In Linnaeus' taxonomic tree, class stood above order in such a way that male parts were given priority in determining the status of the organism in nature.[66]

Adanson was equally critical of Linnaeus' nomenclature and opposed his extensive revision of botanical names, charging that Linnaeus had arbitrarily changed most of the best-known names in botany and medicine. Adanson pointed to the absurdity of Linnaeus' naming a colonial plant *Dillenia* after Oxford University's Johann Dillenius rather than retaining *Sialita*, one of its traditional names. He even refused the name *Adansonia* for the baobab tree and cited "the vanity of botanists" as one of the three causes impeding the progress of the science. Adanson charged that requiring all natural historical names to end only in *ia, um,* or *us* merely revealed Linnaeus' unsophisticated use of Latin and served only to lend his nomenclature a "scientific air."[67] In step with Tournefort, Buffon, Lamarck, and the French philosophes more generally (and contra Linnaeus), Adanson also rejected the use of Latin for his scientific works, choosing instead to publish in his vernacular French.

But Adanson's challenge went beyond ridicule; he also proposed very different naming practices. First, he argued for retaining traditional names, particularly for economically important plants, because nonbotanists—"physicians, apothecaries, and the people who collect herbs in the fields for medical use"—know these names and easily understand one another when using them.[68] Hence, jalap, the four-o'clock flower, should be reinstated for Linnaeus' *Mirabilis*.

For naming new plants, Adanson argued strenuously for adopting vernacular names from any language—"French, English, German, African, American, or Indian"—that were commonly in use, unless they were too long. While botanizing in Senegal, he learned Wolof and recorded numerous names in that language in his great *Familles des plantes* (1763). Adanson thus championed a pragmatic populist approach to naming. When, for example, required to choose between synonyms as in van Reede's *Hortus Malabaricus,* where often both a "Brahman" and Malayalam name was given, Adanson suggested that the shorter and one most easily pronounced by the greatest number of people be chosen. Specifically attacking Linnaeus, who never traveled outside Europe, Adanson wrote, "if dogmatic authors had traveled, they would have seen that in those other countries [Africa, America, or India] our European names are treated as barbarous." For Adanson, nomenclature could not be fixed until a truly natural system was identified. Until that time, nomenclature should be simple and convenient. Adanson's inclusive, decentralized nomenclature cultivated a vision of human unity and diversity that prompted him, like Buffon, to adopt names from many of the world's languages.[69]

Linnaeus, for his part, did not appreciate Adanson's revisions, writing that his competitor's "natural" system was the most unnatural of all. Linnaean aficionados called Adanson's categories *Familia confusarum*. In a letter to Abraham Baeck, Linnaeus took a typically inflated view of his own accomplishments: "Adanson himself has no experience, everything he has written he has compiled from my works, which I can prove."[70]

Why was the Linnean system preferred over Adanson's? In this instance, personal idiosyncrasies and institutional politics came to shape the history of science. From the beginning, Linnaeus self-consciously managed his legacy. He wrote, for example, a history of botany that celebrated his own work as having set "the whole science on a new foundation." Linnaeus also fashioned his garden and position at the University of Uppsala into a vast botanical empire. Traveling little himself, he corresponded with naturalists throughout Europe and sent his many students (twenty-three of whom eventually became professors) with specific instructions in hand to America, Africa, India, Ceylon, Java, Japan, and Australia. Although Linnaeus' son was not the man his father was, he nonetheless carried on the elder Linnaeus' work and consolidated his fame. Thus Linnaeus' methods and nomenclature were passed on and popularized perhaps beyond their intrinsic value.[71]

Adanson, by contrast, never managed to win an academic position. He lost out in the intricate play of politics in the French academic system and was passed over for a position at the Jardin du Roi, the center of eighteenth-century French botany. Consequently he had few students and correspondents. In the spirit of Enlightenment rationalization, Adanson embedded his new botanical nomenclature in a more general language reform that swept away outmoded doubled letters and diphthongs, as in *nommes*, which Adanson rendered *nomes*, or Theophrastus, which he shortened to Teofraste. Adanson's rationalized prose proved enough of an obstacle that, unlike Buffon's wildly popular *Histoire naturelle*, Adanson's books were not widely read. Although his daughter and only child became a working botanist, she, like most women in this era, was barred from occupying an academic position.[72]

Others in Europe were also critical of Linnaeus' system. Sir William Jones, founder of the journal *Asiatick Researches*, objected vehemently to Linnaeus' sexual system of classification, which so "inflame[d]" the imagination as to be completely useless to "well-born and well-educated women." Jones also found "childish" Linnaeus' practice of naming plants for the persons who first described them, a practice Jones suggested ought wholly to be rejected. He found the names *Champaca* and *Hinna* not only to be more elegant, but far more proper for an Indian or an Arabian plant than the Linnaean *Michelia* and *Lawsonia*.

> Nor can I see without pain that the great Swedish botanist considered it as *the supreme and only reward of labour* in this part of natural history, to preserve a name by hanging it on a blossom, and that he declared this mode of promoting and adorning botany, worthy of being *continued with holy reverence*, though so high an honour, he says, *ought to be conferred with chaste reserve, and not prostituted for the purpose of conciliating the good will, or eternizing the memory, of any but his chosen followers; no, not even of saints.*

Jones took issue with one of Linnaeus' hundred and fifty such names in particular, suggesting that the name *Musa* (for banana) did not derive from a proper name but from the Dutch pronunciation of the Arabic word for that fruit. Jones, like Adanson, suggested keeping the common (and often indigenous Indian) name for a plant. In the case of India, however, he preferred Sanskrit (the Latin of India) over names from what he referred to as "vulgar dialects."[73]

Beyond these aesthetic objections, Whitelaw Ainslie, an English surgeon in Madras, raised the practical point that the English names given by European botanists to some of the trees and shrubs of tropical countries were so "obscure and unfamiliar" that he had chosen to substitute common Indian terms so that the plants could be more easily obtained from the native practitioners.[74]

In African and New World colonies, too, voices arose against Linnaeus. Spanish creoles in New Spain, such as the priest and botanist José Antonio de Alzate y Ramírez, complained that the Linnaean system obscured crucial information about a plant's location, environment, and flowering season, along with the soil characteristics required for cultivation. He remarked further that Linnaeus' sexual system failed to capture important characteristics of plants, such as their usages. Jean-Baptiste-Christophe Fusée-Aublet, who worked in both the Isle de France and Cayenne, was another who advocated retaining names indigenous to the places where plants were found. Anti-Linnaean feeling ran so strong in France that in 1772, Louis XV ordered a remaking of the Jardin Royal des Plantes.

Antoine-Laurent de Jussieu, then a professor of botany at the garden, reordered its plants using his and his uncle Bernard's "natural system," and in the process adopted many of Aublet's plant names.[75]

These controversies have been all but forgotten today. Modern practices reinforce Linnaean principles of giving priority to European languages and names in botanical taxonomy and nomenclature. The 1905 International Botanical Congress in Vienna based its code of botanical nomenclature on Candollean Laws, which (following Linnaeus) set the principle of priority as the rule governing the names of taxa. These congresses—held roughly every six years under the rubric "international"—have been pretty much closed European and, increasingly, North American affairs. At the 1959 congress, for example, all but eight of the 118 named delegates came from Europe, North America, Australia, and the Soviet Union. India, China, Africa, and Central and South America, with their ancient botanical traditions, were poorly represented. Further, these congresses have been predominantly male events. At the 1950s meetings, for example, the nineteen members of the taxonomic section included only two women scientists: the cigar-smoking feminist Johanna Westerdijk of Amsterdam, and Constance M. Eardley of Adelaide, Australia. Similarly the proceedings of these meetings are typically published in the languages of the world leaders: English, French, and German, and occasionally Spanish, Russian, or the language of that year's host country (such as Japanese). And the meetings have been held predominantly in first-world cities: Paris (1900), Philadelphia (1904), Vienna (1905), Brussels (1910), London (1924), Ithaca (1926), Cambridge (1930), Amsterdam (1935), Stockholm (1950), Paris (1954), Montreal (1959), Edinburgh (1964), Seattle (1969), Leningrad (1975), Sydney (1981), Berlin (1987), Tokyo (1993), Saint Louis (1999), and Vienna (2005).[76]

Eighteenth-century botanical nomenclature served as an instrument of empire detaching plants from their native cultural moorings and placing them within schema comprehensible first and foremost to Europeans. With the rise of modern botany, a uniquely European system of nomenclature developed that swallowed into itself the diverse geographic and cultural identities of the world's flora.

In the eighteenth century, as we have seen, there were rivals to Linnaeus' nomenclature even within Europe. Had Adanson's system been chosen for the starting point of botanical nomenclature, nomenclature today might be more inclusive of the world's languages. But there are many forms of imperialism. Even Adanson paid little attention to how the Senegalese, Mauritians, and Guianese classified plants or animals. It is difficult today even to know what kinds of native taxonomic and naming systems existed. The rich traditions of the Malabar coast are said to be best represented today in van Reede's *Hortus Malabaricus*. Here and in a number of other instances, non-European knowledge systems have been collected, collated, and preserved in European texts; according to Richard Grove, van Reede's text remains the "only faithful textual record of the accumulated Ezhava botanical knowledge of the seventeenth century." Whatever the dangers of "mummifying" colonial naming systems, it seems that Adanson's expansive naming practices acknowledged global botanical knowledge in a way that Linnaeus' did not.[77]

S. M. Walters of Cambridge University has pointed out that using Linnaeus as the starting point of modern botany has had the added consequence that modern classification and nomenclature is heavily centered around European flora. Roughly two-thirds of the genera in Linnaeus' *Species plantarum* (1753) are European. According to Walters, the selection of the world's flowering plants available to Linnaeus has determined many features of current botanical classification. Walters has demonstrated, in particular, that early taxonomists arranged plants in groups of convenient size so that the size of plant families today are proportional to how long ago the taxon was established: old families tend to be larger and recent families smaller.[78] The relative stability of angiosperm families, he noted, has resulted from the unwillingness of taxonomists to change them, and should not be used as evidence of any essential correctness of this particular taxon's boundaries. William Stearn has added the intriguing argument that "modern science originated outside the tropics and that it is a reasonable assumption that it could not have originated within them." According to Stearn, Linnaeus was able to develop his taxonomic system only because quite early in his career he knew about plants from a few Swedish parishes and Uppsala's poorly stocked garden. He did not become, like later travelers, overwhelmed by the diversity and complexity of nature in the tropics.[79]

There is, in other words, nothing sacred about the Linnaean system of classification and nomenclature. His is in some sense the "QWERTY" universal keyboard of the organic world, whose success owes as much to accidents of founding and dispersion as to, say, inherent natural advantage. That is often the case in history: once things are fixed it is hard to think of them in any other way. And eventually we come to forget that things ever could have been otherwise.

Conclusion: Agnotology

> A large population makes a colony strong and rich; a weak and mediocre population leads to poverty and inaction.
> JEAN-BARTHÉLEMY DAZILLE, 1776

> From ancient times, abortion . . . was common to prevent overpopulation. As soon as it was realized, however, that each individual life is important to the state, it became the most important duty of the state to protect the life of every individual.
> FRANZ GUETNER, 1845

Europeans chose to remain ignorant of West Indian abortifacients. Whose knowledge was it that did not transfer into Europe: Amerindian? African? A hybrid knowledge created by crossing African and Amerindian techniques? What produced Europeans' neglect of abortifacients from abroad and their gradual vilification of induced abortion in their own medical traditions?

There are many forms of ignorance. I am here not interested in the sequestering of knowledge produced through secrecy, such as guild or military secrets, or the secrets of the Spanish, who did not publish the intelligence gathered from their many royal expeditions into the New World so as to retain an advantage over their enemies, or even the secrets of the many colonial slaves who hid their medicines from Europeans. Nor am I interested in ignorance produced by overtly suppressing knowledge considered worthless or dangerous, such as European witchcraft or West Indian obeah. What I am interested in is how in the eighteenth century a confluence of circumstances allowed the cultivation of certain types of knowledge over others. Funding priorities, global strategies, national policies, the structures of scientific institutions, trade patterns, and gender politics all pushed investigation toward certain parts of nature and away from others. Before turning to the case of abortifacients, let me discuss two other distinctive ignorances in eighteenth-century botany.[1]

The distinguished English botanist William Stearn has drawn attention to a fundamental distortion in eighteenth-century taxonomists' knowledge. A burning question for early modern European taxonomists was how similar were the plants of one continent to those of another. John Ray queried Sir Hans Sloane in Jamaica, for instance, whether any species of plants were common to America and Europe, and for exact information concerning plants native to that island. Sloane himself, as we saw earlier, realized that much of the floral uniformity he observed across the Caribbean basin was a result of plants being carried (intentionally and unintentionally) from the South American mainland and elsewhere into the islands, first by the Tainos, then by the Spanish, Dutch, English, and enslaved Africans. The impression of floral uniformity in the tropics was further heightened by the fact that Europeans who collected in these areas before 1753 did so mostly in ports and along coasts, regions highly disturbed from over 200 years of European voyaging and trade. Sacks of produce standing in harbors before being shipped, for instance, often carried soil and seeds of weedy species, which were then inadvertently transported to other countries and naturalized. As a consequence, tropical ports around the globe soon came to host the same ruderal flora. A collector, unaware of this process, might find the same plant in both the East and the West Indies, and assume it to be indigenous to each area where it was found. This human-made uniformity led taxonomists erroneously to assume that tropical flora was highly uniform instead of being, as it was, regionally highly diverse.[2]

Stearn's observation points to a particular type of ignorance. Ignorance of the rich diversity in tropical flora was produced by an unawareness that distinctive cultural practices had cultivated floral uniformity along European trade routes. What distinguishes this type of ignorance from that surrounding abortifacients is that once the error was discovered, it was energetically corrected. Incorrect scientific conclusions were quickly revised when data from Humboldt, Bonpland, Cook, and Bank's voyages revealed the great variety in tropical flora. European botanists and taxonomists were not invested in the notion of uniformity; they accepted revisions as new information became available.

Other misconceptions were created by eighteenth-century technologies of conveyances. In Humboldt's day, for example, plants (and especially small varieties) were better known than stones and minerals. Plants were lighter and more easily transported. Among plants, voyagers gave preference to succulents and bulbs because these were more likely to survive successfully the long and expensive passage back to Europe.[3] To the extent that Europeans consciously made these choices, they were quickly reversed as ships became larger and speedier.

The ignorance surrounding abortifacients was of an entirely different kind. When discussing the agnotology of abortifacients, it is important to understand that knowledge of them was rarely suppressed by decree. Instructions to travelers did not warn bioprospectors against collecting this knowledge. Physicians often cautioned against the dangers of the use of this class of drugs in their practices in Europe, but at the same time they knew and used different abortive techniques. Indeed, the lives of many women depended on this knowledge. When new exotic abortifacients were discovered, as indeed they were repeatedly by naturalists for over a century, knowledge of them was not cultivated. Unlike the two examples above, cultural forces closed Europe's borders to the importation of abortive techniques from abroad. When knowledge became available, it was not embraced.

Whose knowledge was it that was rebuffed? What characterized the chain of knowing and where was it broken? Maria Sibylla Merian reported that both Amerindians and African slaves used the peacock flower as an abortifacient. We do not know where this plant originated, whether it was indigenous to the Caribbean or whether it drifted there by sea or sailed on board a merchant or slave ship. We also have no documents to pin down how the knowledge of the *Poinciana*'s use as an abortifacient originally spread from person to person, culture to culture. We do know, however, that it was in use in Jamaica in 1687 when Sloane voyaged, in Surinam in 1699–1701 when Merian worked there, and in Barbados in the 1750s when Griffith Hughes observed its use as an emmenagogue. We also know that this plant grew in Saint Domingue as early as the 1640s and 1650s when Poincy saw it, and that it was used there as an abortifacient as late as the 1790s, when Descourtilz wrote about it.

A number of scenarios might be developed to explain the presence of this plant and its abortive uses across the Caribbean basin. A first possibility is an unintentional transmission of the plant. The plant that has become known as the *Poinciana* may have been swept from the Guiana coast and Orinoco valley into the Caribbean by the flood waters that were strong enough to divert the South Equatorial current northward. This current is known to have often carried plants and small animals into the Windward Islands. The hardy pod and seeds would have had little trouble traveling on their own, and we know that seeds of this genus can tolerate salt water.[4]

The plant might also have stolen away in the fodder of livestock or soils of plants transported for cultivation from one part of the globe to another. The spread of cultivated plants to new areas has been a constant feature of human history. Seeds and plants of various sorts were deliberately shipped for purposes of commerce, medicine, food, and curiosity. Dutch botanists in Ceylon, for example, shipped chestloads of specimens (often in separate vessels to ensure safe arrival) to Dutch gardens in Holland and the East Indies beginning in the seventeenth century. Europeans carried seeds of dietary staples everywhere they settled; even their "revictualing" stations (at the Cape of Good Hope, Saint Helena, or Mauritius, for example) were often stocked with imported European plants and livestock. Opportunities were abundant for the intentional or unintentional spread of this particular flowering shrub.[5]

Another scenario suggests that the *Poinciana* may be of African origins (or that it was naturalized there earlier via trade to the East) and was subsequently carried to the Caribbean by Europeans. Richard Ligon, voyaging in the seventeenth century, reported having brought seeds of the plant from Saint Jago, in the Cape Verde archipelago off the west coast of Africa, to Barbados in the West Indies. It should be kept in mind, however, that Cape Verde was a shipping crossroads and entrepôt in the seventeenth century. If Ligon carried the plant from Saint Jago, it could have come from anywhere in the world that the Europeans had ports. Moreover, Poincy found the *Poinciana* in Hispaniola at least a decade before Ligon carried seeds to Barbados. The eighteenth-century English master gardener Philip Miller cast further doubt on Ligon's report, saying that it was "very certain that the plant grows naturally in Jamaica, where the late Dr. Houstoun found it in the woods at a great distance from any settlements." Houstoun also located it "growing naturally at La Vera Cruz, and at Campeachy [Compeche], where he also found the two varieties with red and yellow flowers."[6]

Alternatively, the plant might have had African origins and been carried to the Caribbean by slaves. In her *Black Rice,* Judith Carney has documented, for example, that Africans brought not only rice but also the technologies for its cultivation from Africa to the New World. Traders and slaves often carried plant stocks used for foods or medicines with them from Africa. Thomas Dancer, island botanist and keeper of the Bath Botanical Garden in Jamaica, for example, noted that the "Bichey" (an African fruit) had been introduced into Jamaica "by Negroes" before Sloane's time and that "Aka" (ackee, another African fruit) was introduced "by Negroes in some of Mr. Hibbert's ships."[7]

Africans, who had long practiced abortion, may have brought the hearty seeds of the *Poinciana* with them when carried into slavery, though this is unlikely. The *Poinciana* grows in Western Africa, but it is not among the abortifacients widely used there. Slaves may, however, have found plants similar to those they had used back home already growing in the tropical Caribbean. A plant resembling Merian's peacock flower, the *Caesalpiniaceae swartzia madagascariensis,* grows on the west coast of Africa. Its seeds are well known in Senegal (and also Zambia) as an abortifacient. However slaves came to use the *Poinciana* in Surinam, Saint Domingue, and Jamaica, it is clear they knew abortifacients before they entered the Caribbean's fraught sexual economy.[8]

The nineteenth-century Swiss botanist Augustin-Pyrame de Candolle suggested yet another possibility, namely that the peacock flower had its origins in India and was subsequently transported by Europeans to the Caribbean. (A 1991 *Flora of Ceylon* states just the opposite, that it was brought to southwest Asia from the Americas.) The peacock flower grows profusely in India and surrounding areas, and was imported into Europe from the Dutch East Indies as early as the 1670s; Merian most likely saw it in Amsterdam before traveling to Surinam because she called it by its generic East Indian name: "peacock flower." If the peacock flower was used as an abortifacient in the East Indies in the seventeenth or eighteenth centuries, this knowledge was not taken into Europe.[9]

Finally, one might postulate a tropical American origin of the flowering plant, and an aboriginal Amerindian knowledge of its abortive virtues. As we have seen, Spanish documents show that Amerindians were familiar with abortifacients before contact. By the seventeenth century, the *Poinciana* was used as an abortive throughout the Caribbean. This common usage from Surinam up through the French Antilles to Jamaica suggests that the plant and knowledge of its uses was known to the forebears of the Tainos, the Saladoid peoples, and followed their migration out of South America into the islands. Humboldt, writing at the end of the eighteenth century, was impressed by the extensive use of abortifacients (which he did not identify) that he found in the Orinoco basin. The Saladoid peoples expanded first from what is today northeastern Venezuela into the Guianas. As soon as they discovered Grenada, they moved quickly (in less than a century) through the Lesser Antilles and into Puerto Rico (the Greater Antilles show evidence of human habitation by about 4000 B.C.). This quick movement of peoples, knowledges, foods, customs, rituals, and technologies may account for the similarities in the uses of plants found in the region. While it is possible that displaced Africans taught the Tainos the use of the plant now known as the *Poinciana*, it is more likely that the Tainos and Arawaks taught its uses to the newly arrived Africans. Dancer listed the *Poinciana* among the "most rare indigenous plants" grown in Jamaica's Bath Botanical Garden. Arthur Broughton, who catalogued Hinton East's extensive private botanical garden in Jamaica, also noted that the *Poinciana* was indigenous to the Caribbean: a yellow variety was introduced into Jamaica by a Mr. Shakespeare from Honduras in 1782.[10]

The *Poinciana* was not the only abortifacient identified by naturalists in the West Indies. Edward Bancroft wrote of the "gulley-root" *(Petiveria alliacea),* also known as garlic weed or henweed, which is indigenous to the Amazon. The abortive qualities of this plant were known to the indigenous peoples there. By the mid eighteenth century, when Bancroft was writing, knowledge of its many uses (including abortion) had reached Barbados, where Bancroft worked before moving on to Guiana. Although originally Amerindian knowledge, Bancroft accused also slaves of using the plant for abortion, which he considered the ruin of West Indian planters. At some point, slave women learned from the Arawaks (who still resided on nearly every plantation in Guiana) how to prepare this plant as an abortifacient. Although Bancroft and other European physicians learned of the gully-root's uses, bioprospectors did not deem this knowledge worthy of transport back to Europe.[11] There is no mention of abortive medicaments made from gully-root in the *Pharmacopoeia* of London, Paris, Leiden, or Edinburgh.

Bancroft also pointed to the abortive qualities of "ocro" or okra *(Hibiscus esculentus),* a plant of African origins. Geobotanists trace okra to the Abyssinian center of cultivated plants, an area that includes present-day Ethiopia and certain mountainous parts of Eritrea. Okra was introduced to Brazil sometime before 1658, and was known in Surinam by the 1680s. It dries well and is easily carried; slavers may have brought this plant to the New World as food for their cargos, and slaves may even have stowed it away aboard ship. Bancroft reported that female slaves who "intend to procure abortion" found it advantageous to "lubricate the uterine passages by a diet of these pods." They then induced abortion using either the gully-root or the "sensitive plant" *(Mimosa pudica)*, a native of tropical America and most notably Brazil. Here we have an example of slave women transferring their medical techniques to the New World and creatively mixing them with remedies learned from the Amerindians. Women played a key role as healers in West Africa; it is not surprising that they adapted their knowledge to new circumstances. In addition to the African abortifacients used in the West Indies, about sixty other medicinal plants in Jamaica (out of some 160 commonly used there) are known to have served in medicaments in Africa. These include Jamaica senna and the Kola or bissy nut.[12]

In the French island of Saint Domingue, Descourtilz identified several indigenous plants used as abortifacients, as discussed in Chapter 3. The first, which he called by its Linnaean name *Aristolochia bilobata,* also known as the horseshoe creeper or West Indian Dutchman's pipe, seems to have been prescribed by physicians in the Antilles. Descourtilz cautioned them against giving it to pregnant women. Birthwort was already well known to Europeans, and this West Indian variety was assimilated into that taxon. Descourtilz also mentioned the *Eryngium foetidum* (fitweed or long coriander), native to Central America. Despite its origins, it was burdened with a Greco-Latinate name unrelated to the languages in the land of its origin. The plant is still used today in Surinam for fever, diarrhea, vomiting, and flu, and in Puerto Rico it appears in the salsa eaten with chips. The *Veronica frutescens,* also used by slave women, was called *cougari* by the Caribs; African women, we may assume, learned of its use from Amerindians. Finally, Descourtilz noted another *emmenagogue excitante* found in the islands, the *Trichilie à trois folioles* or *arbre à mauvais'gen,* for which he had no Linnaean name.[13]

We can see, then, that knowledge of abortifacients poured into the Caribbean from South America and from Africa. But the flow of that knowledge into Europe was blocked. In this case trade winds of prevailing opinion impeded shiploads of New World abortifacients from reaching Europe.

Many of Descourtilz's abortifacients received Linnaean names, yet none found their way into European *Pharmacopoeia* of the period. Why was this so? As I mentioned earlier, knowledge of exotic abortifacients was not something that was *overtly* suppressed. Agnotology calls for deeper analysis. What we need to know is how societies are structured so that certain knowledges become reviled and their development blocked. Much in the organization of eighteenth-century medical communities, colonial agencies, and governmental policies blocked bioprospectors, who were in the West Indies expressly to find useful and profitable drugs, from importing knowledge of exotic abortifacients into Europe.

As discussed in Chapter 4, European medical communities were decidedly unreceptive to abortifacients: enthusiasm for the new experimental medicine did not extend to these potentially lifesaving drugs. As I have suggested in Chapter 3, there is some evidence that abortion, like most female medicine, belonged to the domain of midwifery. Physicians employed abortifacients in their practices, but they were generally only used when a woman's life was seriously in danger. Physicians may have had little experience with the full range of abortion practices in Europe at the time. Although legal, abortion had never been undertaken lightly: physical danger argued against it even when moral trepidation did not. Physicians, concerned that the low professional standing of midwives not rub off on them as they attempted to develop scientific obstetrics, tended to distance themselves from the less savory aspects of midwifery, including the practice of abortion. With the rise of medical jurisprudence at the turn of the nineteenth century, legal communities across Europe tightened restrictions on abortions, or *fausses-couches forcées* as the French government called them, making it difficult if not impossible to undertake research on abortifacients.[14]

The colonial enterprise was, as we have seen, largely male—most planters and slaves were men, as were colonial administrators, naturalists, and colonial physicians. The voyages of scientific discovery, as part of the colonial enterprise, showed little interest in the female side of life. Indeed, colonial administrators, such as Philippe Lonvilliers de Poincy and Hendrik Adriaan van Reede tot Drakenstein, were most interested in medicines to protect traders, planters, and trading company troops, among whom few women were found. For this reason physicians collected and experimented with drugs, such as Peruvian bark, jalap, and ipecacuanha, useful to the soldiers, sailors, and European settlers in the tropics. The masculine nature of the colonial enterprise did not, in itself, negate the collection of remedies destined specifically for females. As we have seen, humoral medicine in this period still adhered to the importance of understanding how sexual differences affected the course of disease and treatment. Throughout the early modern period, physicians took care to test new remedies in females.

We can imagine now that the massive colonial expansion of Europe might have set the stage for developing anti-fertility drugs. The discovery of the New World touched off a dramatic and unprecedented widespread movement of plants; plants moved everywhere and in every direction. Before and after contact, Tainos and Caribs traded cassava, maize, yams, bananas, tobacco, plantains, pineapples, cotton, and dye-plants across their routes connecting South America and the islands of the Caribbean. Slaves introduced numerous plants from Africa into the New World, including rice and Angolan peas; Ligon confirmed that "Negroes" brought with them seeds of a plant whose large root (perhaps the yam) when dried "tasted well." The English transshipped mulberry trees for raising silkworms from the Far East to their Virginia colonies; they carried Mexican jalap and ipecacuanha via England to Barbados. The Portuguese introduced the South American tomato into Barbados from their ports in Lisbon. Across the span of the sixteenth and seventeenth centuries, Europeans carried lucrative economic plants—wheat, sugarcane, indigo, and coffee from east to west, while transporting corn, potatoes, sweet potatoes, tobacco, and *Cinchona* from west to east. By the end of the eighteenth century, an intercontinental system of some 1,600 botanical gardens connected European holdings around the globe. The garden in Martinique was typical, featuring 360 genera of plants brought from Peru, Europe, Japan, Madagascar, Egypt, China, Indonesia, Surinam, and other farflung territories.[15]

In this ferment of activity, bioprospectors might have collected anti-fertility drugs given different political and gender priorities. Such medicinal plants moved easily with the Tainos from South America into the Caribbean and with West Africans from their homelands into the Greater and Lesser Antilles. Here, however, the chain of knowledge was broken. Despite the general enthusiasm for new commodities in Europe, anti-fertility drugs did not flood those markets.

Developing abortifacients or any drugs used to control fertility in this period worked directly against the interests of mercantilist states. Leaders in the Dutch republic, the constitutional monarchy of England, and the absolute monarchy of France promoted policies designed to increase population, not control births. Mercantilist governments sought to augment national wealth by producing growing and healthy populations that served to increase the production of crops and goods, fill the ranks of standing armies, and pay substantial taxes and rents. The French, in particular, feared their population was in decline, given the celibacy of Catholic churchmen and women, the forced emigration of Protestants after the revocation of the Edict of Nantes, war, the low rates of marriage among domestic servants, the neglect of agriculture, and general debauchery. High on the list also were abortion, infanticide, and infant mortality—all considered a flagrant waste of human life at its very start. As population increasingly became a matter of state, Nina Gelbart has argued, "women's bodies came to be thought of as a kind of national property, somehow coming under the stewardship and use of rights of the state, counted on to ensure the regular fecundity of society. Married women were considered morally obliged, patriotically bound to perform the public function of producing citizens." Observers at the time discussed children as commodities, celebrated as "the wealth of nations, the glory of kingdoms, and the nerve and good fortune of empires." In this climate, women were encouraged "to multiply and rear the human species."[16]

Mercantilists, in other words, were strongly pronatalist. The English botanist and physician Nehemiah Grew, for example, prepared a memorandum for Queen Anne on "The Meanes of a Most Ample Encrease of Wealth and Strength of England" listing the four essential components of England's economy as land, manufactures, foreign trade, and population. Of these, population—to supply labor, soldiers, sailors, and markets—was considered of great import for increasing national wealth. In the same period Colbert took steps to increase the French population in the West Indies. In 1668, he instructed governors in the islands to lower the age of marital eligibility to eighteen for boys and fourteen for girls. An increased population in the islands, Colbert reminded company officials, would increase profits for the company and thus for France.[17]

Mercantilist governments enlisted the aid of physicians, botanists, and midwives in "growing" their populations. Physicians—empirically oriented and publicly engaged—served as government-sponsored medical police, promoting public health to increase the vigor, strength, wealth, and prosperity of the state. City hospitals and lying-in hospitals were expanded and improved in efforts to decrease morbidity and mortality among the poor and working populations. Physicians also developed public health measures such as smallpox inoculation, which by 1803 was perceived as "one very great cause of increasing population." Efforts were also made to improve midwifery, including national systems of exams and licensing under the control of surgical corporations. In France, Angélique Marguerite Le Boursier du Coudray was employed by the royal government to educate midwives throughout France, work she considered a service to the public good *(bien public)*. In her midwifery manual, she repeatedly declared that midwives serve their country by conserving future "subjects of the state," those children who are "precious in the eyes of God, useful to their family, and necessary for the state." Some of the manikins she prepared for teaching purposes reached France's Caribbean colonies.[18]

Physicians in the colonies also considered it part of their mission to engender healthy populations. In the view of many, "population determine[d] the prosperity, wealth, and strength of a colony." Jean-Barthélemy Dazille, a royal physician in Saint Domingue and former army surgeon in Cayenne, expressed these sentiments in his book on how best to cure the special illnesses of "Negroes." Outlining the mercantile system wherein the wealth of a colony flowed back to France, Dazille emphasized that the opulence of a colony lay primarily in an abundant population of slaves because "without Negroes there is no culture, no products, and no riches."[19]

It was the physician's job to keep slaves healthy; it was the botanist's job to feed them cheaply. The most celebrated example of efforts to introduce cheap foods aimed at lowering the cost of feeding plantation slaves was Sir Joseph Banks's naturalization of the Polynesian breadfruit to the West Indies in the 1790s. Realizing the benefits of this prolific foodstuff on his voyage with Captain Cook to Tahiti in 1769, Banks and his colleague, Daniel Solander, urged King George III to introduce breadfruit to the West Indies, and in 1787 the English king dispatched the infamous Captain William Bligh to Tahiti to accomplish that goal. The crew spent six months in Tahiti collecting and propagating plants, then sailed westward across the Pacific Ocean with over a thousand saplings aboard ship. Although the *Bounty* never reached the West Indies, the British admiralty eventually succeeded in bringing the breadfruit to Jamaica and Saint Vincent. Michel-René Hilliard d'Auberteuil, writing from Saint Domingue in the 1770s, noted other sources of cheap and plentiful foods. Observing that the earliest slaves brought to the West Indies had carried rice and manioc (which they made into a tasty bread) with them from their homes in Africa, he calculated that one slave working only two hours per day could cultivate enough of these foods to feed twenty slaves. Better, however, from the planters' point of view, was to find foods for the slaves that grew naturally in the islands. Bananas, limes, potatoes, and sweet potatoes grew everywhere, required little maintenance, and yielded food in all seasons. Hilliard d'Auberteuil estimated that, joining these two systems, 16,000 *carreaux* (an area equal to about 45,000 acres) could feed 290,000 slaves.[20]

Colonial physicians were also enlisted in the effort to increase births and decrease infant mortality. Charles Arthaud wrote concerning the need to train midwives in the French colonies: "women should not have the liberty to sacrifice themselves to ignorance. This is opposed to the interest of the state." He recommended that colonial legislatures test and certify midwives, and that the king's physicians and surgeons actively supervise them. Laws governing reproduction in the colonies were for the most part imported directly from Europe. The French colonies in 1718 reaffirmed Henri II's Ordinance of 1556/7 against hidden pregnancies and infanticide, cautioning that it was "necessary to warn of an extremely loathsome crime, common in our realm, that many women, having conceived an infant by dishonest or other means are persuaded by bad will . . . to disguise and hide their pregnancy . . . and deliver their fruit secretly and then suffocate, kill, or otherwise do away with it—throwing it into a secret place or burying it in profane ground—without having given it the sacrament of baptism." The Ordnance continued: "women who hide their pregnancy and then suffocate or kill their infant will be punished and made an example of to stop this horrendous crime. If they hide their pregnancy and their baby is found dead, it will be assumed that they have murdered it, even if they claim that the child was born dead." The Ordnance was reaffirmed in the French colonies again in 1758, 1765, and 1784. No midwife or surgeon was allowed to deliver a woman secretly.[21]

Midwives in Europe had long been agents of church and state responsible for regulating birthing practices, reporting concealed pregnancies, irregular practices, and discovering the names of fathers of illegitimate offspring (often wrest from women during labor). Colonial medical establishments also sought to enlist midwives in their efforts to end the loss of slaves to infanticide and abortion. Colonial law required that all slaves declare their pregnancies to a midwife, who was in turn to report it to a surgeon, who was required to register it. Since colonial physicians considered plantation midwives responsible for helping women abort or kill their newborn infants, slave midwives were severely punished along with the mother when she miscarried or produced a dead infant. Schemes to train and license midwives included efforts to win over their loyalty. At the same time, planters offered slave women trinkets, more food, and time to cultivate it as incentives to bear more children.

Foundling homes were also established in both Europe and its colonies in efforts to preserve more children for the benefit of the state. In colonial orphanages, the economics of race determined children's fate. Laws in Martinique, for instance, declared that all foundlings were to be taken to the Hospital of the White Ladies *(Hôpital des Dames Blanches)* where, in the presence of the royal prosecutor, they were to be declared either white or colored by the royal physician and royal surgeon. If a child were declared white, this was recorded in the child's baptismal certificate. If the child's color was in doubt, he or she was to be kept one year at the expense of the king and then reexamined. If a child was declared "colored," this was to be recorded in his or her baptismal certificate and the child would be sold for the king's profit. If a child was declared black and sold, and his or her mother came at some later date to claim the child or to prove that he or she was a free person, the mother was required to pay all expenses incurred for the child's care.[22]

European desires to "grow" populations in the colonies often overrode other state policies. Hilliard d'Auberteuil noted that the revocation of the Edict of Nantes in 1685, which drove Protestants from French soil and forbade Jews from settling in French colonies, was, in fact, rarely enforced. One finds, he wrote, people from "all countries and all religions" in the colonies. If settlers were prudent and hard working, no one inquired into their place of birth or the nature of their religious beliefs. Everything, he added, was calculated in terms of their usefulness to the colony. In Jamaica, during the reign of William III, the island council requested that the crown expel all the Jews from the island, but the king, noting their industriousness, did not.[23]

In such climates agents of botanical exploration—trading companies, scientific academies, and governments—had little interest in expanding Europe's store of anti-fertility pharmacopoeia. Alexander von Humboldt, writing at the turn of the nineteenth century, revealed how and why European scientific men did not collect contraceptives and abortifacients from the New World. When difficulties with language impeded Humboldt's ability to identify and understand how particular herbs were used for these purposes, he was able to enlist the aid of a Jesuit priest, Gili, who served as confessor to the indigenous peoples living along the Orinoco River for fifteen years and boasted an intimate knowledge of *i segreti delle donne maritate* (the secrets of married women). Humboldt expressed his surprise about the safety of the abortifacients the Amerindians employed and discussed the need for such drugs in Europe, sympathizing with the young mothers there who were "afraid of having children, because they know not how to feed, clothe, and provide for them." Yet he refused to transmit information about these efficacious herbs to Europe. While he was well aware that Europeans had a working knowledge of abortifacients (he listed savin, aloe, and the essential oils of cinnamon and clove), he feared that the introduction of New World abortives and contraceptives into Europe would increase "the depravity of manners in towns, where one quarter of the children see the light only to be abandoned by their parents." He also feared that the new substances might prove too powerful for delicate European constitutions. "The robust constitution of the savage, in whom the different systems are more independent of each other," he wrote, "resists better and for a longer time an excess of stimulants, and the use of deleterious agents, than the feeble constitution of civilized women."[24]

More important, Humboldt made clear that his reluctance to collect such medical knowledge had to do with mercantilist concerns about population growth. Listing the causes of depopulation among the Indians of the Orinoco, Humboldt dismissed smallpox, which had so ravaged other peoples (according to him the disease had not yet penetrated inland to this remote area). Humboldt highlighted instead the Amerindian's "repugnance" for the Christian missions, the unhealthy hot and damp climate, the poor food they received, the devastating children's diseases, and, last but not least, women's control of their own fertility. These mothers, he wrote, were well known for preventing pregnancy by the use of herbs. Being aware that contraception and abortion were becoming taboo subjects among learned European men, Humboldt concluded by adding, "I thought it necessary to enter into these pathological details, far from agreeable as they are, because they make known a part of the causes, which in the rudest state of our species as well as in a high degree of civilization, render the progress of population almost imperceptible."[25]

Abortion practices, in sum, were deeply embedded in colonial struggles. Gender relations in eighteenth-century Europe and its colonies were such that European bioprospectors did not collect information about abortion or seek to understand this aspect of Caribbean medical practices. Merian's *flos pavonis* thus participated in both a revolution in the history of botany and a transformation in the history of the body. At a time of the rapid expansion of science more generally, European knowledge of anti-fertility agents waned. Gender politics lent recognizable contours not to a distinctive body of knowledge but, in this instance, to a distinctive body of ignorance. The agnotology of abortives among Europeans was not for want of knowledge collected in the colonies; it resulted from protracted struggles over who should control women's fertility. Bodies of ignorance, in turn, came to mold the lived experience of European women. European women's loss of easy access to contraceptives and abortifacients curbed their reproductive and often professional freedoms. An image of upper- and middle-class European women developed that celebrated them as both angels in the home and fecund beings hopelessly subservient to the beck and call of nature. The curious history of the *flos pavonis* reveals how voyagers selectively culled from the bounty of nature knowledge responding to national and global policies, patterns of patronage and trade, developing disciplinary hierarchies, personal interests, and professional imperatives. In the process, much useful knowledge was lost.

Knowledge that did not travel to Europe remained alive nonetheless in the Caribbean. In my travels to Belize, Jamaica, Costa Rica, Martinique,

Guadeloupe, Dominica, the Dominican Republic, and so forth, I have queried numerous people concerning the use of abortifacients today. In Costa Rica I was told by a male guide of Spanish heritage that "everyone" knows these remedies and still uses them today. In a village near where we were hiking, he told me that "a little virgin" had recently aborted a child conceived out of wedlock. In Dominica, I had an animated, two-hour conversation about Carib history and culture with one of the approximately 3,000 ethnic Caribs who have survived in the Caribbean basin.

Figure C.1. A merchant selling simples. Postcard from Martinique, turn of the twentieth century.

Feeling comfortable with this very open and interesting woman, I eventually turned to the topic of birth control. She launched into her answer, then shot me a glance and said quietly, "but it's secret." I did not press the issue. After a moment's reflection, she called to her husband and together they picked a plant growing at their backdoor step. She told me that to prevent conception, after intercourse a woman ingests a tea made from the plant and also washes herself with it.

Further along in Dominica, perhaps five miles from my first encounter, an unlicensed country doctor (who distinguished himself from the "bush doctors" practicing for pay in the towns) provided me with the names of three abortifacients. One of these, which he called "Shadow Benny" *(Eryngium foetidum)*, was an abortifacient identified by Descourtilz as actively used in Saint Domingue in the 1790s. Since abortion is illegal in Dominica, the herbalist was anxious that I not reveal his name.[26] While he spoke openly about abortion, this man of African heritage flatly denied that the plant the Carib woman had given me (which I showed him) worked as a contraceptive. It was unclear to me how to interpret this disagreement.

We see in these casual meetings specters of the eighteenth-century encounters between European bioprospectors and the peoples of the Caribbean. There is the language problem: the Carib woman and I conversed in English (her native language has died out in Dominica and she works with Awaraks in Surinam in a project to revive it); nonetheless, we did not share the same names for plants. I could not ask her if she knew the peacock flower, Barbados Pride, *Poinciana pulcherrima*, because I had not yet been in the country long enough to discover their local term for Merian's flowering bush. Then there was the problem of secrecy and fear because abortion in this largely Catholic country is illegal. One wonders what easy, safe, and effective methods of birth control and abortion have been lost to European and American women because of state politics that enmesh innocent plants in their web.

Notes

Introduction

1. Merian, *Metamorphosis*, commentary to plate 45.
2. Guillot, "La vraie 'Bougainvillée.'" Shteir, *Cultivating Women*.
3. In addition to individual works cited below, see John MacKenzie, ed., *Imperialism and the Natural World* (Manchester: University of Manchester, 1990); N. Jardine, J. Secord, and E. Spary, eds., *Cultures of Natural History: From Curiosity to Crisis* (Cambridge: Cambridge University Press, 1995); Yves Laissus, ed., *Les Naturalistes français en Amérique de Sud* (Paris: Édition du CTHS, 1995); Tony Rice, *Voyages: Three Centuries of Natural History Exploration* (London: Museum of Natural History, 2000).
4. Robert Proctor, *Cancer Wars: How Politics Shapes What We Know and Don't Know about Cancer* (New York: Basic Books, 1995), 8; Robert Proctor, "Agnotology: A Missing Term to Describe the Study of the 'Cultural Production of Ignorance'" (manuscript).
5. See, e.g., Steven Shapin and Simon Schaffer, *Leviathan and the Air-Pump: Hobbes, Boyle, and the Experimental Life* (Princeton: Princeton University Press, 1985); Thomas Laqueur, *Making Sex: Body and Gender from the Greeks to Freud* (Cambridge, Mass.: Harvard University Press, 1990); Schiebinger, *Nature's Body;* Nelly Oudshoorn, *Beyond the Natural Body: An Archeology of Sex Hormones* (New York: Routledge, 1994); Mario Biagioli, ed., *The Science Studies Reader* (New York: Routledge, 1999).
6. Pratt, *Imperial Eyes,* 6. Balick and Cox, *Plants, People, and Culture,* 29–30.
7. Crosby, *Columbian Exchange;* Lucile Brockway, "Plant Science and Colonial Expansion: The Botanical Chess Game," in *Seeds and Sovereignty: The Use and Control of Plant Resources,* ed. Jack Kloppenburg, Jr. (Durham: Duke University Press, 1988), 49–66; Sidney Mintz, *Sweetness and Power: The Place of Sugar in Modern History* (New York: Viking, 1985). Guerra, "Drugs from the Indies." Mackay, *In the Wake of Cook,* 123–143.
8. Henry Hobhouse, *Seeds of Change: Five Plants that Transformed Mankind* (London: Sidgwick & Jackson, 1985); Clifford Foust, *Rhubarb: The Wondrous Drug* (Princeton: Princeton University Press, 1992); Jarcho, *Quinine's Predecessor;* Larry Zuckerman, *The Potato: How the Humble Spud Rescued the*

Western World (Boston: Faber and Faber, 1998); Susan Terrio, *Crafting the Culture and History of French Chocolate* (Berkeley: University of California Press, 2000); Henry Hobhouse, *Seeds of Wealth: Four Plants that Made Men Rich* (London: Macmillan, 2003).
9. Miller, *Gardener's Dictionary*, s.v. "Poinciana (pulcherrima)."
10. Stearn, "Botanical Exploration," 175. Much of the history of botany has been written as the rise of systematics. See, for example, Julius von Sachs, *Geschichte der Botanik vom XVI. Jahrhundert bis 1860* (Munich, 1875); Edward Lee Greene, *Landmarks of Botanical History*, 2 vols. (Washington, D.C.: Smithsonian Institution, 1909).
11. Daubenton, *Histoire naturelle*, ix. Haggis, "Fundamental Errors"; Jaramillo-Arango, *Conquest;* John Dixon Hunt, ed., *The Dutch Garden in the Seventeenth Century* (Washington, D.C.: Dumbarton Oaks, 1990). Harold J. Cook, "The Cutting Edge of a Revolution? Medicine and Natural History near the Shores of the North Sea," in *Renaissance and Revolution: Humanists, Scholars, Craftsmen and Natural Philosophers in Early Modern Europe,* ed. J. Field and F. James (Cambridge: Cambridge University Press, 1993), 45–61; Steven Harris, "Long-Distance Corporations, Big Sciences, and the Geography of Knowledge," *Configurations* 6 (1998): 269–304.
12. Theodor Fries, ed., *Bref och skrifvelser,* 9 vols. (Stockholm: Aktiebolaget Ljus, 1907–1922), I, vol. 8, 27; cited in Koerner, *Linnaeus,* 104. Thomas Dancer, a physician in Jamaica, described colonial botanical gardens as "not now, as formerly, considered merely an appendage to a college or a university, but . . . object[s] of general concern with enlightened men of every description, even the mercantile class, in maritime and manufacturing towns" (*Some Observations,* 3–4). Bourguet and Bonneuil, ed., *Revue,* 14. Jacob Bigelow, *American Medical Botany, Being a Collection of the Native Medicinal Plants of the United States* (Boston, 1817–1820), vii. Gascoigne, *Science in the Service of Empire;* Drayton, *Nature's Government;* Spary, *Utopia's Garden;* Olarte, "Remedies for the Empire"; François Regourd, "Sciences et colonisation sous l'ancien régime: Le Cas de la Guyane et des Antilles Françaises" (Ph.D. diss., Université de Bordeaux 3, 2000).
13. Smellie, ed., *Encyclopaedia Britannica,* s.v. "Botany." Diderot and d'Alembert, eds., *Encyclopédie,* s.v. "Botanique." Nicolas, "Adanson et le mouvement colonial," 447.
14. Stearn in Linnaeus, *Species plantarum,* 68. See also Koerner, *Linnaeus,* 6–7, 43, 48–49. For the close relationship between medicine and botany, see also Philippe Pinel, *The Clinical Training of Doctors,* ed. Dora Weiner (1793; Baltimore: Johns Hopkins University Press, 1980).
15. Koerner, *Linnaeus,* 121.
16. De Beer, *Sloane,* 72–73. There is an earlier mention of milk chocolate by an American physician in 1672, but Sloane was the first to popularize the drink in England. He brought his recipe with him from Jamaica and sold it to an apothecary who marketed it as "Sir Hans Sloane's milk chocolate." Hot chocolate was produced by Cadbury's until 1885.

17. Director of the Madrid Botanical Garden cited in Olarte, "Remedies," 46. Diderot and d'Alembert, eds., *Encyclopédie*, s.v. "Amérique."
18. Thunberg, *Travels*, vol. 1, ix. Guerra has argued that the search for "spices" (a term referring to all seasoning, medicinal drugs, perfumes, and dyestuffs) opened over-water trade routes ("Drugs from the Indies"). By the eighteenth century, botanical exploration followed trade routes. Adanson, *Voyage*, 318. Stroup, *Company of Scientists;* Mackay, "Agents of Empire," 39.
19. Aublet, "Observations sur la culture du café," bound with Benjamin Moseley, *Traité sur les propriétés et les effets du café* (Paris, 1786), 100–104. See also Jean Tarrade, *Le Commerce colonial de la France à la fin de l'Ancien Régime*, 2 vols. (Paris: Presses universitaires de France, 1972), vol. 1, 34.
20. MacKay, "Agents of Empire," 6, 38–57. Latour, *Science in Action*, chap. 6. Long, *History*, vol. 2, 590. Daniel Headrick, *The Tools of Empire* (New York: Oxford University Press, 1981).
21. McClellan, *Colonialism and Science*, 148. Duval, *King's Garden*, 69, 99. MacKay, *In the Wake of Cook*, 17.
22. McClellan, *Colonialism and Science*, 63; Hulme, *Colonial Encounters*, 4; Philip Curtin, *The Rise and Fall of the Plantation Complex* (Cambridge: Cambridge University Press, 1990). Lafuente and Valverde, "Linnaean Botany"; Jorge Cañizares-Esguerra, *How to Write the History of the New World: Histories, Epistemologies, and Identities in the Eighteenth-Century Atlantic World* (Stanford: Stanford University Press, 2001).
23. *Petit Robert*, s.v. "Botaniste"; *Oxford English Dictionary*, s.v. "Botany."
24. For mortality rates among soldiers and sailors, see Francisco Guerra, "The Influence of Disease on Race, Logistics and Colonization in the Antilles," *Biological Consequences of European Expansion, 1450–1800*, ed. Kenneth Kiple and Stephen Beck (Aldershot: Ashgate, 1997), 161–173, esp. 164–167.
25. Thunberg, *Travels*, vol. 1, viii. Aublet, *Histoire*, vol. 1, xvii; Daubenton, *Histoire naturelle*, x.
26. Trapham, *Discourse*, 28, 30. Ligon, *History*, 2, 99. [Bourgeois], *Voyages*, 503. Humboldt (and Bonpland), *Personal Narrative*, vol. 5, 209.
27. Blair, *Pharmaco-Botanologia*, v. Knowledge of Native Americans, Africans, and other peoples of the Third World is today often improperly termed "indigenous knowledge" to distinguish it from "true," "universal" science. Furthermore, when "indigenous" is used (as it often is) merely as a synonym for "Third World," it homogenizes a broad spectrum of culturally diverse knowledges and practices. "Indigenous" means simply native to a place; Europeans had (and still have) their own indigenous knowledge just as do non-European peoples. Achoka Awori, "Indigenous Knowledge, Myth or Reality?" *Resources: Journal of Sustainable Development in Africa* 2 (1991): 1; Peter Meehan, "Science, Ethnoscience, and Agricultural Knowledge-Utilization," in *Indigenous Knowledge Systems and Development*, ed. David Brokensha, D. Warren, and Oswald Werner (Washington, D.C.: University Press of America, 1980), 379. See also, Mary Alexandra Cooper, "Inventing the Indigenous: Local Knowledge and Natural History" (Ph.D. diss., Harvard University, 1998).

28. [Schaw], *Journal*, 114.
29. Varro Tyler, "Natural Products and Medicine," in *Medicinal Resources of the Tropical Forest*, ed. Michael Balick, Elaine Elisabetsky, and Sarah Laird (New York: Columbia University Press, 1996), 3–10, esp. 7.
30. Merson, "Bio-prospecting," *Nature and Empire*, ed. MacLeod, 284.
31. Cox, "The Ethnobotanical Approach." J. W. Harshberger, "The Purposes of Ethno-Botany," *Botanical Gazette* 21 (1896): 146–154.
32. Environmental Policy Studies Workshop, 1999, School of International and Public Affairs, *Access to Genetic Resources* (New York: Columbia University School of International and Public Affairs, Environment Policy Studies, Working Paper #4, 1999), 3–16, 18–23.
33. The line of demarcation was negotiated ten degrees farther West by the treaty of Tordesillas in June, 1494 to recognize Portuguese interests in Brazil. Walvin, *Fruits of Empire*, 2–7.
34. On gender and race, see Schiebinger, *Nature's Body;* Kathleen Wilson, *The Island Race: Englishness, Empire and Gender in the Eighteenth Century* (London: Routledge, 2003); Felicity Nussbaum, *The Limits of the Human: Fictions of Anomaly, Race, and Gender in the Long Eighteenth Century* (Cambridge: Cambridge University Press, 2003).
35. Seneca, cited in Noonan, *Contraception*, 27n33. Lewin, *Fruchtabtreibung*. Boord, *Breviarie of Health*, 7. I thank Andrew Wear for calling this passage to my attention. Also Astruc, *Traité*, vol. 5, 326–327.
36. For a discussion of Caribbean slave women's resistances and tactics for survival, see Jenny Sharpe, *Ghosts of Slavery: A Literary Archaeology of Black Women's Lives* (Minneapolis: University of Minnesota Press, 2003), xiv–xxiii.
37. Maehle, *Drugs on Trial*.
38. Freind, *Emmenologia*, 68–69, 73. Sloane, *Voyage*, vol. 1, cxliii.
39. Louis Montrose, "The Work of Gender in the Discourse of Discovery," *Representations* 33 (1991): 1–41, esp. 8. De Beer, *Sloane*, 38–41; Sloane, *Voyage*, vol. 1, v; Goslinga, *The Dutch in the Caribbean and in the Guianas*, 268.
40. McNeill, "Latin," 755.
41. Olarte, "Remedies," 116. See also Engstrand, *Spanish Scientists*.
42. Tarcisco Filgueiras, "In Defense of Latin for Describing New Taxa," *Taxon* 46 (1997): 747–749.
43. McNeill, "Latin," 755.
44. [Schaw], *Journal*, 27–28, 31, 50. Sloane, *Voyage*, vol. 2, 346. Thiery de Menonville, *Traité*. F. Richard de Tussac, *Flore des Antilles*, 4 vols. (Paris, 1808–1827), vol. 1, 9–10.
45. Aublet, *Histoire*, vol. 1, xvii.

1. Voyaging Out

1. Stafleu, *Linnaeus*, 145. Linnaeus cited in Koerner, *Linnaeus*, 115.
2. Lafuente and Valverde, "Linnaean Botany."

3. On voyaging priests, see P. Fournier, *Voyages et découvertes scientifiques des missionnaires naturalistes français* (Paris: Paul Lechevalier, 1932). On Jesuit naturalists, see Steven Harris, "Long-Distance Corporations, Big Sciences, and the Geography of Knowledge," *Configurations* 6 (1998): 269–304. One sees this transition in the appointment of Pierre Barrère as royal botanist in Guiana. The crown first sought a man of God who would also run the hospital in Cayenne. Barrère was appointed when none could be found, but he refused to treat the soldiers for free, charging them one pistol per visit. He claimed he was paid as a botanist, not as a physician. Lacroix, *Figures de savants*, vol. 3, 31–35, esp. 32.
4. De Beer, *Sloane*, chap. 1. Stagl, *History of Curiosity*, 85. Koerner, *Linnaeus*, 56.
5. Ray and Lister cited in de Beer, *Sloane*, 26–28. Men above the rank of baron could become members of the Royal Society without the scrutiny given other applicants.
6. De Beer, *Sloane*, 30–31. Marcus Rediker, *Between the Devil and the Deep Blue Sea: Merchant Seamen, Pirates, and the Anglo-American Maritime World, 1700–1750* (Cambridge: Cambridge University Press, 1987), 124. Reede, *Hortus;* Heniger, *Hendrik Adriaan van Reede*, 269. Cannon, "Botanical Collections," esp. 141; Grainger, *Essay*, iv; Moseley, *A Treatise on Sugar.* See also Walvin, *Fruits of Empire.*
7. De Beer, *Sloane*, 32–42.
8. [Thomas Birch], "Memoirs relating to the Life of Sir Hans Sloane, formerly President of the Royal Society," British Library, Manuscripts Collection, Add. 4241, 25. De Beer, *Sloane*, 101; MacGregor, "The Life," 15, 23.
9. Sloane, *Voyage*, vol. 1, preface.
10. Ibid., preface, xlvi.
11. Ibid., preface.
12. Thunberg, *Travels*, vol. 2, 132. Maria Riddell, *Voyages to the Madeira, and Leeward Caribbean Isles with Sketches of the Natural History of these Islands* (Edinburgh, 1792), preface.
13. Among her many works, see T. E. Bowdich and Sarah Bowdich, *Excursions in Madeira and Porto Santo* (London, 1825). See also D. J. Mabberley, "Robert Brown of the British Museum: Some Ramifications," in Alwyne Wheeler and James Price, eds., *History in the Service of Systematics* (London: Society for the Bibliography of Natural History, 1981), 101–109, esp. 103–104; Pycior, Slack, and Abir-Am, eds., *Creative Couples.* For nineteenth-century women travelers, see, e.g., Shteir, *Cultivating Women;* and Susan Morgan, *Place Matters: Gendered Geography in Victorian Women's Travel Books about Southeast Asia* (New Brunswick: Rutgers University Press, 1996).
14. Trapham, *Discourse*, 5. [Bourgeois], *Voyages*, 438. Thunberg, *Voyages*, vol. 2, 281.
15. Johann Blumenbach, *On the Natural Varieties of Mankind* (1795) trans. Thomas Bendyshe (1865; New York: Bergman, 1969), 212n2. Blumenbach codified notions long current in the culture. Pouppé-Desportes, *Histoire des*

maladies, vol. 1, 57. Marie de Rabutin-Chantal, marquise de Sévigné, *Correspondance,* ed. Roger Duchêne (Paris: Gallimard, 1972), vol. 1, 370.
16. Davis, *Women on the Margins,* 169–171.
17. Young boys from central Europe were also among those who, displaced by famine and in search of work in the bustling port cities, were kidnapped and shipped out to serve on company vessels. Pimentel, "The Iberian Vision," 23.
18. Sandrart, cited in Elisabeth Rücker, "Maria Sibylla Merian," *Fränkische Lebensbilder* 1 (1967): 225; Schiebinger, *Mind,* 26–27, chap. 3; Davis, *Women on the Margins;* Wettengl, "Maria Sibylla Merian"; and Segal, "Merian as a Flower Painter," 84. Archives nationales, Paris, AJ XV 510, no. 331.
19. Merian, *Metamorphosis,* "An den Leser."
20. Rücker, *Merian,* 17, 19, and 21; Segal, "Merian as a Flower Painter," 86; Wettengl, "Maria Sibylla Merian," 18.
21. James Anderson, *Correspondence for the Introduction of Cochineal Insects from America* (Madras, 1791), 18–19.
22. Merian, *Metamorphosis,* commentary to plate 52. See also Sörlin, "Ordering the World for Europe," 52; Merian to Petiver, 27 April, British Library, Manuscripts Collection, Sloane 4064, f. 70; English translation Sloane 3321, f. 176.
23. Merian, *Metamorphosis,* "An den Leser," commentary to plates 4, 5, 10, 11, 21, 27, 32, 36, 42, 44, 48, 49, 59. Davis, *Women on the Margins,* 177.
24. Merian, *Metamorphosis,* "An den Leser," commentary to plate 35; Merian to Volkammer, 8 Oct. 1702, in Rücker, *Merian,* 22–23.
25. Stewart Mims, *Colbert's West India Policy* (New Haven: Yale University Press, 1912), 81. Cole, *Colbert,* vol. 2, 1–55.
26. James E. McClellan III and François Regourd, "The Colonial Machine: French Science and Colonization in the Ancien Régime," in *Nature and Empire,* ed. MacLeod, 31–50, esp. 32; Gascoigne, *Science in the Service of Empire;* Drayton, *Nature's Government,* 66–67.
27. Opposition by the Faculté de médecine delayed the signing of the royal edict until 1635. Stroup, *A Company of Scientists,* 169–179; Spary, *Utopia's Garden.*
28. Duval, *King's Garden,* 12, 19, 45, 48. Joseph Pitton de Tournefort, *Relation d'un voyage du Levant,* 2 vols. (Paris, 1717).
29. Duval, *King's Garden,* 36–37.
30. La Condamine, "Sur l'arbre du quinquina," 326; *Colloque International "La Condamine"* (Mexico: IPGH, 1987). La Condamine, *Relation abrégée,* 26–27. See also Roger Hahn, *The Anatomy of a Scientific Institution: The Paris Academy of Science 1666–1803* (Berkeley: University of California Press, 1971).
31. Thiery de Menonville, *Traité,* civ. Mackay, *In the Wake of Cook,* 182. McClellan, *Colonialism and Science,* 152–156.
32. Jeremy Baskes, *Indians, Merchants, and Markets* (Stanford: Stanford University Press, 2000), 9–15.

33. Thiery de Menonville, *Traité,* vol. 1, 5–6, 43.
34. Ibid., 59–60.
35. Ibid., 61.
36. Ibid., vol. 2, 39–40, 44–46.
37. Ibid., vol. 1, 137–138, 144–145, 184, 190–191.
38. Ibid., 208–209.
39. Ibid., xcvi, civ. *Affiches américaines* 3 (1780), supplement. McClellan, *Colonialism and Science,* 154; Pluchon, "Le Cercle des philadelphes."
40. Mackay, *In the Wake of Cook,* 182; Koerner, *Linnaeus,* 150; Rushika Hage, *Cochineal* (Minneapolis: James Ford Bell Library, 2000).
41. Thiery de Menonville, *Traité,* vol. 1, civ, 260. Emphasis added.
42. Merson, "Bio-prospecting," *Nature and Empire,* ed. MacLeod, 284. See also Kerry ten Kate and Sarah Laird, *The Commercial Use of Biodiversity: Access to Genetic Resources and Benefit-Sharing* (London: Earthscan, 1999).
43. Duval, *King's Garden,* 12, 18.
44. Patrice Bret, in "Le Réseau des jardins coloniaux: Hypolite Nectoux (1759–1836) et la botanique tropicale, de la mer des Caraïbes aux bords du Nil," *Les naturalistes français,* ed. Laissus, 185–216, esp. 187. Leblond, *Voyage;* Lacroix, *Figures de savants,* vol. 3, 78.
45. John Woodward, *Brief Instructions for Making Observations in all Parts of the World,* ed. V. A. Eyles (1696; London: Society for the Bibliography of Natural History, 1973), 12–13. Joseph Banks, *The Endeavour Journal of Joseph Banks (1768–1771),* ed. J. P. Sandford, 2 vols. (Sydney: State Library of New South Wales, 1998), vol. 1, 157–158; Drayton, *Nature's Government,* 67.
46. Taillemite, *Bougainville,* vol. 1, 349. Étienne Taillemite, director of conservation at the French National Archives, published the four journal accounts (from Bougainville, Fesche, de Rochefort, and de Versailes) associated with Bougainville's voyage.
47. Ibid., vol. 1, 350. Commerson cited from manuscript reproduced in Monnier et al., *Commerson,* 99. The ordinance of 15 April 1689, book IV, title III, article xxv. The ordinance of 25 March 1765, article MLXIII, reaffirmed the 1689 ordinance. Reprinted in Taillemite, *Bougainville,* vol. 1, 90.
48. Monnier et al., *Commerson,* 97.
49. Guillot, "La vraie 'Bougainvillée,'" 38.
50. Commerson's, Vivez's, and Bougainville's accounts reprinted in Taillemite, *Bougainville,* vol. 1, 349; vol. 2, 101, 237–241, 485.
51. Ibid., vol. 2, 238.
52. Ibid., vol. 1, 349, vol. 2, 240. Also Denis Diderot, *Supplément au voyage de Bougainville,* ed. Gilbert Chinard (Paris: Droz, 1935), 131–132.
53. Taillemite, *Bougainville,* vol. 1, 349–350.
54. Ibid., vol. 1, 349; vol. 2, 241.
55. Ibid., vol. 2, 101.
56. Ibid., vol. 1, 89, 349.
57. Lizabeth Paravisini-Gebert, "Cross-Dressing on the Margins of Empire," in *Women at Sea,* ed. Lizabeth Paravisini-Gebert and Ivette Romero-Cesareo

(New York: Palgrave, 2001), 59–98; Catalina de Erauso, *Lieutenant Nun: Memoir of a Basque Transvestite in the New World,* trans. Michele Stepto and Gabriel Stepto (Boston: Beacon Press, 1996); Brigitte Eriksson, "A Lesbian Execution in Germany, 1721: The Trial Records," *Journal of Homosexuality* (1980–1981): 27–40. Sophie Germain, *Oeuvres philosophiques,* ed. Hippolyte Stupuy (Paris, 1896), 271. Kenneth Manning, "The Complexion of Science," *Technology Review* (Nov./Dec. 1991): 63. Elizabeth Blackwell, *Pioneer Work in Opening the Medical Profession to Women* (1895; New York: Schocken, 1977), vii. Laissus, "Voyageurs naturalistes," 316.

58. Bougainville in John Reihold Forster's English translation of Bougainville's journal, *A Voyage Round the World* (London, 1772), 300. Monnier et al., *Commerson,* 99, 109–111.
59. Taillemite, *Bougainville,* vol. 1, 443. Pycior, Slack, and Abir-Am, eds., *Creative Couples.*
60. Laissus, "Voyageurs naturalistes," 316; Jean Chaïa, "Jean-Baptiste Patris: Medecin botaniste à Cayenne," *95ᵉ Congrès national des sociétés savantes* 2 (1970): 189–197.
61. Lafuente, "Enlightenment in an Imperial Context," 161. Lafuente limits "creole" to Europeans born in the Americas (excluding from this concept American-born Africans).
62. McClellan, *Colonialism and Science,* part III; Drayton, *Nature's Government,* 64–65; Nassy, *Essai,* 164. The Cercle des Philadelphes enjoyed one woman member, Mlle. Lemasson Le Golf, du Havre (Pluchon, "Le Cercle des philadelphes," 168). *Mémoires du cercle des philadelphes* 1 (1788)—this seems to have been the only volume published.
63. Jacques Michel, "La Guyane sous l'Ancien Régime," *G. H. C.* 18 (1990): 178.
64. Aublet, *Histoire,* preface.
65. Ibid.
66. Louis Malleret, ed., *Un Manuscrit inédit de Pierre Poivre: Les Mémoires d'un voyageur* (Paris: Publications de l'Ecole Française d'Extréme-Orient, 1968). See also Emma Spary, "Of Nutmegs and Botanists: The Colonial Cultivation of Botanical Identity," in *Colonial Botany,* ed. Schiebinger and Swan. Balick and Cox, *Plants, People, and Culture,* 135.
67. Chaïa, "A Propos de Fusée-Aublet."
68. Aublet, *Histoire,* preface. Henri Froidevaux, "Les Recherches scientifiques de Fusée Aublet à la Guyane Française," *Bulletin de géographie historique et descriptive* (1897): 425–469. Mark Plotkin, Brian Boom, and Malorye Allison, *The Ethnobotany of Aublet's Histoire des plantes de la Guiane Françoise, 1775* (St. Louis: Missouri Botanical Garden, 1991).
69. Aublet, *Histoire,* preface.
70. Chaïa, "A Propos de Fusée-Aublet," 61–62.
71. Blunt, *Compleat Naturalist,* 117. Latour, *Science in Action,* chap. 6.
72. Marie-Noëlle Bourguet, "Voyage et histoire naturelle (fin XVIIᵉᵐᵉ siècle—

début XIXème siècle)," *Le Muséum au premier siècle de son histoire,* ed. Claude Blanchaert, Claudine Cohen, Pietro Corsi, Jean-Louis Fischer (Paris: Muséum National d'Histoire Naturelle, 1997), 163–196, esp. 177. See also Daubenton, *Histoire naturelle,* viii. Koerner, *Linnaeus,* chap. 5.
73. Drayton, *Nature's Government,* Part I. Stearn, ed., *Humboldt.*
74. Sörlin, "Ordering the World for Europe," 64.
75. Sloane, *Catalogus plantarum.* Dandy, *Sloane Herbarium;* MacGregor, "The Life," 22–24, 28; Cannon, "Botanical Collections."
76. Aublet, *Histoire,* 104.
77. Schiebinger, *Mind,* chap. 2. There are some exceptions, such as Lady Margaret Cavendish Bentinck, Duchess of Portland, who kept extensive gardens at Bulstrode, Buckinghamshire. See David Allen, *The Naturalist in Britain* (1976; Princeton: Princeton University Press, 1994), 24–25.
78. "Lists of Plants in the Garden at Badminton, chiefly made by Mary Somerset" and "Lists of Seeds and Plants Belonging to Mary Somerset," British Library, Manuscripts Collection, Sloane 3343, 4070–4072. The most complete history of the duchess's gardens is Douglas Chambers's excellent "'Storys of Plants': The Assembling of Mary Capel Somerset's Botanical Collection at Badminton," *Journal of the History of Collections* 9 (1997): 49–60. In the nineteenth century, one of the family's women invented the game of badminton (court dimensions were taken from the entrance hall of the Badminton House). Cottesloe and Hunt, *The Duchess of Beaufort's Flowers,* 10. Sir Hans Sloane's Collection, Natural History Museum, London, H. S. 131–142. Sherard cited in Phyllis Edwards, "Sir Hans Sloane and His Curious Friends," in *History in the Service of Systematics,* ed. Alwyne Wheeler and James Price (London: Society for the Bibliography of Natural History, 1981), 27–35, esp. 30. See also Dandy, ed., *Sloane Herbarium,* 209–215.
79. Cottesloe and Hunt, *The Duchess of Beaufort's Flowers,* 19. William Aiton, *Hortus Kewensis* (London, 1789).
80. Sir Hans Sloane's Collection, Natural History Museum, London, H. S. 66. Cited in Dandy, *Sloane Herbarium,* 210. Nothing more is known of this woman.
81. Cottesloe and Hunt, *The Duchess of Beaufort's Flowers,* 16.
82. Segal, "Merian as a Flower Painter," 79–82.
83. Merian to Petiver, 4 June 1703, British Library, Manuscripts Collection, Sloane 4063, f. 201; Sloane 4067, f. 51; the same letter is sent in French, Merian to Petiver, April 1704, Sloane 4064, f. 5. "History of Surinam Insects," abbreviated and methodized by J. Petiver, British Library, Manuscripts Collection, Sloane 3339, ff. 153–160b.
84. Merian to Petiver, 27 April 1705, British Library, Manuscripts Collection, Sloane 4064, f. 70; English translation Sloane 3321, f. 176; Sloane 4064, f. 70; English translation Sloane 3321, f. 176.
85. Merian to Petiver, 27 April 1705, British Library, Manuscripts Collection, Sloane 3321, f. 176.

86. Humboldt (and Bonpland), *Personal Narrative,* vol. 5, 389. Gunnar Broberg, "*Homo sapiens:* Linnaeus' Classification of Man," in *Linnaeus: The Man and His Work,* ed. Tore Frängsmyr (Berkeley: University of California Press, 1983), 185–186. Seymour Phillips, "The Outer World of the European Middle Ages," in *Implicit Understandings,* ed. Stuart Schwartz (Cambridge: Cambridge University Press, 1994), 23–63; Anthony Grafton, *New Worlds, Ancients Texts* (Cambridge, Mass.: Harvard University Press, 1992).
87. La Condamine, *Relation abrégée,* 111, 52, 106. Josine Blok, *The Early Amazons* (Leiden: E. J. Brill, 1995).
88. La Condamine, *Relation abrégée,* 104, 111.
89. Ibid., 103–109.
90. Ibid., 105–108.
91. Humboldt (and Bonpland), *Personal Narrative,* vol. 5, 387–394.
92. Ibid. Sir Everard Ferdinand Im Thurn, *Among the Indians of Guiana* (London, 1883), 385.
93. Mary Terrall, "Gendered Spaces, Gendered Audiences: Inside and Outside the Paris Academy of Sciences," *Configurations* 3 (1995): 207–232, esp. 219–223. Scientists today, such as volcanologists, admit to a "gallant dare devilishness" (William Rose, "Volcanic Irony," *Nature* 411 [2001]: 21). Spary, *Utopia's Garden,* 83–84. Linnaeus, *Critica Botanica,* no. 238.
94. Humboldt, *Vues des Cordillères,* commentary to plate 5.
95. La Condamine, *Relation abrégée,* 25.
96. Adanson, *Voyage,* 131.
97. Merian to Johan Volkammer, 8 October 1702, Trew-Bibliothek, Brief-Sammlung Ms. 1834, Merian no. 1, Universitätsbibliothek Erlangen, reprinted in Rücker, *Merian,* 22. [Schaw], *Journal,* 19–78.
98. "Letter to M. De la Condamine from M. Godin des Odonais," in *A General Collection,* ed. Pinkerton, vol. 14, 256–269.
99. Philip Curtin, "The White Man's Grave: Image and Reality, 1780–1850," *Journal of British Studies* 1 (1961): 94–110. Thunberg, *Travels,* vol. 2, 280. See also Boxer, *The Dutch Seaborne Empire,* 243; Philip Curtin, *Disease and Empire: The Health of European Troops in the Conquest of Africa* (Cambridge: Cambridge University Press, 1998), 3–4. Thunberg, *Travels,* vol. 1, 99. Trapham, *Discourse,* 70. Trevor Burnard, "'The Countrie Continues Sicklie': White Mortality in Jamaica, 1655–1780," *Social History of Medicine* 12 (1999): 45–72. Humboldt (and Bonpland), *Personal Narrative,* vol. 5, 244–245.
100. Richard Hakluyt, "A Short and Brief Narration of the Navigation made . . . to the Islands . . . called New France," in *A General Collection,* ed. Pinkerton, vol. 12, 659–660.
101. "Letter of M. De La Condamine, written in 1773, to M. ***; giving an account of the fate of those astronomers who participated in the requisite operations for the measurement of the earth, begun in 1735," ibid., vol. 14, 257–258.

102. Thiery de Menonville, *Traité*, vol. 1, 147. Adanson, *Voyage*, 336. Duval, *King's Garden*, 77.
103. Risse, "Transcending Cultural Barriers," 32.

2. Bioprospecting

1. Long, *History*, vol. 1, 6.
2. Nicolas Culpeper, *A Physicall Directory* (London, 1649), "To the Reader," A1. Pierre-Henri-Hippolyte Bodard, *Cours de botanique médicale comparée*, 2 vols. (Paris, 1810), vol. 1, xviii, xxx. Eli Heckscher, *Mercantilism*, 2 vols., trans. Mendel Shapiro (London: George Allen and Unwin, 1935); Robert Ekelund and Robert Tollison, *Politicized Economies: Monarchy, Monopoly, and Mercantilism* (College Station: Texas A&M University Press, 1997).
3. [Bourgeois], *Voyages*, 460. Long listed duties Great Britain received in the 1770s on various economic and medicinal plants imported into England from Jamaica. These plants included sugar (whose medicinal merits Long discussed at length), rum, pimento, ginger, cotton, coffee, indigo, sago, mahogany, aloes, cassia, guaiacum, winter's bark, jalap, sarsaparilla, tamarinds, vanilla, and cacao (*History*, vol. 1, 590). Descourtilz, *Flore pittoresque*, vol. 1, 41.
4. [Bourgeois], *Voyages*, 459; also Long, *History*, vol. 2, 590. Fermin estimated that drugs lost three-quarters of their virtue in the voyage across the Atlantic (*Description générale*, vol. 1, 83–84).
5. Drayton, *Nature's Government*, 92. Barrera, "Local Herbs," 174. Harold Cook, "Global Economies and Local Knowledge," *Colonial Botany*, ed. Schiebinger and Swan; Grove's claim is that Reede collected more than mere "information," or relatively raw data, and that the systems of classification in Reede's book relied on Ezhava "knowledge," or systematized thought (*Green Imperialism*, 78, 89–90). See also Mark Harrison, "Medicine and Orientalism: Perspectives on Europe's Encounters with Indian Medical Systems," *Health, Medicine and Empire: Perspectives on Colonial India*, ed. Biswamoy Pati and Mark Harrison (New Delhi: Orient Longman, 2001), 37–87. For the information vs. knowledge distinction see Peter Burke, *A Social History of Knowledge from Gutenberg to Diderot* (Cambridge: Polity Press, 2000), 11.
6. Sloane, *Voyage*, vol. 1, preface; Sloane, *Catalogus plantarum*. Merian, *Metamorphosis*, commentary to plate 45.
7. Keegan, "The Caribbean," 1262. Barrera, "Local Herbs," 166–167.
8. Thomas Walduck to James Petiver, Oct. 29, 1710: "By the Spaniards' own account, they destroyed in a few years one million and 200 thousand souls [Indians] upon [illegible], 3 millions upon Hispaniole, 6 millions in New Spain, 600 thousand upon Jamaica" (British Library, Manuscripts Collection, Sloane 2302). Historians' estimates of the number of Tainos living on the island of Hispaniola when Columbus arrived in 1492 vary from 60,000 to 4

million, with 1 million being a commonly agreed-upon figure. By 1514 only about 26,000 remained, according to a Spanish census. The numbers fell to 18,000 in 1518 and to 2,000 in 1542. Noble David Cook, *Born to Die: Disease and New World Conquest, 1492–1650* (Cambridge: Cambridge University Press, 1998), 23–24. By 1524, the Taino ceased to exist as a separate population. Rouse, *Tainos,* 169; David Henige, "On the Contact Population of Hispaniola: History as Higher Mathematics," *Hispanic American Historical Review* 58 (1978): 217–237.

The Caribbean Sea was named by the Spanish in the sixteenth century after the warlike Carib Indians, who inhabited primarily the Lesser Antilles when the Spanish first arrived. The Spanish gave the sea this name even though they came into contact primarily with Tainos, who populated Hispaniola, Cuba, Jamaica, and Puerto Rico. The Spanish called these Indians Taino, meaning "good" or "noble," words that the island-dwellers used to indicate to Columbus that they were not Island-Caribs, although they had no specific name for themselves. Irving Rouse has argued that the Spanish tended to call any aggressive natives "Caribs," and that as the Tainos came to resist Spanish cruelties, they too were often taken to be Caribs (*Tainos,* 5, 23, 155). Peter Hulme has pointed out that the term Arawak (the Tainos are of South American Arawakan heritage) was coined by Fray Gregorio Batela in 1540, whereas Carib, or Kalina, was something the native peoples called themselves. Hulme's larger point is that the ethnic map of the Caribbean area as described by early European settlers was itself the product of colonization insofar as Europeans sometimes made distinctions that the Amerindians themselves did not and vice versa (*Colonial Encounters,* 60, 67). See also Henri Stehlé, "Évolution de la connaissance botanique et biologique aux antilles françaises," *Comptes rendues du congrès des Sociétés Savantes* 1 (1966): 275–290, esp. 281; and David Watts, *The West Indies: Patterns of Development, Culture and Environmental Change since 1492* (Cambridge: Cambridge University Press, 1987).

9. Pouppé-Desportes, *Histoire des maladies,* vol. 3, 59. Pierre Barrère also provided Latin, French, and "Indian" names for plants, but he did not call his listing a *Pharmacopoeia (Essai).*
10. Pouppé-Desportes, *Histoire des maladies,* vol. 3, 59.
11. [Bourgeois], *Voyages,* 67. [Anon.], *Histoire des désastres,* 47. The Bibliothèque nationale de France suggests this book, published in 1795, was written by Michel-Étienne Descourtilz, but he did not go out to Saint Domingue until 1799.
12. Craton, *Searching,* 54. Stedman, *Stedman's Surinam,* 63.
13. [Bourgeois], *Voyages,* 458, 470. For others who admired slave medicine, see Grainger, *Essay,* and Sheridan, *Doctors and Slaves,* 80–82.
14. Judith Carney, "African Traditional Plant Knowledge in the Circum-Caribbean Region," *Journal of Ethnobiology* 23 (2003): 167–185. [Bourgeois], *Voyages,* 470.
15. [Bourgeois], *Voyages,* 468, 470.

16. De Beer, *Sloane*, 41–42. Knight is cited in Sheridan, *Doctors and Slaves*, 81.
17. Descourtilz, *Flore pittoresque*, vol. 1, 16–17. Adelon et al., eds., *Dictionaire*, vol. 14, s.v. "femme," 654.
18. Barrère, *Nouvelle relation*, 204. He was paid 2,000 livres per year. Nassy, *Essai historique*, 64. See also Goslinga, *The Dutch in the Caribbean and in the Guianas*, 359–360; Robert Cohen, *Jews in Another Environment: Surinam in the Second Half of the Eighteenth Century* (Leiden: Brill, 1991). Lacroix, *Figures de savants*, vol. 3, 31–35. Sloane, *Voyages*, vol. 1, xiii–xiv.
19. La Condamine, "Sur l'arbre du quinquina," 330. On the tradition of animals using herbs to heal themselves, see L. A. J. R. Houwen, "'Creature, Heal Thyself': Self-Healing Animals in the *Hortus sanitatis*" (unpublished lecture, Department of English, Ruhr-Universität Bochum, Germany).
20. Pouppé-Desportes, *Histoire des maladies*, vol. 3, 81. Long, *History*, vol. 2, 380. James, *Medicinal Dictionary*, vol. 1, preface.
21. Long, *History*, vol. 2, 381.
22. Tzvetan Todorov, *The Conquest of America: The Question of the Other*, trans. Richard Howard (New York: Harper & Row, 1984); Stephen Greenblatt, *Marvelous Possessions: The Wonder of the New World* (Chicago: University of Chicago Press, 1991); Latour, *Science in Action*, chap. 6; and Anke te Heesen, "Accounting for the Natural World," in *Colonial Botany*, ed. Schiebinger and Swan. Spary, *Utopia's Garden*, 84.
23. Pratt, *Imperial Eyes*, 6–7.
24. Long, *History*, vol. 2, 287; Thunberg, *Travels*, vol. 1, 73–75. Thunberg wrote of these practices: "For as the Company is often in want of men, and does not care to give better pay, they are obliged to overlook the methods used by these infamous traders in human flesh to procure hands." Merian, *Metamorphosis*, commentary to plate 45; see also plates 7, 25, and 13.
25. Barker, Hulme, and Iversen, eds., *Colonial Discourse*, 7.
26. Bancroft, *Essay*, 3.
27. Pierre Pelleprat, *Introduction à la langue des Galibis* (Paris, 1655), 3; see also Pimentel, "The Iberian Vision," 26. La Condamine, "Sur l'arbre du quinquina," 340. La Condamine, *Relation abrégée d'un voyage* (Paris, 1745), 53–55. Humboldt (and Bonpland), *Personal Narrative*, vol. 3, 301–303.
28. Humboldt (and Bonpland), *Personal Narrative*, vol. 5, 431.
29. Leblond, *Voyage*, 138.
30. Cited in Blackburn, *New World Slavery*, 281.
31. Campet, *Traité*, 55.
32. Rochefort, *Histoire naturelle*, 449.
33. Risse, "Transcending Cultural Barriers," 32; but see also Estes, "The European Reception," 12; and George Foster, *Hippocrates' Latin American Legacy* (Langhorne, Penn.: Gordon and Breach, 1994). Spary, *Utopia's Garden*,

87; see also Anthony Pagden, *European Encounters with the New World* (New Haven: Yale University Press, 1993), 21.
34. Jerome Handler, "Slave Medicine and Obeah in Barbados, ca. 1650 to 1834," *New West Indies* 74 (2000): 57–90. See also Renny, *History of Jamaica*, 171.
35. Thomson, *Treatise*, 9–10.
36. Moseley, *A Treatise on Sugar*, 190–205.
37. Jerome Handler and Kenneth Bilby, "On the Early Use and Origin of the Term 'Obeah' in Barbados and the Anglophone Caribbean," *Slavery and Abolition* 22 (2001): 87–100; Fuller, *New Act of Assembly*, xl.
38. Humboldt (and Bonpland), *Personal Narrative*, vol. 5, 256.
39. Ibid., 132. La Condamine, *Relation abrégée*, 74–75.
40. La Condamine, *Relation abrégée*, 74–75.
41. Humboldt (and Bonpland), *Personal Narrative*, vol. 5, 132. In 1660 France and England guaranteed Saint Vincent island to the Caribs. In 1700 the governor of Martinique divided Saint Vincent into two halves—one for the "red Caribs" and one for the "black Caribs" (descendants of maroon Africans). Pouliquen, "Introduction," in Leblond, *Voyage*, 11. *Code de la Martinique* (Saint Pierre, 1767), 457–458. Sloane, *Voyage*, vol. 1, xviii.
42. [Bourgeois], *Voyages*, 487. Fermin, *Traité des maladies*, preface.
43. Grainger, *Essay*, 70.
44. Monardes, *Joyfull Newes*, vol. 1, 136–137; Estes, "The European Reception," 10. Alonso de Ovalle, "An Historical Relation of the Kingdom of Chile," *A General Collection*, ed. Pinkerton, vol. 14, 38. La Condamine, "Sur l'arbre du quinquina," 329.
45. [Bourgeois], *Voyages*, 487. Fermin, *Description générale*, vol. 1, 209.
46. Thiery de Menonville, *Traité*, vol. 1, 14. Sloane, *Voyages*, vol. 1, liv–lv.
47. Edward Ives, *Voyage from England to India, in the Year 1756* (London, 1773), 462.
48. William Eamon, *Science and the Secrets of Nature: Books of Secrets in Medieval and Early Modern Culture* (Princeton: Princeton University Press, 1994), 4–5.
49. Jaramillo-Arango, *Conquest*, 79.
50. Thunberg, *Travels*, vol. 2, 286. Nicolas, "Adanson et le mouvement colonial," 440. Smith, *Wealth*, vol. 1, 69. Guerra, "Drugs from the Indies," 29. Developing a new drug takes approximately twelve years. Merck estimates that for every 10,000 substances that are evaluated, twenty are selected for animal testing. Of these twenty, ten will be tested in humans and only one approved by the U.S. Food and Drug Administration for sale. Robert Pear, "Research Cost for New Drugs Said to Soar," *New York Times* (1 December 2001): C1, 14. Chadwick and Marsh, eds., *Ethnobotany*, 21, 42, 88
51. Koerner, "Women and Utility," 251. Mackay, *In the Wake of Cook*, 15. R. G. Latham, ed., *The Works of Thomas Sydenham*, 2 vols. (London: Sydenham Society, 1848–1850), vol. 1, 82.

52. Lowthrop, *Philosophical Transactions*, vol. 3, 252–255.
53. Ibid. For a later episode in the development of styptic, see Harold Cook, "Sir John Colbatch and Augustan Medicine," *Annals of Science* 47 (1990): 475–505.
54. Sloane, *Account*, 1. British Library, Manuscripts Collection, Medical Receipts Seventeenth Century, Sloane. 3998 ff. 1-34, 50–58, 60 b-75.
55. Sloane, *Account*, 5.
56. Ibid., 13–14.
57. S. W. Zwicker, *Breviarium apodemicum methodice concinnatum* (Danzig, 1638), cited in Stagl, *History of Curiosity*, 78. Margaret Hannay, "'How I These Studies Prize': The Countess of Pembroke and Elizabethan Science," in *Women, Science and Medicine*, ed. Hunter and Hutton, 109–113, 67–76.
58. Schiebinger, *Mind*. Carolus Clusius, *Rariorum aliquot Stirpium, per Pannoniam, Austriam, et vicinas . . . Historia* (Antwerp, 1583), 345; see also Jerry Stannard, "Classici and Rustici in Clusius' Stirp. Pannon. Hist. (1583)," in *Festschrift anlässlich der 400 jährigen Widerkehr der wissenschaftlichen Tätigkeit von Carolus Clusius (Charles de l'Escluse) im pannonischen Raum*, ed. Stefan Aumüller (Eisenstadt: Burgenländischen Landesarchiv Sonderheft V, 1973), 253–269. Sydenham cited in de Beer, *Sloane*, 25.
59. Her cure was published in "Mrs. Stephen's Cure for the Stone," *London Gazette*, June 16, 1739, n.p. See also Stephen Hales, *An Account of Some Experiments and Observations on Mrs. Stephen's Medicines for dissolving the Stone* (London, 1740); James Parsons, *A Description of the Human Urinary Bladder . . . to which are added Animadversions on Lithontriptic Medicines, particularly those of Mrs. Stephens* (London, 1742); Arthur Viseltear, "Joanna Stephens and the Eighteenth Century Lithontriptics: A Misplaced Chapter in the History of Therapeutics," *Bulletin of the History of Medicine* 42 (1968): 199–220; Maehle, *Drugs on Trial*, chap. 2. It was common for governments to buy the secrets of useful cures for "public utility" in this period; Brockliss and Jones, *Medical World*, 622–623. Cope, *Cheselden*, 24–25.
60. David Hartley, *A Supplement to a Pamphlet Entitled, A View of the Present Evidence for and against Mrs. Stephen's Medicines* (London, 1739), 37–38.
61. Ibid., 49–51.
62. Ibid., 39–40, 52.
63. Woodville, *Medical Botany*, vol. 1, 139.
64. Withering, *Foxglove*, 2–10. Cox, "The Ethnobotanical Approach," 26.
65. Schiebinger, *Mind*, 237–238. Koerner, "Women and Utility," 250–251.
66. John Douglas, *A Short Account of the State of Midwifery in London, Westminster* (London, 1736), 19.
67. Margaret Pelling, "Thoroughly Resented? Older Women and the Medical Role in Early Modern London," in *Women, Science and Medicine*, ed. Hunter and Hutton, 63–88, esp. 67, 72. Blair, *Pharmaco-Botanologia*. J. Burnby,

"The Herb Women of the London Markets," *Pharmaceutical Historian* 13 (1983): 5–6.
68. But see Genevieve Miller, "Putting Lady Mary in Her Place," *Bulletin of the History of Medicine* 55 (1981): 3–16.
69. Miller, *Adoption of Inoculation*, 63n57.
70. Ibid., 55–63.
71. Emanuel Timonius, "An Account, or History, of Procuring the Small Pox by Incision, or Inoculation," *Philosophical Transactions of the Royal Society of London* 29 (1714): 72–82; Sloane, "An Account of Inoculation." Miller, *Adoption of Inoculation*, 76–79.
72. John Andrew, *The Practice of Inoculation Impartially Considered* (Exeter, 1765), vii.
73. Anita Desai, intro., *The Turkish Embassy Letters* by Lady Mary Wortley Montagu (London: Virago, 1994), xv. Maitland, *Account*, 7–8.
74. Condamine, *History*, 9. Miller, *Adoption of Inoculation*, 60.
75. Schiebinger, *Nature's Body*, 126–134. La Condamine, *History*, 3, 9, 10–12. Charles Arthaud in Saint Domingue claimed that inoculation was discovered in the climate where despotism places a high price on beauty (*Memoire*, 8).
76. Wagstaffe, *A Letter to Dr. Freind*. Boylston, *Historical Account*.
77. Lord Wharncliffe, ed., *The Letters and Works of Lady Mary Wortley Montagu*, 2 vols. (London, 1893), vol. 1, 309, 352–353. British Library, Manuscripts collection, Add. 34327, folio 7.

3. Exotic Abortifacients

1. Goslinga, *The Dutch in the Caribbean and in the Guianas*, 268.
2. Hilliard d'Auberteuil, *Considérations*, 65.
3. James, *Medicinal Dictionary*, vol. 3, s.v. "Poinciana." *Caesalpinia pulcherrima* is listed in Norman Farnsworth's large study as an abortifacient and emmenagogue—"Potential Value of Plants as Sources of New Antifertility Agents I," *Journal of Pharmaceutical Sciences* 64 (1975): 535–598, esp. 565 where the leaf and seed are noted as abortifacients—and in R. Casey, "Alleged Anti-Fertility Plants of India," *Indian Journal of Medical Science* 14 (1960): 590–600, esp. 593. Julia Morton, *Atlas of Medicinal Plants of Middle America* (Springfield, Ill.: Charles Thomas, 1981), 284–285; John Watt and Maria Breyer-Brandwijk, *The Medicinal and Poisonous Plants of Southern and Eastern Africa* (Edinburgh: Livingstone, 1962), 564–565; Walter Lewis and Memory Elvin-Lewis, *Medical Botany: Plants Affecting Man's Health* (New York: John Wiley, 1977), 42.
4. Merian, *Metamorphosis*, commentary to plate 45. Sloane, *Voyage*, vol. 2, 49–50; Sloane cites Merian's work in an addendum to his text (*Voyage*, vol. 2, 384). Merian's book was published in 1705; Sloane's in 1707 (vol. 1) and 1725 (vol. 2). Descourtilz, *Flore pittoresque*, vol. 1, 27–30. The plant induces uterine contractions.

5. Tournefort, *Élémens,* vol. 1, 491–492; vol. 3, plate 391. See also Du Tertre, *Histoire,* vol. 1, 125–126. Chevalier, *Lettres,* 111. Descourtilz, *Flore pittoresque,* vol. 1, 27–30. On Descourtilz, see Leon Rulx, "Descourtilz," *Conjonction* 39 (1952): 40–48.
6. Stedman, *Stedman's Surinam,* 26, 271–272. Goslinga detailed some of these punishments in Surinam: In 1741 two slaves each lost a leg; in 1765 three legs were amputated and one tendon cut. In 1772 two tendons were cut and three legs lost. Between 1765 and 1787 at least sixteen blacks had a leg amputated (*The Dutch in the Caribbean and the Guianas,* 382, 399). *Le Code noir* (Paris, 1685), article 38; Long, *History,* vol. 2, 440; Sloane, *Voyage,* vol. 1, lvii. Dalling's testimony in *House of Commons,* ed. Lambert, vol. 72, 433.
7. Sloane, *Voyage,* vol. 1, lvii, cxliii; vol. 2, 50. One assumes that by "flour," Sloane means this plant can be grown as a beautifully flowered fence. "Flour" could also, however, refer to women's "flowers," "flours," or "fleurs"—words that in English and French commonly denoted menstruation. Other great "dissemblers" of illness, according to Sloane, were servants, "both Whites and Blacks." Shorter, *Women's Bodies,* 181; see also McLaren, *History,* 160. Charles-Louis-François Andry also warned physicians to make sure a woman was not pregnant before prescribing an emmenagogue. Too frequently, he wrote, young women who wish to save their honor seek to trick a doctor and in so doing commit a terrible crime in order to hide a human weakness (*Matière médicale,* vol. 2, 22). Dimsdale wrote, "I have not inoculated any woman whom I knew to be pregnant; but some who concealed their pregnancy have been inoculated by me, who, I fancy, hoped for an event that did not happen, I mean miscarriages" (*Present Method,* 21–22). Rublack, "The Public Body," 64.
8. James, *Medicinal Dictionary,* vol. 1, s.v. "abortus, or aborsus." See also Adrian Wilson, in "William Hunter and the Varieties of Man-Midwifery," *William Hunter and the Eighteenth-Century Medical World,* ed. W. F. Bynum and Roy Porter (Cambridge: Cambridge University Press, 1985), 343–369, esp. 350–351. Sloane, *Voyage,* vol. 1, cxliii; also Shorter, *Women's Bodies,* 190. Ersch and Gruber, *Encyclopädie,* s.v. "Abtreibung."
9. Riddle, *Contraception;* Riddle, *Eve's Herbs;* Shorter, *Women's Bodies.* This is also Susan Klepp's view; see her "Lost, Hidden, Obstructed," 71–73. J. Thomsen in the nineteenth century assumed that this knowledge belonged to the "secret region of women's lives" and it was passed along through "wise women" *(weise Frauen)* ("Ein Fall von Abtreibung der Leibesfrucht," *Vierteljarsschrift für gerichtliche und öffentliche Medicin* 1 [1864]: 315–328, esp. 316). Monica Green argues, however, that in medieval Europe women's health care was not exclusively handled by women (*Women's Healthcare in the Medieval West* [Aldershot, Hamp.: Ashgate, 2000]). Dazille, *Observations sur les maladies des negres,* vol. 2, 56; Williamson, *Medical . . . Observations,* vol. 2, 206–207. See also Sheridan, *Doctors and Slaves,* 95, 268–291; Moitt, *Women and Slavery,* 63–68; and Karol Weaver, "The Enslaved Healers of

Eighteenth-Century Saint Domingue," *Bulletin of the History of Medicine* 76 (2002): 429–460. In addition, large plantations had a yaws hut attendant and often a separate attendant for infants and children. In Jamaica and Antigua, too, plantation hospitals were run day to day typically by an "aged Negress." She was supervised by a European male physician. Renny, *History*, 179; Adair, *Unanswerable Arguments*, 118. Paul Brodwin, *Medicine and Morality in Haiti: The Contest for Healing Power* (Cambridge: Cambridge University Press, 1996), 28–32.

10. Dancer, *Medical Assistant*, 200. Fermin, *Traité des maladies*, 13, 98–100. In Europe, too, women were often asked to address their complaints to the wife of a particular male medical practitioner (Jacques Barbeu du Bourg, *Gazette d'Épidaure, ou Recueil de nouvelles de médecine*, 4 vols. [Paris, 1762], vol. 3, 30).

11. Zvi Loker, "Professionnels medicaux dans la colonie de Saint Domingue au XVIIIème siècle," *Revue de la société haïtienne d'histoire, de géographie et de géologie* 39 (1981): 5–33. The *Almanach historique de Saint Domingue* ([Cap François, 1779], 111) listed a Becht, as royal physician and "*accoucheur* for the colony." No female midwives were listed in this almanac. *Affiches américaines* (7 August 1769), no. 31, 266. Arthaud, *Observations*, 78. "Quaker Records: 'At a Meeting of the Midwives in Barbadoes II.XII.1677,'" *Journal of the Barbados Museum and Historical Society* 24 (1957): 133–134.

12. Riddle, *Eve's Herbs*, 11, 180–181; *The Case of Mary Katherine Cadière, against the Jesuite Father John Baptist Girard* (London, 1731), 13. See also Rublack, "The Public Body," 62. Mauriceau, *Traité*, 191–192. Sloane, *Voyage*, vol. 1, 13.

13. Schiebinger, *Has Feminism Changed Science?*

14. *Oxford English Dictionary*, s.v. "aborted." See Schiebinger, *Nature's Body*, chap. 2.

15. Mauriceau, *Traité*, 187. This was also the midwife Angélique Marguerite Le Boursier du Coudray's understanding of *fausse-couche* (*Abrégé*, 44); Charles-Chretien-Henri Marc, the author of the entry on "fausse-couche" in the *Dictionaire des sciences médicales* also considered this term to refer to something issuing from a "fausse-grossesse" (Adelon et al., eds., *Dictionaire*, vol. 14, s.v. "fausse-couche"). James, *Medicinal Dictionary*, vol. 1, s.v. "abortus, or aborsus." Boord, *Breviarie of Health*, 7. Smellie, *Encyclopedia Britannica*, s.v. "midwifery."

16. Burton, "Human Rights," 427–428. Grimm's *Deutsches Wörterbuch* (1854) lists under "*Fehlgeburt*" both "*abortus*" and "*fausse-couche.*" For fine distinctions surrounding miscarriage, see Ersch and Gruber, *Encyclopädie*, s.v. "Fehlgebären." Zedler, *Universal Lexicon*, s.v. "Abortus," "Abtreiben."

17. *Oxford English Dictionary*, s.v. "abortion." Owsei Temkin and Lilian Temkin, eds., *Ancient Medicine: Selected Papers of Ludwig Edelstein* (Baltimore: Johns Hopkins Press, 1967), 9, 13; Zedler, *Universal Lexicon*, s.v. "Abortus"; Adelon et al., eds., *Dictionaire*, vol. 2, s.v. "avortement," 492–494. See also

Chauncey Leake, ed., *Percival's Medical Ethics* (Baltimore: Williams and Wilkins, 1927), 134–135.
18. Stukenbrock, *Abtreibung,* 19–20. See also Robert Jütte, ed., *Geschichte der Abtreibung: Von der Antike bis zur Gegenwart* (Münich: C. H. Beck, 1993); Sibylla Flügge, *Hebammen und heilkundige Frauen: Recht und Rechtswirklichkeit im 15. und 16. Jahrhundert* (Frankfurt am Main: Stroemfeld Verlag, 1998); Günter Jerouschek, "Zur Geschichte des Abtreibungsverbot," *Unter anderen Umständen: Zur Geschichte der Abtreibung,* ed. Gisela Staupe and Lisa Vieth (Dresden: Deutsches Hygiene-Museum, 1993), 11–26. Church law cited in Riddle, *Eve's Herbs,* 131, 158. Common law cited in John Keown, *Abortion, Doctors and the Law* (Cambridge: Cambridge University Press, 1988), 4–5, 173n39. William Blackstone, *Commentaries on the Laws of England,* 4 vols. (Oxford, 1765), vol. 1, 129–130; François André Isambert, ed., *Recueil général des anciennes lois Françaises,* 29 vols. (Paris, 1821–1833), s.v. "infanticide"; "grossesse." Adelon et al., eds., *Dictionaire,* vol. 2, s.v. "avortement," 495–496; Burton, "Human Rights," 431. Law codes in French colonies were generally the same as in France except where differences in place, persons, and goods required attention to local customs; this was true also for the French provinces. *Code de la Martinique* (Saint Pierre, 1767), preface.
19. Frank, *System,* vol. 2, 61. Stukenbrock, *Abtreibung,* 20.
20. Mauriceau, *Traité,* 191–192. Also Frank, *System,* vol. 2, 84–122; Tardieu's *Étude.* Mauriceau cited in James, *Medicinal Dictionary,* s.v. "abortus, or aborsus."
21. Duden, *Disembodying Women,* 79–82.
22. Dancer, *Medical Assistant,* 267. Signs of pregnancy are discussed in Albrecht von Haller, *Vorlesungen über die gerichtliche Arzneiwissenschaft,* 2 vols. (Bern, 1782), vol. 1, 52–61; abortifacients are discussed in vol. 2 of Frank, *System.* See also Gottlieb Budaeus, *Miscellanea medico-chirurgia, practica et forensia* (Leipzig, 1732–1737); William Cummin, *The Proofs of Infanticide Considered* (London, 1836); Tardieu, *Étude.* Freind, *Emmenologia,* 5–7. Smellie, *Encyclopedia Britannica,* s.v. "midwifery."
23. *Encyclopedia Britannica,* 11th ed. (New York: Encyclopaedia Britannica, 1910), s.v. "medical jurisprudence."
24. Stofft, "Avortement criminal," 79. Adelon et al., *Dictionaire,* vol. 2, s.v. "avortement," 497.
25. Diderot and d'Alembert, eds., *Encyclopédie,* s.v. "fausse-couche." Mauriceau, *Traité,* 191–192.
26. Norman Himes, *Medical History of Contraception* (Baltimore: Williams and Wilkins, 1936); Noonan, *Contraception.* Lewin, *Fruchtabtreibung;* Leibrock-Plehn, *Hexenkräuter.* McLaren, *History.* Shorter, *Women's Bodies;* Riddle, *Contraception* and *Eve's Herbs.* Gunnar Heinsohn and Otto Steiger, *Die Vernichtung der Weisen Frauen* (Herbstein: März, 1985). Rublack, "The Public Body," 65.

27. Riddle, *Contraception;* Riddle, *Eve's Herbs;* Leibrock-Plehn, *Hexenkräuter.* On herbals, see Wilfrid Blunt and Sandra Raphael, *The Illustrated Herbal* (New York: Thames and Hudson, 1979); Jerry Stannard, *Herbs and Herbalism in the Middle Ages and Renaissance* (Aldershot, Hampshire: Ashgate, 1999). Nicholas Culpeper, *Pharmacopoeia Londinensis; or, The London Dispensatory* (London, 1669), 36. Gerard, *Herbal,* s.v. "Of Savin." Some herbals provide recipes for "provoking the flowers in women," see Lucile Newman, "Ophelia's Herbal," *Economic Botany* 33 (1979): 227–232.

28. See, e.g., Jean Donnison, *Midwives and Medical Men: A History of Inter-Professional Rivals and Women's Rights* (London: Heinemann, 1977); Riddle, *Contraception;* Hilary Marland, ed., *The Art of Midwifery: Early Modern Midwives in Europe* (London: Routledge, 1993); Gelbart, *King's Midwife;* Evenden, *Midwives.*

29. See Alexander Hamilton, *A Treatise of Midwifery* (Edinburgh, 1785). Remedy in Beryl Rowland, *Medieval Woman's Guide to Health: The First English Gynecological Handbook* (Kent, Ohio: Kent State University Press, 1981), 97. Talbot, *Natura Exenterata,* 193–194.

30. Bourgeois, "Instructions," 120. *Statuts et Reiglemens ordonnez pour toutes les matronnes, ou saiges femmes* (Paris, 1587), 6; Wendy Perkins, *Midwifery and Medicine in Early Modern France: Louise Bourgeois* (Exeter: University of Exeter Press, 1996), 4. James Aveling records this prohibition for English midwives from 1567 onward (*English Midwives* [1872; London: Elliott, 1967]). See also Evenden, *Midwives,* 205–208. A Strasbourg ordinance of 1650 prohibited midwives from inducing abortions (Shorter, *Women's Bodies,* 189). Rublack, "The Public Body," 68.

31. Louise Bourgeois, *Recueil de secrets* (Paris, 1710), 84–87. Bourgeois, "Instructions," 121. Nicholas Culpeper, *A Directory for Midwives* (London, 1760), 61. One plant Sharp suggested pregnant women avoid was alpine snakeroot *(Eryngium alpinum),* which has abortive qualities. Riddle, *Eve's Herbs,* 154. Leibrock-Plehn, *Hexenkräuter,* 157; on Siegemund, see also Waltrund Pulz, "Gewaltsame Hilfe? Die Arbeit der Hebammen im Spiegel eines Gerichtskonflikts (1680–1685)," *Rituale der Geburt,* ed. J. Schlumbohm, B. Duden, J. Gélis, and P. Veit (Munich: Beck Verlag, 1998). Le Boursier Du Coudray, *Abrégé,* 44–51. Only in the nineteenth century is there evidence that midwives were the ones primarily in charge of abortions. Three-quarters of the 1,145 persons accused of criminal abortion in France between 1851 and 1865 were women, and the majority of them were midwives. Of the hundred accused in the years between 1846 and 1850, thirty-seven were midwives, nine were physicians, one was a pharmacist and herbalist, two were "charlatans," and two were matrons. Tardieu, *Étude,* 12, 22.

32. Gelbart, *King's Midwife,* 27–28; Stofft, "Avortement criminel," 76. Jürgen Schlumbohm, "'The Pregnant Women Are Here for the Sake of the Teaching Institution,'" *Social History of Medicine* 14 (2001): 59–78, esp. 66.

33. *Lettres de Gui Patin*, ed. J. H. Reveillé-Parise, 3 vols. (Paris, 1846), vol. 3, 225–226. Stofft, "Avortement criminal," 79. Pierre Dionis, Parisian master surgeon, also assumed it was midwives who gave women abortives. He mentioned the fatal consequences for the midwife if a woman of high standing were to die as a result (*Traité*, 419).
34. Rublack, "The Public Body."
35. Hermann, *Materia medica*, 130–131, 214–216, 279–281, index. Carl Linnaeus, *Materia medica* (Amsterdam, 1749).
36. Boccaccio, cited in Leibrock-Plehn, *Hexenkräuter*, 163. Among these means might have been the juice of unripe grapes and rue. Mme. Claude Gauvard, *"De Grace especial": Crime, état et société en France à la fin du Moyen Age*, 2 vols. (Paris: Publications de la Sorbonne, 1991), vol. 1, 316–317. Ben Jonson, *Epicoene, or, The Silent Woman* (London, 1620), 58–62. Donatien-Alphonse-François, marquis de Sade, *The Bedroom Philosopher* (Paris: Olympia Press, 1957), 53.
37. Mary Wollstonecraft, *The Wrongs of Woman* (1798; Oxford: Oxford University Press, 1976), 109. Daniel Defoe's Moll Flanders used a midwife to rid herself of a child after she was delivered (*Moll Flanders*, ed. James Sutherland [Cambridge, Mass.: Riverside Press, 1959], 148–150). Lewin, *Fruchtabtreibung*, 377. Surgical abortions became more common in the nineteenth century. There was the sensational story of Eliza Wilson, thirty-two years old. She paid four shillings to an abortionist, one Mrs. Dryden, who operated with her instruments twice. When that had no effect, she paid 2 pounds 10 shillings to a Mrs. Spencer Linfield, whom the neighbors testified was often visited by "ladies of fashion" for such reasons. Wilson died from these last attempts. *Full Account of the Extraordinary Death of Eliza Wilson, by Mrs. Linfield, Midwife at Walworth, Caused by Abortion* (London, 1845). Löseke, *Materia medica*, 387–388. Modern tests (on pigs and rats) show savin an effective, if sometimes fatal, abortive. N. Page et al., "Teratological Evaluation of *Juniperus sabina* Essential Oil in Mice," *Planta Medica* 55 (1989): 144–146; Brondegaard, "Sadebaum," 335.
38. Shorter noted the top four European abortifacients as penny-royal, sage, thyme, and rosemary, with ergot, rue, and savin following closely behind (*Women's Bodies*, 183); my study, however, shows savin as most often used. J. H. Dickson and W. W. Gauld, "Mark Jameson's Physic Plants: A Sixteenth Century Garden for Gynaecology in Glasgow?" *Scottish Medical Journal* 32 (1987): 60–62. Monica Green, "Constantius Africanus and the Conflict between Religion and Science," in *The Human Embryo*, ed. G. R. Dustan (Exeter: University of Exeter Press, 1990), 47–69; Woodville, *Medical Botany*, vol. 2, 256; Riddle, *Contraception*, 160; Linnaeus cited in Brondegaard, "Sadebaum," 340, 344.
39. Woodville, *Medical Botany*, vol. 2, 256. Zedler, *Universal Lexicon*, s.v. "Sadebaum." Riddle, *Eve's Herbs*, 54.
40. Professor Klose zu Breslau, "Vermischte Bemerkungen aus dem Gebiet der

practischen Medicin," *Journal der practischen Heilkunde* (1820): 3–18, esp. 5–6.
41. Cited in Lewin, *Fruchtabtreibung*, 328. Brondegaard, "Sadebaum," 341, 342.
42. Stofft, "Avortement criminel." G.-R. Le Febvre de Saint-Ildephont and L.-A. de Cézan, ed., *État de médecine, chirurgie et pharmacie en Europe pour l'année 1776, présenté au Roi* (Paris, 1776), 231. McLaren has provided various estimates of abandonment to foundling hospitals: in Rouen between 1670 and 1862, over 65,000 were abandoned, and in Paris over 4,000 per year for the same period (*History*, 162). Diderot and d'Alembert, eds., *Encyclopédie*, s.v. "fausse-couche."
43. Smith, *Wealth*, vol. 1, 88.
44. Diderot and d'Alembert, eds., *Encyclopédie*, s.v. "fausse-couche."
45. Bancroft, *Essay*, 371–372.
46. Bartolomé de Las Casas, *Historia de las Indias*, 3 vols. (Mexico: Fondo de Cultura Económica, 1951), vol. 2, 206. Girolamo Benzoni, *La Historia del Mundo Nuovo* (1572; Caracas: Academia Nacional de la Historia, 1967), 94.
47. Humboldt (and Bonpland), *Personal Narrative*, vol. 5, 28–32. See also N. Y. Sandwith, "Humboldt and Bonpland's Itinerary in Venezuela," *Humboldt*, ed. Stearn, 69–79.
48. Humboldt (and Bonpland), *Personal Narrative*, vol. 5, 31–32. Thomas Jefferson, *Notes on the State of Virginia*, ed. Thomas Abernethy (New York: Harper and Row, 1964), 58.
49. Labat, *Nouveau voyage*, vol. 2, 122, 126.
50. Craton, *Searching*, 87.
51. Fuller, *New Act of Assembly*, vi.
52. Stedman, *Stedman's Surinam*, 136.
53. Barbara Bush discusses the role of women in this insurrection (*Slave Women*, 65–73). Stedman, *Stedman's Surinam*, 130, 266. Poisons used included *le jus de la canne de Madere, le Mancanilier, le Laurier Rose*, and *la graine de Lilas*. Hilliard d'Auberteuil, *Considérations*, vol. 2, 139.
54. Stedman, *Stedman's Surinam*, 272. Ligon cited in Bush, *Slave Women*, 121.
55. Du Tertre, *Histoire*, vol. 2, 505. Long, *History*, vol. 2, 440. Women were freed more often than men, at least in French holdings in New Orleans, because they were less expensive and perceived as less of a threat to whites. L. Virginia Gould, "Urban Slavery–Urban Freedom," in *More than Chattel*, ed. Gaspar and Hine, 298–314, esp. 306. See also Goslinga, *The Dutch in the Caribbean and in the Guianas*, 529. William Beckford in Jamaica objected to masters who freed slaves simply so that they did not have to feed them in their old age (*Remarks*, 23, 96).
56. Rouse, *Taino*, 145, 151, 158; C. R. Boxer, *Women in Iberian Expansion Overseas, 1415–1815* (New York: Oxford University Press, 1975), 35–36.
57. Craton and Walvin, *A Jamaican Plantation*, 14–19. Gautier, *Soeurs*, 31, 33. Moitt, *Women and Slavery*, 10. Blackburn, *New World Slavery*, 291. Du

Tertre, *Histoire,* vol. 2, 455. Oddly enough many European women in the colonies were unmarried or widowed. Only half the French women in Saint Domingue were married. These unmarried women may have been the spunky widows of whom Stedman spoke: the women are in good health because of their healthy living. "I have known many wives to outlive four husbands but never a man to wear out two wives" (*Stedman's Surinam,* 22); Goslinga, *The Dutch in the Caribbean,* 279; McClellan, *Colonialism and Science,* 56–57. Long noted the population of European men and women in Jamaican towns for the year 1673 (*History,* vol. 1, 376). Direction des Archives de France, ed., *Voyage,* 47–50. Trevor Burnard, "European Migration to Jamaica, 1655–1780," in *The William and Mary Quarterly* 53 (1996): 769–796.

58. B. W. Higman, *Slave Populations of the British Caribbean, 1807–1834* (Baltimore: John Hopkins University Press, 1984), 100, 115–119; Geggus, "Slave and Free Colored Women," 259–260; Moitt, *Women and Slavery,* 12–30; Gautier, *Soeurs,* 33; Beckles, *Centering Woman,* 3, 7.

59. McClellan, *Colonialism and Science,* 57; Cauna, "L'État sanitaire des esclaves," 50. See also Trevor Burnard, "'The Countrie Continues Sicklie': White Mortality in Jamaica, 1655–1780," *Social History of Medicine* 12 (1999): 45–72. Ramsay, *Essay,* 83.

60. Ward, *British West Indian Slavery,* 176.

61. Stedman, *Stedman's Surinam,* xxx, 20. It is worth quoting Stedman in some detail: "I must describe this Custom which I am convinced will be highly censured by the sedate European Matrons but which is, nevertheless, as common as it is almost necessary to the bachelors who live in this climate. These gentlemen, all without exception, have a female slave (mostly a Creole) in their keeping, who preserves their linens clean and decent, dresses their victuals with skill, carefully attends them (they being most excellent nurses) during the frequent illnesses to which Europeans are exposed in this country, prevents them from keeping late hours, knits for them, sews for them &c. These girls who are sometimes Indians, sometimes Mulattos, and often Negroes, naturally pride themselves in living with a European, whom they serve with as much tenderness, and to whom they are generally as faithful as if he were their lawful Husband to the great Shame of so many fair ladies, who break through ties more sacred, and indeed bound with more solemnity. Nor can the above young women be married in any other way, being by their state of servitude entirely debarred from every Christian privilege and ceremony, which makes it perfectly lawful on their side, while they hesitate not to pronounce as harlots those who do not follow them (if they can) in this laudable example, in which they are encouraged, as I have said, by their nearest relations and friends." See also Ann Stoler, *Capitalism and Confrontation in Sumatra's Plantation Belt, 1870–1979* (New Haven: Yale University Press, 1985).

62. Stedman, *Stedman's Surinam,* 186. Dr. Jackson's testimony in *House of Com-*

mons, ed. Lambert, vol. 82, 56. Thunberg, *Travels,* vol. 1, 137–138, 303. Geggus, "Slave and Free Colored Women," 265.

63. Labat, *Nouveau voyage,* vol. 2, 126.
64. Ibid., 122–126, 128; Gautier, *Soeurs,* 31; Beckles, *Centering Woman,* 27–32; Garrigus, "Redrawing the Color Line," 29. The 1685 *Code noir* prescribed that when a free man married a slave, she be manumitted and her children born free. Free men of European origin, however, rarely married their slave concubines. If one did, he was to suffer a heavy fine (though this was rarely enforced). Sue Peabody, "Négresse, Mulâtrese, Citoyenne: Gender and Emancipation in the French Caribbean, 1650–1848," in *Gender and Emancipation in the Atlantic World,* ed. Pamela Scully and Diana Paton (Raleigh: Duke University Press, 2004).
65. Labat, *Nouveau voyage,* vol. 2, 128–132; Beckles, *Centering Woman,* 74.
66. Stedman, *Stedman's Surinam,* 133. Goslinga, *The Dutch in the Caribbean,* 357–379.
67. Goslinga, *The Dutch in the Caribbean,* 358.
68. A little later Menonville accosted a man following him whom Menonville assumed to be a spy. The man, it turned out, was seeking Menonville's sexual favors. As Menonville put it, this gentleman was no other than what at court, "where all things are painted in their fairest colors," is termed the prince's friend (*l'ami du prince*). Thiery de Menonville, *Traité,* vol. 1, 69–70, 128.
69. David de Isaac Cohen Nassy, *Historical Essay on the Colony of Surinam,* trans. Simon Cohen (1788; Cincinnati: American Jewish Archives, 1974), 41. Goslinga, *The Dutch in the Caribbean,* 376. Bancroft, *Essay,* 375.
70. Aublet, "Observations sur le traitement des négres," Manuscript NHB 452, published as "Observations sur les négres esclaves" *Histoire,* vol. 2, 111ff. Laissus, "Voyageurs naturalistes," 316; Direction des Archives de France, ed., *Voyage,* 64; *Nouvelle biographie universelle* (Paris, 1852–1866), s.v. "Aublet"; "Extrait d'un manuscript de Robert Paul Lamonon, déposé à la Bibliothéque nationale," *Magasin Encyclopédique,* ed. A. L. Millin, 12 (1802): 365–367. See also Trevor Burnard, "The Sexual Life of an Eighteenth-Century Jamaican Slave Overseer," *Sex and Sexuality in Early America,* ed. Merril Smith (New York: New York University Press, 1998), 163–189; Douglas Hall, *In Miserable Slavery: Thomas Thistlewood in Jamaica, 1750–1786* (London: Macmillan, 1989).
71. Monique Pouliquen, "Introduction," in Leblond, *Voyage,* 5–19; Pouliquen, ed., *Voyages,* 39; Monique Pouliquen, "Que sont devenus les manuscrits de Jean-Baptiste Leblond?" *G. H. C. Bulletin* 79 (1996): 1532–1535.
72. Barbara Bush, "White 'Ladies,' Coloured 'Favourites' and Black 'Wenches': Some Considerations on Sex, Race and Class Factors in Social Relations in White Creole Society in the British Caribbean," *Slavery and Abolition* 2 (1981): 244–262, esp. 249; Hazel Carby, *Reconstructing Womanhood: The Emergence of the Afro-American Woman Novelist* (New York: Oxford University Press, 1987), 20–39; Evelyn Brooks Higginbotham, "African-American

Women's History and the Metalanguage of Race," *Signs: Journal of Women in Culture and Society* 17 (1992): 251–274, esp. 262–266; Nussbaum, *Torrid Zones*. Sloane, *Voyage*, vol. 1, 248–249. Like most of the drugs naturalists discussed in this era, the *Caraguata-acanga* had many uses; it was also considered good for "fevers and cures worms," making wine, and healing mouth ulcers. See also Long, *History*, vol. 3, 738. *Bromelia pinguin* was still used in Jamaica in 1955 as an abortifacient. G. Asprey and Phyllis Thornton, "Medicinal Plants of Jamaica," *West Indian Medical Journal* 4 (1955): 68–82, 145–168.

73. Long, *History*, vol. 2, 436. [Schaw], *Journal*, 112–113. [Edward Trelawny], *An Essay Concerning Slavery, etc.* (London, 1746), 35–36. In the East Indies, too, Europeans reported that indigenes practiced abortion. In the early seventeenth century Pieter Both, the first Dutch governor-general, advised against sending any more Dutch women to the Moluccas or Ambon because they led scandalous and unedifying lives, "to the great shame of our nation." He advocated, instead, that his men marry indigenous women, but not Muslims since they deliberately aborted any babies they conceived by Christian men. Boxer, *The Dutch Seaborne Empire*, 216.

74. McClellan, *Colonialism and Science*, 59.

75. Bancroft, *Essay*, 371–372.

76. Descourtilz, *Flore pittoresque*, vol. 8, 284.

77. Ibid., 306, 317.

78. Stedman, *Stedman's Surinam*, 22, 148. Jackson testimony from 1791 in *House of Commons*, ed. Lambert, vol. 82, 54–55. Susan Socolow, "Economic Roles of the Free Women of Color of Cap Français," in *More than Chattel*, ed. Gaspar and Hine, 279–297, esp. 288. The population of Saint Domingue was divided roughly into three groups: Europeans, either directly from Europe or born on the island; free people of color, including either freed Africans or "mulatos, mestizos, or quadrons"; and slaves, who, tallying 450,000 in the late eighteenth century, were seven times more numerous than the other two classes combined.

79. Sheridan, *Doctors and Slaves*, 224; McClellan, *Colonialism and Science*, 53. David Geggus, "Une Famille de La Rochelle et ses Plantations de Saint-Domingue," in *France in the New World*, ed. David Buisseret (East Lansing: Michigan State University Press, 1998), 119–138, esp. 127. Hilliard d'Aubertueil, *Considérations*, vol. 1, 65; [Collins], *Practical Rules*, 151. Grainger, *Essay*, 5. Beckford, *Remarks*, 26. Similarly, the cost of curing slaves was calculated against the cost of replacing them. The preamble to a 1646 act in Virginia stated that medical attention, despite its uncertain and often detrimental effects, was so much more expensive than the replacement costs of slaves and servants that the planters found it more humane and economical to let their slaves die. Cowen, "Colonial Laws."

80. In the 1780s and 1790s, as the slave trade was threatened, it became common for slave mothers to be given incentives "to encourage the population."

These measures eventually entered into island laws. In Martinique a law of 1786 prescribed reduced labor for pregnant slave women (Geneviève Leti, *Santé et société esclavagiste à la Martinique* [Paris: Editions L'Harmattan, 1998], 116). In Jamaica, a law of 1792 prescribed that owners pay the overseer three pounds to be divided equally between the slave mother and midwife (this was repealed in 1827). Henrice Altink, "Representations of Slave Women in Discourses of Slavery and Abolition, 1780–1838" (Ph.D. diss., University of Hull, 2002), chap. 1. Stedman, *Stedman's Surinam*, 272–273; Aublet, *Histoire*, vol. 2, 120; Thomas Norbury Kerby, native of Antigua, testimony concerning the slave trade in 1790 in *House of Commons*, vol. 72, 303. Marietta Morrissey, *Slave Women in the New World: Gender Stratification in the Caribbean* (Lawrence: University Press of Kansas, 1989), 101–102.

81. Records from the Worthy Park plantation books are cited in Craton and Walvin, *Jamaican Plantation*, 134. Blackburn, *New World Slavery*, 291; Gautier, *Soeurs*, 122–123; Geggus, "Slave and Free Colored Women," 267.

82. Thomas testimony in *House of Commons*, ed. Lambert, vol. 71, 252. John Williamson, a physician in Jamaica from 1798 to 1812, also commented that abortion was "common" and "frequent" among the slaves (*Medical . . . Observations*, vol. 1, 198, 200). [Collins], *Practical Rules*, 153. Adair, a physician in Antigua from 1777 to 1783, made similar observations (*Unanswerable Arguments*, 121). M. Cassan, *Considérations sur les rapports qui doivent existe entre les colonies et les métropoles* (Paris, 1790), 125–127. See also Ward, *British West Indian Slavery*, 165–189; Bush, *Slave Women*, chap. 7.

83. Arthaud, *Observations*, 75; Debien, *Esclaves*, 363–366; Cauna, "L'État sanitaire des esclaves," 52; Ramsay, *Essay*, 90.

84. Thomson, *Treatise*, 111. Long, *History*, vol. 2, 433. See also Sheridan, *Doctors and Slaves*, 224–245.

85. [Le Père Nicolson], *Essai sur l'histoire naturelle de l'isle de Saint-Domingue* (Paris, 1776), 55. This was a common theme; see Williamson, *Medical . . . Observations*, vol. 2, 200. Campet, *Traité*, 58–59; Hilliard d'Auberteuil, *Considérations*, vol. 2, 66. [Collins], *Practical Rules*, 157. See also [Anon.], *Negro Slavery; or, a View of some of the more Prominent Features of that State of Society* (London, 1823), 75–76: "Abortion is so frequent [in Jamaica] as to lead to an opinion, that means are taken to procure it, on account of ill disposition to their masters, and other barbarous reasons, for which there can be no excuse."

86. [Anon.], *Histoire des désastres*, 89–90.

87. Jean-Barthélemy Dazille, *Observations sur le tétenos* (Paris, 1788), 216–217. Dazille, *Observations sur les maladies des negres*, vol. 2, 75–77. Michel-Étienne Descourtilz, *Voyages d'un naturaliste*, 3 vols. (Paris, 1809), vol. 3, 119.

88. 22.1.1788 Sieur Tourtain au Comte de la Luzerne, AN, Col. F3 90, fol. 237, cited in Gautier, *Soeurs*, 114. See also Debien, *Esclaves*, 365.

89. Jackson testimony in *House of Commons*, ed. Lambert, vol. 82, 58. Jackson's

observations were also true of the French islands. See Cauna, "L'État sanitaire des esclaves," 21. Stephen Fuller, *Two Reports from the Committee of the Honourable House of Assembly of Jamaica* (London, 1789).
90. Médéric-Louis-Élie Moreau de Saint-Méry, *Description topographique, physique, civile politique et historique de la partie française de l'isle Saint-Domingue*, 3 vols. (1797; Paris: Librairie Larose, 1958), vol. 1, 61. Klepp, "Lost, Hidden, Obstructed," 101. Thomson, *Treatise*, 113.
91. Bernard Moitt, "Slave Women and Resistance in the French Caribbean," in *More than Chattel*, ed. Gaspar and Hine, 245.

4. The Fate of the Peacock Flower in Europe

1. Long, *History*, vol. 3, 852–853, which forms part of his "Synopsis of Vegetable and other Products of this Island, proper for Exportation, or Home Use and Consumption."
2. [Denis Joncquet], *Hortus Regius* (Paris 1666), 3; *Traitté des plantes par Bohin Fasius et celles du Jardin Roual, leurs vertu et leurs qualities* (1694), Ms. 1906, Bibliothèque centrale, Muséum national d'histoire naturelle (MNHN), Paris; *Catalogus plantarum [Horti Regii Parisiensis]* (1766), 228; René Desfontaines, *Catalogus plantarum Horti Regii Parisiensis*, 3rd ed. (Paris, 1829), 303; Bernard de Jussieu, *Hortus Regius Parisiensis ordine alphabetico, conscriptus 1728*, Ms. 1364, Bibliothèque centrale, MNHN, Paris, 194; "Catalogue des plantes apportées en France par le capitaine de vaisseau Milius commandant le Lys," Ms. 305, Bibliothèque centrale, MNHN, Paris. Like Poincy, Jean-Baptiste du Tertre attempted to grow the *Poinciana* in France. Although the seeds he brought with him from the Antilles developed one small tree the "height of a finger," it perished during the first cold spell (*Histoire*, vol. 2, 154). Herman Boerhaave, *Index alter plantarum quae in Horto Academico Lugduno-Batavo* (Leiden, 1720), part 2, 57; Breyne, *Exoticarum*, 61–64; Wijnands, *The Botany of the Commelins*, 59; Carl Linnaeus, *Hortus Upsaliensis, exhibens plantas exoticas* (Horto Upsaliensis Academiae, Stockholm, 1748), vol. 1, 101.
3. Miller, *Gardener's Dictionary*, s.v. "Poinciana (pulcherrima)."
4. Herman Boerhaave, *Historia plantarum, quae in Horto Academico Lugduni-Batavorum crescunt cum earum characteribus, & medicinalibus virtutibus* (Rome, 1727), 488–489. Boerhaave did not mention the peacock flower in his *Materia medica* (London, 1741) that he intended to serve as a compendium of ancient and modern medicine. His correspondence shows that he was familiar with Merian's work. *Boerhaave's Correspondence*, ed. G. A. Lindeboom (Leiden: Brill, 1962), part 1, 78. Herman Boerhaave, *Traité de la vertu des médicamens*, trans. M. de Vaux (Paris, 1729), 391. *Catalogue des plantes du Jardin de Mrs. les apoticaires de Paris* (Paris, 1759). Charles Alston, *Index plantarum, praecipue officinalium, quae in Horto Medico Edinburgensi* (Edinburgh, 1740); Alston, *Lectures*.
5. Dancer, *Medical Assistant*, 380; [Bourgeois], *Voyages*, 465.

6. Woodville, *Medical Botany*, vol. 1, vii; Thomson, *Treatise*, 144; Cullen, *Treatise*, vol. 1, vi; Alston, *Lectures*, vol. 1, 2.
7. Bernard, *Introduction*, 101; Maehle, *Drugs on Trial;* Jean Astruc, *Doutes sur l'inoculation de la petite verole* (Paris, 1756); Hildebrandt, *Versuch*, 86.
8. Störck, *Essay*, 12–13.
9. Felice Fontana, *Traité sur le vénin de la vipère, sur les poisons Americains, sur le laurier-cerise et sur quelques autres poisons végétaux* (Florence, 1781); Melvin Earles, "The Experimental Investigation of Viper Venom," *Annals of Science* 16 (1960): 255–269. Roger French, *Dissection and Vivisection in the European Renaissance* (Aldershot: Ashgate, 1999), 207. See also Maehle, "Ethical Discourse," 218, 225.
10. Fermin, *Description générale*, vol. 1, 70.
11. La Condamine, *Relation abrégée*, 208–210. See also Wolfgang-Hagen Hein, "The History of Curare Research," *Botanical Drugs of the Americas in the Old and New Worlds*, ed. Wolfgang-Hagen Hein (Stuttgart: Wissenschaftliche Verlagsgesellschaft, 1984), 43–49.
12. Condamine, *Relation abrégée*, 208–210.
13. Hérissant, "Experiments," 77–78.
14. Ibid. Before and even after the development of chloroform in the 1890s, curare was used in experiments to render animals motionless. Anti-vivisectionists of the period argued that although the animal could not move, it remained conscious and aware of pain. Maehle, "Ethical Discourse."
15. Johann Friedrich Gmelin, *Allgemeine Geschichte der Gifte*, 3 vols. (Leipzig, 1776), vol. 1, 34.
16. Julia Douthwaite, *The Wild Girl, Natural Man, and the Monster: Dangerous Experiments in the Age of Enlightenment* (Chicago: University of Chicago Press, 2002), 72. Johann Ritter, cited in Stuart Strickland, "The Ideology of Self-Knowledge and the Practice of Self-Experimentation" (Max-Planck-Institut für Wissenschaftsgeschichte, preprint 65, 1997), 25.
17. Albrecht von Haller, "Abhandlung über die Wirkung des Opiums auf den menschlichen Körper," *Berner Beiträge zur Geschichte der Medizin und der Naturwissenschaften* 19 (1962): 3–31. Simon Schaffer, "Self Evidence," *Critical Inquiry* 18 (1992): 327–362, esp. 336.
18. Boylston, *Historical Account*, vi. This continued into the modern period. Sir Douglas Black noted in 1973, "It's much easier to get the volunteer's cooperation for a study if you sit by the bedside and say that you have done the same experiment to yourself and are none the worse for it. And if you do it on yourself first, it gives you a wonderful chance to iron out any bugs." Lawrence Altman, *Who Goes First: The Story of Self-Experimentation in Medicine* (New York: Random House, 1987), 12.
19. Rolf Winau, "Experimentelle Pharmakologie und Toxikologie im 18. Jahrhundert" (Mainz: Habil. Schrift, 1971). Störck, *Essay*, 12–14.
20. Thomson, *Treatise*, 145–146.
21. Hérissant, "Experiments," 79.

22. Bourgeois, "Instructions," 119. Alathea Talbot published a book of 1720 "receipts and experiments" in 1655; reports of trials in patients are not included *(Natura Exenterata)*. Marie de Maupeou Fouquet was another to publish a book of recipes based, as she claimed, on experiments (*Recueil de receptes choisies* [Villefranche, 1675]). Lynette Hunter, "Women and Domestic Medicine: Lady Experimenters, 1570–1620," *Women, Science and Medicine*, ed. Hunter and Hutton, 89–107. [Anon.], *A Sovereign Remedy for the Dropsy* (London, 1783).
23. Maehle, *Drugs on Trial;* Brockliss and Jones, *Medical World*. Ryan complained that ethics are "totally neglected" in medical schools. Physicians are taught "the mysteries of medicine, without the slightest allusion to the duties they owe each other or the public" (*Manual*, 37–38). *Lettres à M. Moreau contre l'utilité de la transfusion* (Paris, 1667).
24. Hildebrandt, *Versuch*, 85. Risse, *Hospital Life*, 5. See also Guy, *Practical Observations*, xii; Susan Lawrence, *Charitable Knowledge: Hospital Pupils and Practitioners in Eighteenth-Century London* (Cambridge: Cambridge University Press, 1996), 237; Johanna Geyer-Kordesch, "Medizinische Fallbeschreibungen und ihre Bedeutung in der Wissensreform des 17. und 18. Jahrhunderts," *Medizin, Gesellschaft und Geschichte* 9 (1990): 7–19. Brockliss and Jones, *Medical World*, 730–782.
25. Paula Findlen, "Controlling the Experiment: Rhetoric, Court Patronage and the Experimental Method of Francesco Redi," *History of Science* 31 (1993): 35–64.
26. Miller, *Adoption of Inoculation*, 226; *Gentleman's Magazine* cited in Razzell, *Conquest*, viii. Charles-Marie de La Condamine, *Memoires pour servir à l'histoire de l'inoculation* (Paris, 1768), 71. Arthaud, *Memoire*, 11.
27. Mary Fissell, "Innocent and Honorable Bribes: Medical Manners in Eighteenth-Century Britain," in *The Codification of Medical Morality*, ed. Robert Baker, Dorothy Porter, and Roy Porter (Dordrecht: Kluwer, 1993), 19–46; also Gianna Pomata, *Contracting a Cure: Patients, Healers, and the Law in Early Modern Bologna* (Baltimore: Johns Hopkins University Press, 1998).
28. Dimsdale, *Present Method*, 4.
29. Withering, *Foxglove*, 3–4.
30. Hildebrandt, *Versuch*, 77. Rolf Winau, "Vom kasuistischen Behandlungsversuch zum kontrollierten klinischen Versuch," *Versuche mit Menschen in Medizin, Humanwissenschaft und Politik*, ed. Hanfried Helmchen and Rolf Winau (Berlin: Walter de Gruyter, 1986), 83–107. Denis Dodart, *Mémoires pour servir à l'histoire des plantes* (Paris, 1676), 10. Christian Wolff, *Disputatio philosophica de moralitate anatomes circa animalia viva occupatae* (Leipzig, 1709), 28–40.
31. Maupertuis, *Lettre*, section 11, "Utilités du supplice des criminels."
32. Francis Bacon, *Sylva sylvarum* (Leiden, 1648), book iv, experiment 400. Claude Bernard reported that in the sixteenth century the Grand Duke of Tuscany offered a condemned criminal to Gabriello Fallopius, professor of

anatomy at Pisa and, later, Padua, with permission to kill or dissect him at pleasure. Fallopius killed the man with an overdose of opium. Another criminal, known as the "archer of Meudon," successfully withstood a nephrotomy and was pardoned (*Introduction*, 100).

33. J. B. Denis, *Lettre écrite à Monsieur de Montmort* . . . (Paris, 1667); Farr, "The First," 160.
34. Maupertuis, *Lettre*. Hildebrandt reported a study of *Aconitum*, an extremely poisonous buttercup, on criminals in the late sixteenth century (*Versuch*, 77). Bynum, "Reflections," 32. Maitland, *Account*, 8. Andrew, *Practice*, vii. Anton von Störck also noted that smallpox induced by inoculation was "very advantageous" to the *beau sexe* because, unlike the natural smallpox, it did not mark their faces (*Traité*, 111).
35. See Nicholas Orme and Margaret Webster, *The English Hospital: 1070–1570* (New Haven: Yale University Press, 1995). John Leake, *An Account of the Westminster New Lying-in Hospital* (London, 1765), 1. See Amit Rai, *Rule of Sympathy: Sentiment, Race, and Power, 1750–1850* (New York: Palgrave, 2002), 33.
36. Risse, *Hospital Life*, 21–22. Brockliss and Jones, *Medical World*, 673–700.
37. Sigrun Engelen, "Die Einführung der Radix Ipecacuanha in Europa" (Ph.D. diss., Universität Düsseldorf, Institut für Geschichte der Medizin, 1967), 38–46. Monro, *Treatise*, vol. 3, 201. Charles Talbot, "America and the European Drug Trade," in *First Images of America: The Impact of the New World on the Old*, ed. Fredi Chiappelli, 2 vols. (Berkeley: University of California Press, 1976), vol. 2, 833–851, esp. 840. Lafuente, "Enlightenment in an Imperial Context," 161–162. See also Risse, "Transcending Cultural Barriers." John Hume, "An Account of the True Bilious, or Yellow Fever," in *Letters*, [Monro, ed.], 195–264. Pierre Pluchon, ed., *Histoire des médecins et pharmaciens de marine et des colonies* (Toulouse: Privat, 1985); McClellan, *Colonialism and Science*, 92–94, 128–129, 133–134.
38. Donald Hopkins, *Princes and Peasants: Smallpox in History* (Chicago: University of Chicago Press, 1983), 82, 224–225.
39. Schiebinger, *Has Feminism Changed Science?* chap. 6.
40. Brockliss and Jones, *Medical World*, 411. Wagstaffe, *Letter to Dr. Freind*, 4. Harold Cook makes the important point that military hospitals already in the seventeenth century were used to develop medical procedures and drugs that could be used in large populations without observing the Galenic niceties of lengthy preparations for individual constitutions ("Practical Medicine and the British Armed Forces after the 'Glorious Revolution,'" *Medical History* 34 [1990]: 1–26). William Bynum has similarly emphasized that historians must distinguish at least three types of therapeutics in the ancient and early modern worlds: the gentler, highly individualized diagnoses and therapies for the well-to-do, the quick and "resolutive" treatment for poorer free men, and mass-market treatments for slaves, especially in the ancient world ("Reflections on the History of Human Experimentation," in *The Use of Human Be-*

ings in Research, ed. Stuart Spicker et al. [Dordrecht: Kluwer, 1988], 29–46, esp. 32).

41. It remains unclear who initiated the Newgate experiments. Here I follow Sloane's "An Account of Inoculation," 517. Richard Mead also reported that the experiments were done "by order of his Sacred Majesty, both for the sake of his own family, and of his subjects" (*Works,* vol. 2, 145). Emanuel Timonius, "An Account, or History, of Procuring the Small Pox by Incision, or Inoculation," *Philosophical Transactions of the Royal Society of London* 29 (1714): 72–82, esp. 72. The notion that men are the subjects of medical trials is so pervasive that many historians have reported the Newgate prisoners to be six men, missing completely the fact that the experiment was designed to test equal numbers of men and women.

42. James Parsons, *A Description of the Human Urinary Bladder* (London, 1742); Jean-Dominique-Luc Ambialet, *Essai sur l'usage et l'abus du quinquina* (Montpellier, 1801), 31. "Some New Experiments of Injecting Medicated Liquors into Veins," *Philosophical Transactions of the Royal Society of London* 30 (1667): 564–565. Lowthrop, *Philosophical Transactions,* vol. 3, 234.

43. Lowthrop, *Philosophical Transactions,* vol. 3, 234.

44. Thomas Fowler, *Medical Reports of the Effects of Tobacco* (London, 1785), 72–79.

45. Philippe Pinel, *The Clinical Training of Doctors,* ed. and trans. Dora Weiner (1793; Baltimore: Johns Hopkins University Press, 1980), 78–79.

46. Fanny Burney, *Selected Letters and Journals,* ed. Joyce Hemlow (Oxford: Oxford University Press, 1986).

47. Störck, *Essay,* case III. Breast cancers were curiously sized at this time in terms of birds' eggs. We know that British women are still sometimes referred to as "birds" and that the emblem of feminine physical proportions in the eighteenth century was the ostrich, remarkable for is capacious pelvis and long, willowy neck. Schiebinger, *Mind,* chap. 7.

48. Störck, *Essay,* case XI, 49–52.

49. Guy, *Practical Observations,* xiii, 36–39. For belladonna cancer cures, see James, *Modern Practice of Physic;* and M. Marteau, "Observation sur la guérison d'un cancer à la mammelle," *Journal de médecine, chirurgie, pharmacie* 14 (1761): 11–27.

50. La Condamine, *History,* 15–16, 28.

51. Kay Dickersin and Yuan-I Min, "Publication Bias: The Problem that Won't Go Away," *Annals of the New York Academy of Sciences* 703 (1993): 135–148. Guy, *Practical Observations,* 6.

52. Cope, *Cheselden,* 24–25. Leake, *Lectures,* preface. On manikins, Gelbart, *King's Midwife.*

53. *The Jamaica Physical Journal* 1 (1834): 1. In the nineteenth century, Jamaican physicians such as Dancer and Thomson published their medical treatises in Jamaica. After 1765, students from the British West Indies sometimes

studied at the Medical School, College of Philadelphia (later University of Pennsylvania). K. R. Hill and I. S. Parboosingh, "The First Medical School of the British West Indies and the First Medical School of America," *West Indian Medical Journal* 1 (1951): 21–25. The Cercle des Philadelphes in Saint Domingue investigated all branches of science, including medicine. Their early publications include studies especially of the therapeutic use of mineral waters.

54. Thomson, *Treatise*, 151–156.
55. Quier inoculated 700 persons in 1768 and 146 in 1774. He also inoculated in 1773 but did not record the number (Monro, ed., *Letters*, 8, 64).
56. Sheridan, *Doctors and Slaves*, 66, 254–255; Heinz Goerke, "The Life and Scientific Works of Dr. John Quier," *West Indian Medical Journal* 5 (1956): 23–27.
57. Sheridan, *Doctors and Slaves*, 252.
58. [Monro, ed.], *Letters*, 43, 56. Soldiers were another population often inoculated by command, not choice (Paul Kopperman, "The British Army in North America and the West Indies, 1755–1783," manuscript).
59. [Monro, ed.], *Letters*, 8, 23–24, 65.
60. Ibid., 13, 17, 25. Mead remarked that most children, even West Indian slaves, were not inoculated until age five (*Works*, vol. 2, 146). James Grainger, a physician in Saint Kitts, risked inoculating his daughter in 1761 while she was still teething. John Nichols, *Illustrations of the Literary History*, 8 vols. (London, 1858), vol. 7, 277.
61. Dimsdale, *Present Method*, 21–22. *Memoir of the Late William Wright, M.D.* (Edinburgh, 1828), 340–341.
62. [Monro, ed.], *Letters*, 11, 54–56.
63. Ibid., 46, 67–69.
64. On slave midwives, see Moreau de Saint-Méry, *Description*, vol. 1, 22; Robert Thomas, surgeon in the islands of Saint Christopher and Nevis, testimony in *House of Commons*, ed. Lambert, vol. 71, 248. Beame cited in Bush, *Slave Women*, 139.
65. Michael Craton, *Searching for the Invisible Man* (Cambridge, Mass.: Harvard University Press, 1978), 218, 259–264.
66. Letter IV from Dr. Thomas Fraser, M.D. of Antigua to Dr. D. Monro, London, May 22, 1756 in [Monro, ed.], *Letters*, 110, 106–107.
67. Wagstaffe, *A Letter to Dr. Freind*.
68. On the issue of the interchangeability of black and white bodies, see Londa Schiebinger, "Human Experimentation in the Eighteenth Century: Natural Boundaries and Valid Testing," in *The Moral Authority of Nature*, ed. Lorraine Daston and Fernando Vidal (Chicago: University of Chicago Press, 2003), 384–408. Thomson, *Treatise*, 3.
69. Sloane, *Voyage*, vol. 1, x. Stedman cited in Bush, *Slave Women*, 142; Bancroft, *Essay*, 371–372; Descourtilz, *Flore pittoresque*, vol. 8, 284, 306, and 317; James, *Medicinal Dictionary*, s.v., "abortus"; Horatio Wood, *A Treatise on Therapeutics* (London, 1874), 537; Riddle, *Eve's Herbs*, 231.

70. See, for example, *Pharmacopoea Londinensis* (1618); *Pharmacopoea Amstelredamensis, senatus auctoritate munita, & recognita* (1636); *Codex Medicamentarius Parisiensis* (1638). The London pharmacopoeia of May 1618, for example, included 712 compound remedies arranged in pharmaceutical classes: simple waters, decoctions, electuaries, lozenges, ointments, etc. It also included 680 simples, classed according to type, for example roots, leaves, gums and resins, animal parts, salts, and metals. M. P. Earles, *The London Pharmacopoeia Perfected* (London: Chameleon Press, 1985), 13, 15; George Urdang, "Pharmacopoeias as Witnesses of World History," *Journal of the History of Medicine and Allied Sciences* 1 (1946): 46–70; John Abraham, *Science, Politics, and the Pharmaceutical Industry: Controversy and Bias in Drug Regulation* (New York: Saint Martin's, 1995), 38; Georges Dillemann, "Les Remèdes secrets et la réglemention de la pharmacopée Française," *Revue d'histoire de la pharmacie* 23 (1976): 37–48; David Cowen, "Colonial Laws Pertaining to Pharmacy," *American Pharmaceutical Association* 23 (1934): 1236–1243. [Chandler], *Frauds Detected*, 1–2, 5–11.
71. Cowen, *Pharmacopoeias*, 5. Monro, *Treatise*, vol. 2, 444–446; Joseph Pitton de Tournefort, *Materia Medica; or, a Description of Simple Medicines Generally us'd in Physick* (London, 1708); Georg Ernst Stahl, *Materia medica* (Dresden, 1744); James, *Modern Practice of Physic*, summarizing the work of Van Swieten, Hoffman, Boerhaave; Andry, *Matière médicale;* Löseke, *Materia medica*. Griffith Hughes, *The Natural History of Barbados* (London, 1750), 201. Long, *History*, vol. 3, 815–816. Descourtilz, *Guide sanitaire*, 166.
72. Chevalier, *Lettres*, 111–117.
73. *Memoir of the Late William Wright, M.D.* (Edinburgh, 1828), 85, 183, 270.
74. Sloane, *Voyage*, vol. 1, xx.
75. *Registres du Comité de Librairie* (March 1763), vol. 1, 122.
76. Jürgen Schlumbohm, "'The Pregnant Women are Here for the Sake of the Teaching Institution': The Lying-In Hospital of Göttingen University, 1751–1830," *Social History of Medicine* 14 (2001): 59–78.
77. Antoine Arnault, *Existe-t-il des agents emménagogues?* (Paris, 1844). Today women regulate their periods at various stages of life with birth control pills, contraceptive injections, hysterectomy, and other drugs and medical procedures. Physicians are unable to say what proportion of women employ menstrual regulators because each one has several uses (birth control pills regulate bleeding as well as pregnancy; hysterectomy ends excessive bleeding but also serves as a cancer treatment). Gianna Pomata, "Menstruating Men: Similarity and Difference of the Sexes in Early Modern Medicine," in *Generation and Degeneration: Tropes of Reproduction in Literature and History from Antiquity through Early Modern Europe*, ed. Valeria Finucci and Kevin Brownlee (Durham: Duke University Press, 2001), 109–152. Dancer, *Medical Assistant*, 263; Monardes, *Joyfull Newes*, 13–14.
78. Descourtilz, *Flore pittoresque*, vol. 8, 276. This "long train of evils," from Descourtilz, *Flore*, vol. 8, 276; J. B. Chomel, *Abrégé de l'histoire des plantes*

usuelles, 3 vols. (Paris, 1738), vol. 1, 146; and Freind, *Emmenologia*, 78. Jussieu, *Traité*, 339–356; Smellie, *Encyclopedia Britannica*, s.v. "emmenagogues."

79. Angus McLaren, *Reproductive Rituals: The Perception of Fertility in England from the Sixteenth Century to the Nineteenth Century* (London: Methuen, 1984), 102–106. Riddle, *Contraception* and *Eve's Herbs;* Shorter, *Women's Bodies*.
80. Diderot and d'Alembert, eds., *Encyclopédie*, s.v. "emmenagogue"; Zedler, *Universal Lexicon*, s.v. "emmenagoga"; Carl Linnaeus, *Materia medica* (Leipzig, 1782), 225–226, 248. See also Hermann, *Materia medica* and Gerard's *Herbal*.
81. Dionis, *Traité*, 419.
82. On emmenagogues, see Jussieu, *Traité*, 339–356. Monardes, *Joyfull Newes*, 13–14. Guenter Risse, "Medicine in New Spain," in *Medicine in the New World*, ed. Ronald Numbers (Knoxville: University of Tennessee Press, 1987), 12–63. Sloane, *Voyage: Sesamum veterum* (vol. 1, 161), vanilla (vol. 1, 180), a certain pepper (vol. 1, 242), *Lobus echinatus* (vol. 2, 41), and aloe (vol. 2, 379). Alibert, *Nouveaux élémens*, vol. 3, 69. *Mémoire sur les plantes médicinales de Saint-Domingue*, Ms. 1120, Bibliothèque centrale, MNHN, Paris. Woodville, *Medical Botany*, vol. 1, 22; Ainslie, *Materia medica*, 4; *Pharmacopoeia Collegii Regalis Medicorum Londinensis* (London, 1836).
83. William Buchan, *Domestic Medicine; or, the Family Physician* (Philadelphia, 1774), 393; Étienne Geoffroy, *A Treatise on Foreign Vegetables* (London, 1749), 81.
84. Freind, *Emmenologia*, 68–69, 73.
85. Ibid., 179–184.
86. Ibid., 190–195.
87. Ibid., 128–131. Freind provided recipes for the four different emmenagogues he gave her.
88. Ibid., 143–145.
89. Cullen, *Treatise*, vol. 2, 365–366, 566–587. John O'Donnell, "Cullen's Influence on American Medicine," in *William Cullen and the Eighteenth Century Medical World*, ed. A. Doig, J. P. S. Ferguson, I. A. Milne, and R. Passmore (Edinburgh: Edinburgh University Press, 1993), 234–251, esp. 241.
90. Francis Home, *Clinical Experiments, Histories, and Dissections* (London, 1782), 410–421. See also Woodville, *Medical Botany*, vol. 2, 256–258.
91. William Lewis, *An Experimental History of the Materia Medica* (London, 1784), 548; Monro, *Treatise*, vol. 3, 243; Henry Beasley, *The Book of Prescriptions* (London, 1856), 441.
92. Diderot and d'Alembert, eds., *Encyclopédie*, s.v. "savine." Adelon et al., eds., *Dictionaire*, vol. 2, s.v. "avortement," 489. Ryan, *Manual*, 154; Löseke, *Materia medica*, 387–388.
93. Zedler, *Universal Lexicon*, s.v. "Sadebaum." Cullen, *Treatise*, vol. 1, 161.

Brande, *Dictionary*, 357. See also Anthony Todd Thomson, *The London Dispensatory* (London, 1815), 251–252; *Medical Botany: Or, History of Plants in the Materia Medica of the London, Edinburgh, & Dublin Pharmacopoeias* (London, 1821), 100. Alibert, *Nouveaux élémens,* vol. 3, 71–72. Smellie cited in Cotte, *Considérations médico-légales.*

94. Astruc, *Traité,* vol. 5, 326–327. Tardieu, *Étude,* 2.
95. Frank, *System,* vol. 2, 64–67. Gabriel-François Venel, *Précis de matière médicale,* 2 vols. (Paris, 1787), vol. 1, 299. Philippe Vicat, *Matière médicale tirée de Halleri,* 2 vols. (Bern, 1776), vol. 2, 282–284.
96. Hélie, "De l'Action vénéneuse de la rue," 184–185.
97. John Burns, *Observations on Abortion* (London, 1806), 59–65; Jean Romain, *Dissertation sur l'avortement ou fausse-couche* (Montpellier, 1819), 5–7.
98. Ollivier d'Angers, "Mémoire et consultation médico-légale sur l'avortement provoqué," *Annales d'hygiène publique et de médecine légale* 22 (1839): 109–133. Tardieu, *Étude* (1864 ed.), 8. Hélie, "De l'Action vénéneuse de la rue," 217.
99. Tardieu, *Étude,* avertissement. Article 317 of the Napoleonic Code stated that anyone who induced an abortion in a pregnant woman with or without her consent by means of food, drugs, medications, violence, or other means shall be punished in prison or forced labor. Johann Andreae Murray, *Apparatus Medicaminum,* 6 vols. (Göttingen, 1793). Brande, *Dictionary,* 470–471. Hélie, "De l'Action vénéneuse de la rue," 183.
100. Johann Peter Frank and others had argued in favor of abolishing this notion already in the mid eighteenth century (*System,* vol. 2, 84–122). Günter Jerouschek, "Zur Geschichte des Abtreibungsverbots," *Unter anderen Umständen: Zur Geschichte der Abtreibung,* ed. Gisela Staupe and Lisa Vieth (Dresden: Deutsches Hygiene-Museum, 1993), 11–26. Burton, "Human Rights," 427–438. *Encyclopaedia Britannica,* 11th ed. (1910), s.v. "abortion." The Offences Against the Person Act of 1861 (section 58) made it a felony punishable with life imprisonment for anyone, including the woman herself, unlawfully to procure an abortion. Riddle, *Eve's Herbs,* 209, 224.
101. Ersch and Gruber, *Encyclopädie,* s.v. "Abtreibung."
102. Adelon et al., eds., *Dictionaire,* vol. 2, s.v. "avortement," 502–503.
103. W. S. Glyn-Jones, *The Law Relating to Poisons and Pharmacy* (London: Butterworth, 1909), 172.
104. Cotte, *Considérations médico-légales,* 3–4. Shorter, *Women's Bodies,* 209. *Plantes medicinales et phytotherapie* 23 (1989): 186–192.
105. Tanfer Emin, "Technological Change in Pregnancy Termination, 1850–1980," SUNY Stony Brook, Dept. of History, manuscript.
106. See *Pharmacopoea Amstelredamensis* (Amsterdam, 1651); *Pharmacopoea Leidensis* (Leiden, 1718); *Pharmacopoeia Collegii Regalis Medicorum Londinensis* (London, 1721); *Pharmacopoeia Collegii Regii Medicorum Edinburgensis* (Edinburgh, 1722); Peter Shaw, *The Dispensatory of the Royal College of Physicians in Edinburgh* (London, 1727); *The Dispensatory of the*

Royal College of Physicians in London, 2nd ed. (London, 1727); *Pharmacopoea Amstelredamensis* (Amsterdam, 1731); *Codex Medicamentarius, seu Pharmacopoea Parisisensis* (Paris, 1732); *Pharmacopoeia Augustana renovata* (Augustae, 1734); *The British Dispensatory* (London, 1747); *Codex Medicamentarius, seu Pharmacopoea Parisiensis* (Frankfurt am Main, 1760); *Pharmacopoeia Collegii Regalis Medicorum Londinensis* (Paris, 1788); *Pharmacopoea Amstelodamensis Nova* (Amsterdam, 1792); *Pharmacopoea Austriaco-Provincialis* (Milan, 1794); John Thomson, *The Pharmacopoeias of the London, Edinburgh, and Dublin Colleges* (Edinburgh, 1815); William Barton, *Vegetable Materia Medica of the United States; or Medical Botany*, 2 vols. (Philadelphia, 1817); *Codex medicamentarius, sive pharmacopoea gallic* (Paris, 1818); *Medical Botany: Or, History of Plants in the Materia Medica of the London, Edinburgh, & Dublin Pharmacopoeias* (London, 1821); and *Pharmacopoeia Collegii Regalis Medicorum Londinensis* (London, 1836).

5. Linguistic Imperialism

1. Linnaeus, *Critica botanica*, preface, no. 213.
2. Stafleu, *Linnaeus;* John Heller, *Studies in Linnaean Method and Nomenclature* (Frankfurt: Verlag Peter Lang, 1983); Tore Frängsmyr, ed., *Linnaeus: The Man and His Work* (Berkeley: University of California Press, 1983); G. Perry, "Nomenclatural Stability"; Dirk Stemerding, *Plants, Animals, and Formulae: Natural History in the Light of Latour's Science in Action and Foucault's The Order of Things* (Enschede: School of Philosophy and Social Sciences, University of Twente, 1991); Schiebinger, *Nature's Body*, chap. 1; Koerner, *Linnaeus*, chap. 2.
3. Craton and Walvin, *Jamaican Plantation*, 148–149; Jerome Handler and JoAnn Jacoby, "Slave Names and Naming in Barbados, 1650–1830," *William and Mary Quarterly* 53 (1996): 685–728; Trevor Burnard, "Slave Naming Patterns: Onomastics and the Taxonomy of Race in Eighteenth-Century Jamaica," *Journal of Interdisciplinary History* 31 (2001): 325–346.
4. Garrigus, "Redrawing the Color Line," 38.
5. Swartz reclassified the *Poinciana* under *Caesalpinia* in the 1790s; both names are used today (*Observationes*, 165–166).
6. Linnaeus, Letter to Baeck, cited in Nicolas, "Adanson, the Man," *Adanson*, ed. Lawrence, 51.
7. Humboldt (and Bonpland), *Personal Narrative*, vol. 5, 208. Barrère, *Nouvelle Relation*, 39.
8. Jorge Cañizares-Esguerra, "Spanish America: From Baroque to Modern Colonial Science," in *Science in the Eighteenth Century*, ed. Roy Porter (Cambridge: Cambridge University Press, in press), 729; Lafuente and Valverde, "Linnaean Botany and Spanish Imperial Biopolitics."
9. Michel Foucault, *The Order of Things: An Archaeology of the Human Sciences*

(1966; New York: Vintage, 1973), 63–67. Foucault states, "the sign is the pure and simple connection between what signifies and what is signified (a connection that may be arbitrary or not . . .)." On this point, see Staffan Müller-Wille, *Botanik und weltweiter Handel: Zur Begründung eines natürlichen Systems der Pflanzen durch Carl von Linné (1707–1778)* (Berlin: VWB, 1999), chap. 5. See also Gordon McOuat, "Species, Rules and Meaning: The Politics of Language and the Ends of Definitions in 19th Century Natural History," *Studies in the History and Philosophy of Science* 27 (1996): 473–519. Jackson, "New Index."

10. Stearn, "Background," 5. See also Edward Lee Greene, *Landmarks of Botanical History* (Washington, D.C.: Smithsonian, 1909); Linnaeus, *Critica botanica*, no. 256.
11. Gerard, *Herbal*, 843–845. Christian Mentzelius' *Index nominum plantarum multilinguis* (Berlin, 1682). See Jerry Stannard, "Botanical Nomenclature in Gersdorff's Feldtbüch der Wundartzney," in *Science, Medicine, and Society in the Renaissance*, ed. Allen Debus (New York: Science History Publication, 1972), 87–103; Brian Ogilvie, "The Many Books of Nature: How Renaissance Naturalists Created and Responded to Information Overload," History of Science Society Meeting, Vancouver, 2000.
12. McVaugh, *Botanical Results*, 19. See also Simon Varey, ed., *The Mexican Treasury: The Writings of Dr. Francisco Hernández* (Stanford: Stanford University Press, 2000). Rochefort, *Histoire naturelle*, 104–106. Charles Plumier, *Description des plantes de l'Amérique* (Paris, 1693). Reede, *Hortus*. Barrère, *Essai*. Pouppé-Desportes, *Historie des maladies*.
13. Linnaeus to Haller, cited in *Critica botanica*, vii–viii; no. 218, 229. In reference to Merian, Linnaeus wrote: "How absurd is the procedure of many naturalists in regard to insects, who give an account of them and show us pictures, but do not give any names . . . Wherefore I much prefer the barbarous names of the 'Hortus Malabaricus' to the absence of names in Merian's account of the plants of Surinam." Linnaeus' insistence on Greek and Latin roots stands in contradistinction to more recent international codes of botanical nomenclature that allow "the genus name . . . [to] be taken from any source whatever." *International Code of Botanical Nomenclature*, ed. W. Greuter (Konigstein: Koeltz Scientific Books, 1988).
14. Linnaeus, *Critica botanica*, no. 229; Stearn, *Botanical Latin*, 6–7.
15. Linnaeus, *Critica botanica*, no. 241.
16. Ibid., no. 240. Schiebinger, *Nature's Body*, chap. 2.
17. Linnaeus, *Critica botanica*, nos. 238, 240.
18. For Linnaeus' list of "names commemorating distinguished botanists," see ibid., no. 238, 11. Linnaeus' knowledge of several insect species came exclusively from Merian's work. Olof Swartz, *Flora Indiae Occidentalis*, 2 vols. (London, 1797–1806), s.v. "*Meriania purpurea.*" Wettengl, "Maria Sibylla Merian," 13.

19. Linnaeus, *Critica botanica*, no. 238.
20. Ibid.
21. Schiebinger, *Mind*. Linnaeus, *Critica botanica*, nos. 218, 229, 238.
22. Ibid., nos. 237, 238.
23. Ibid., no. 236.
24. John Briquet, *Règles internationales de la nomenclature botanique* (Jena: Fischer Verlag, 1906). Jackson, "New Index," 151.
25. Stearn, "Background," 5. Savage cited in Linnaeus, *Species plantarum*, Stearn's intro., 39–40. See also Aldo Pesante, "About the Use of Personal Names in Taxonomical Nomenclature," *Taxon* 10 (1961): 214–221.
26. Sloane, *Voyage*, preface.
27. Ronald King in Robert Thornton, *The Temple of Flora* (1799; Boston: New York Graphic Society, 1981), 9; Heinz Goerke, *Linnaeus*, trans. Denver Lindley (New York: Scribner, 1973), 108.
28. E. G. Voss, ed., *International Code of Botanical Nomenclature* (Utrecht: Bohn, Scheltema, and Holkema, 1983), 1.
29. Perry, "Nomenclatural Stability," 81.
30. S. L. Van Landingham, "The Naming of Extraterrestrial Taxa," *Taxon* 12 (1963): 282.
31. Sloane, *Voyage*, vol. 1, xlvi.
32. Peter Kolb, *The Present State of the Cape of Good Hope*, trans. Guido Medley (London, 1731). Merian, *Metamorphosis*, intro. 38.
33. This Latin term was used in both the Dutch first edition and the Latin translation.
34. Heniger, *Hendrik Adriaan van Reede*, 162; Breyne, *Exoticarum*, 61–64. Van Reede associated the *crista pavonis* and *tsjétti-mandáru*, but Commelin noted that they are somewhat different. Wijnands, *The Botany of the Commelins*, 59.
35. Fermin, *Description générale*, vol. 1, 218. Roger, *Buffon*, 275–278.
36. Sloane, *Catalogus plantarum*, 149. The *Tacoxiloxochitl* is identified today as the *Calliandra anomala*. Pouppé-Desportes, *Historie des maladies*, vol. 3, 207.
37. Merian, *Metamorphosis*, plate 45. Reede, *Hortus*, vol. 6, 1–2. This book was edited by Caspar Commelin's uncle Jan from 1678 until his death in 1692; Caspar himself prepared an analytical index, which he published separately as *Flora Malabarica sive Horti Malabarici Catalogus* (Leiden, 1696). Hermann, *Horti academici*, 192. See Dan Nicolson, C. Suresh, and K. Manilal, *An Interpretation of van Rheede's Hortus Malabaricus* (Königstein: Koeltz, 1988), 126. On the scripts used in the *Hortus Malabaricus*, see Heniger, *Hendrik Adriaan van Reede*, 148–149.
38. The name *Poincyllane* is recorded by du Tertre and its derivative brought into systematic botany by Tournefort. Du Tertre judged it the most beautiful flower in the French islands; it was also called "Fleur de Saint Martin" (*Histoire*, vol. 2, 154); *Hortus Regius* (Paris, 1666), 3; Tournefort, *Élémens*, vol.

1, 491–492; vol. 3, plate 391. Carl Linnaeus, *Hortus Cliffortianus* (Amsterdam, 1737), 158. Swartz, *Observationes,* 166. Linnaeus ranked van Reede's *Hortus* as one of the two greatest works contributing to his systematics (the other was the Oxford botanist Johann Jakob Dillenius' *Hortus Elthamensis*). Despite this accolade, the wealth of culturally local knowledges embodied in van Reede's project was not embraced by European academic botanists. Carl Linnaeus, *Genera plantarum,* 5th ed. (Stockholm, 1754), xii.
39. Merian, *Metamorphosis,* intro., 38.
40. Jean-Pierre Clement, "Des noms de plantes," *Nouveau monde et renouveau de l'histoire naturelle,* ed. Marie-Cécile Bénassy-Berling, 3 vols. (Paris: Presses de la Sorbonne nouvelle, 1986–1994), vol. 2, 85–109.
41. Olarte, *Remedies,* 116–118. Lopez Ruiz and José Pavón, *Florae Peruvianae et Chilensis* (Madrid, 1794).
42. Donal McCracken, *Gardens of Empire: Botanical Institutions of the Victorian British Empire* (London: Leicester University Press, 1997), 160.
43. D. J. Mabberley, "The Problem of 'Older' Names," in *Improving the Stability of Names,* ed. Hawksworth, 123–134.
44. Stedman, *Stedman's Surinam,* 301. Marcel Dorigny and Bernard Gainot, *La Société des amis des noirs, 1788–1799* (Paris: Unesco/Edicef, 1998). Linnaeus, *Amoenitates academicae,* s.v. "Carl Magnus Blom, *Lignum Quassiae,* 1763." In a letter to Baeck, Linnaeus wrote that Dahlberg overestimated his botanical contributions (*Bref och skrifvelser af och till Carl von Linné utgifna af Upsala Universitet,* part V [Stockholm: Regia Academia Upsaliensis, 1911], 127). Rolander went mad while in Surinam (Koerner, *Linnaeus,* 86).
45. Cullen cited in Woodville, *Medical Botany,* vol. 2, 215–217. See *Pharmacopoea Amstelodamensis Nova* (Amsterdam, 1792).
46. Fermin, *Description générale,* vol. 1, 212–213. Stedman, *Narrative,* 581–582.
47. Nassy, *Essai historique,* 73.
48. Linnaeus, *Amoenitates academicae,* s.v. "Cortice Peruviano," Pt. I Job. Christ. Pet Petersen, 1758. Clements Markham, *A Memoir of the Lady Ana de Osorio, Countess of Chinchon and Vice-Queen of Peru* (London, 1874); Haggis, "Fundamental Errors."
49. La Condamine, "Sur l'arbre du quinquina," 336.
50. Jussieu, *Description.* Jean-Étienne Montucla, *Recueil de pièces concernant l'inoculation de la petite vérole et propres à en prouver la sécurité et l'utilité* (Paris, 1756), 148.
51. La Condamine, "Sur l'arbre du quinquina"; Jussieu, *Description.* Jaramillo-Arango, *Conquest,* 34; Jarcho, *Quinine's Predecessor;* Maehle, *Drugs on Trial.*
52. La Condamine, "Sur l'arbre du quinquina." Stéphanie Félicité, comtesse de Genlis, *Zuma: ou, La Découverte du quinquina* (Paris, 1818).
53. Thomas Skeete, *Experiments and Observations on Quilled and Red Peruvian Bark* (London, 1786), 2.
54. Schiebinger, *Nature's Body,* chap. 1. Gianna Pomata, "Close-Ups and Long

Shots: Combining Particular and General in Writing the Histories of Women and Men," in *Geschlechtergeschichte und Allgemeine Geschichte,* ed. Hans Medick and Anne-Charlott Trepp (Göttingen: Wallstein, 1998), 101–124. Schiebinger, *Mind,* chap. 2.
55. Haggis, "Fundamental Errors."
56. Thunberg, *Travels,* vol. 2, 132. See also Shteir, *Cultivating Women,* 48–50.
57. Linnaeus cited in Blunt, *Compleat Naturalist,* 224.
58. Yves Laissus, "Catalogue des manuscrits de Philibert Commerson," *Revue d'histoire des sciences* 31 (1978): 131–162; Monnier et al., *Commerson,* 99, 109–111.
59. John Edward Smith, *A Grammar of Botany* (London, 1826), 51. C. Váczy, "Hortus Indicus Malabaricus and its Importance for Botanical Nomenclature," in *Botany and History of Hortus Malabaricus,* ed. K. Manilal (Rotterdam: Balkema, 1980), 25–34, esp. 30–31. According to Manilal's count, sixty-one generic and seventy-eight specific names used in botanical nomenclature today have Malayalam roots ("Malayalam Plant Names from Hortus Malabaricus in Modern Botanical Nomenclature," ibid., 70–76). See William Roxburgh, *Plants of the Coast of Coromandel* (London, 1795–1819).
60. Linnaeus, *Critica botanica,* no. 231.
61. Descourtilz, *Flore pittoresque,* vol. 8, 334.
62. Thomas Martyn, cited in David Allen, *The Naturalist in Britain: A Social History* (Harmondsworth: Penguin, 1978), 39. Adanson, *Familles des plantes,* vol. 1, iii–cii.
63. Georges-Louis Leclerc, comte de Buffon, *L'Histoire naturelle, générale et particulière* (Paris, 1749), vol. 1, 23–25. See Roger, *Buffon,* 275–278. Buffon, *Histoire naturelle,* vol. 1, 8, 13–14, 16–18, 26.
64. Roger, *Buffon,* 275.
65. Ibid., 275–278. Félix Vicq-d'Azyr, *Traité d'anatomie et de physiologie* (Paris, 1786), vol. 1, 47–48.
66. Schiebinger, *Nature's Body,* 13–18.
67. Adanson, *Familles des plantes,* vol. 1, xl–xli, cxlix–clii, clxxiii–clxxiv.
68. Ibid., clxxiii.
69. Ibid., cxxiii, cxlix; vol. 2, 318. Stafleu, "Adanson and the 'Familles des plantes,'" 187. Adanson, *Familles des plantes,* vol. 1, clxxi, clxxiii. Nicolas, "Adanson, the Man," 30.
70. Nicolas, "Adanson, the Man," 57. Baeck cited in Stafleu, "Adanson and the 'Familles des plantes,'" 176.
71. Linnaeus, *Amoenitates academicae,* "Incrementa botanices," Jac. Bjuur. Wman., 1753. Linnaeus, *Species plantarum,* Stearn's intro., 11.
72. Adanson was not alone in attempting to reform human language. Volney, as Leibniz before him, wanted to simplify the existing languages by reducing them to a more reasonable, simple grammar and a generally applicable system of writing (Stagl, *History of Curiosity,* 288). Nicolas, "Adanson, the Man," 102.

73. Sir William Jones, "The Design of a Treatise on the Plants of India," *Asiatick Researches* 2 (1807): 345–352. Italics in original.
74. Ainslie, *Materia Medica*, preface.
75. Lafuente and Valverde, "Linnaean Botany and Spanish Imperial Biopolitics." Drayton, *Nature's Government*, 77; J. Fr. Michaud, *Biographie Universelle* (Graz: Akademische Druck, 1966), s.v. "Aublet."
76. Frans Stafleu, "Fifty Years of International Biology," *Taxon* 20 (Feb. 1971): 141–151, esp. 146. Sydney Gould and Dorothy Noyce, *Authors of Plant Genera* (Saint Paul, Minn.: H. M. Smyth, 1965).
77. Grove, *Green Imperialism*, 78, 89–90. Albert Memmi, *The Colonizer and the Colonized*, trans. Howard Greenfeld (New York: Orion Press, 1965), 98.
78. Walters has presented a graph showing that large genera (more than twenty-five species) originated before 1800, medium genera (six to twenty-five species) took shape between 1800 and 1850, and small genera (two to five species) came into being around 1850–1900 ("Shaping," 78). See also Cain, "Logic and Memory."
79. Stearn, "Linnaeus's Acquaintance," 777.

Conclusion: Agnotology

1. Robert Proctor has distinguished three types of ignorance: "ignorance as native state" (the blank slate or absence of knowledge), "ignorance as lost realm" (knowledge selectively lost or repressed), and "ignorance as active construct" (intentional production of doubt, as in the case of the tobacco industry's "doubt [about the hazards of smoking] is our product"). Robert Proctor, "Agnotology: A Missing Term to Describe the Study of the 'Cultural Production of Ignorance,'" (manuscript). Engstrand, *Spanish Scientists*, 3.
2. Stearn, "Linnaeus's Acquaintance."
3. Mary Gunn and L. E. Codd, *Botanical Exploration of Southern Africa* (Cape Town: A. A. Balkema, 1981), 25.
4. Rouse, *Tainos*, 3–4. Private communication, Lee Ann Newsom, Department of Anthropology, Pennsylvania State University. See also Charles Gunn and John Dennis, *World Guide to Tropical Drift Seeds and Fruits* (New York: Quadrangle, 1976).
5. Stearn, "Botanical Exploration," 193; Heniger, *Hendrik Adriaan van Reede tot Drakenstein*, 76–77.
6. Ligon, *History*, 15; Miller, *Gardener's Dictionary*, s.v. *Poinciana (pulcherrima)*.
7. Judith Carney, *Black Rice: The African Origins of Rice Cultivation in the Americas* (Cambridge, Mass.: Harvard University Press, 2001). Many African foods were brought aboard slave ships as food for the passage. Long notes that, in addition to slaves, "African traders" often carried "some valuable drugs" from Africa to the New World (*History*, vol. 1, 491). Dancer also

noted that "persons of the Jewish Nation" introduced dates and palm oil into the island *(Catalogue of Plants)*. Broughton, "Hortus Eastensis."

8. Edward Ayensu, *Medicinal Plants of West Africa* (Algonac, Mich.: Reference Publications, 1978); Maurice Iwu, *Handbook of African Medicinal Plants* (Boca Raton: CRC Press, 1993); Beb Oliver-Bever, *Medicinal Plants of Tropical West Africa* (Cambridge: Cambridge University Press, 1986). Slaves, of course, came from very different parts of Africa, each with its own customs and medical traditions. On the origins of slaves in Surinam, see Stedman, *Stedman's Surinam,* 96; for the French West Indies, see Debien, *Esclaves,* 39–68. Hans Neuwinger, *African Ethnobotany* (London: Chapman & Hall, 1996), 321–324. Barbara Bush discusses some continuities between abortive practices in Africa and Caribbean slave societies in "Hard Labor: Women, Childbirth, and Resistance in British Caribbean Slave Societies," in *More than Chattel,* ed. Gaspar and Hines, 193–217, esp. 204–206. Debien, *Esclaves,* 364–365.

9. Augustin-Pyrame de Candolle, *Prodromus systematis naturalis regni vegetabilis,* 17 vols. (Paris, 1824–1873), vol. 2, 484; M. D. Dassanayake and F. R. Fosberg, eds., *Flora of Ceylon,* 8 vols. (New Delhi: Amerind Publishing, 1980–1994), vol. 7, 46–48. It is today known as an abortifacient in the Dutch East Indies. John Watt and Maria Breyer-Brandwijk, *The Medicinal and Poisonous Plants of Southern and Eastern Africa* (Edinburgh: Livingstone, 1962), 564.

10. Keegan, "The Caribbean," 1269–1271. Dancer, *Catalogue of Plants.* Broughton, "Hortus Eastensis," 481.

11. Bancroft, *Essay,* 371–374.

12. Ibid., 52–53, 371–372. Sheridan, *Doctors and Slaves,* 95.

13. Descourtilz, *Flore pittoresque,* vol. 8, 279–284, 304–306, 317.

14. Debien, *Esclaves,* 364.

15. David Watts, *Man's Influence on the Vegetation of Barbados, 1627–1800* (Hull, Yorkshire: University of Hull Occasional Papers in Geography, 1966), 46. Hilliard d'Aubertuil, *Considérations,* vol. 2, 50; Ligon, *History,* 99. "Of a Letter, sent lately to Robert Moray out of Virginia," *Philosophical Transactions of the Royal Society of London* 1 (1665–1666), 201–202. James Grainger wrote that every owner of an estate ought to have the following medicines sent to him annually from England: "Spanish Flies, Castor, Calcined Hartshorn, Spirit of Hartshorn, Sal Volatile Drops, Cloves, Oil of Cinnamon, Ipecacuana, Jalap, Opium, Nutmegs, Rhubarb, Spirit of Lavender, Tinctura Thebaic, Alum, Common Caustic, Crude Mercury, Corrosive Sublimate, Oil of Turpentine, Plaster, common, Turner's Cerate, Verdigrease, Vitriol" (*Essay,* 95). McClellan, *Colonialism and Science,* 148.

16. Gelbart, *King's Midwife,* 91. Joseph Raulin, *De la conservation des enfans,* 3 vols. (Paris, 1768), vol. 1, "épitre au roi."

17. Grew cited in Miller, *Adoption of Inoculation,* 226. See also Cole, *Colbert,*

vol. 2, 41–45; Eli Heckscher, *Mercantilism*, 2 vols. (London: George Allen & Unwin, 1935), vol. 2, 160–161.
18. Hilliard d'Auberteuil, *Considérations*, vol. 2, 50. *Gentleman's Magazine* cited in Razzell, *Conquest*, viii. Le Boursier Du Coudray, *Abrégé*, ii, viii, 3, 13.
19. Dazille, *Observations sur les maladies des negres*, vol. 1, 1–2.
20. Mackay, *In the Wake of Cook*, 123–143. Hilliard d'Auberteuil, *Considérations*, vol. 2, 58. For the value of the *carreau*, see McClellan, *Colonialism and Science*, xvii.
21. Arthaud, *Observations*, 74, 78–79. Jacques Gelis, "Obstétrique et classes sociales en milieu urbain aux XVIIe et XVIIIe siècles: Évolution d'une pratique," *Histoire des sciences médicales* 14 (1980): 425–433, esp. 429. Ordnance cited in Médéric-Louis-Élie Moreau de Saint-Méry, *Loix & constitutions des colonies Françoises de l'Amérique sous le vent*, 6 vols. (Paris, 1784–1790), vol. 1, 4–6, vol. 4, 222–223, 837.
22. *Second Supplément au Code de La Martinique* (Saint-Pierre, 1786), 357–358.
23. Hilliard d'Auberteuil, *Considérations*, vol. 2, 52. Long, *History*, vol. 1, 570; vol. 2, 293.
24. Humboldt (and Bonpland), *Personal Narrative*, vol. 5, 28–32. Frank also reported that women killed their infants because they feared poverty (*System*, vol. 2, 13).
25. Humboldt (and Bonpland), *Personal Narrative*, vol. 5, 28–32.
26. The roots of the *Trichilia hirta*, known as "jobobán" in the Dominican Republic, are also used today as an abortive. Private communication, Lee Ann Newsom.

Bibliography

Adair, James Makittrick. *Unanswerable Arguments against the Abolition of the Slave Trade.* London, [1790].
Adanson, Michel. *Familles des plantes.* 2 vols. Paris, 1763.
——— *A Voyage to Senegal.* London, 1759.
Adelon, Nicolas Philibert, et al., eds. *Dictionaire des sciences médicales.* 60 vols. Paris, 1812–1822.
Ainslie, Whitelaw. *Materia Medica of Hindoostan, and Artisan's and Agriculturist's Nomenclature.* Madras, 1813.
Aiton, William. *Hortus Kewensis.* London, 1789.
Alibert, Jean-Louis-Marie. *Nouveaux élémens de thérapeutique et de matière médicale.* 3 vols. Paris, 1826.
Alston, Charles. *Lectures on Materia Medica.* 2 vols. London, 1770.
Andry, Charles-Louis-François. *Matière médicale extraite des meilleurs auteurs.* 3 vols. Paris, 1770.
[Anon.]. *Histoire des désastres de Saint-Domingue.* Paris, 1795.
Aristotle. *History of Animals.* Ed. and trans. D. M. Balme. Cambridge, Mass.: Harvard University Press, 1991.
Astruc, Jean. *Traité des maladies des femmes.* 6 vols. Paris, 1761–1765.
Arthaud, Charles. *Memoire sur l'inoculation de la petite vérole.* Cap-Français, 1774.
——— *Observations sur les lois.* Cap-Français, 1791.
Aublet, Jean-Baptiste-Christophe. *Histoire des plantes de la Guiane Françoise, rangées suivant la méthode sexuelle.* 4 vols. London and Paris, 1775.
Balick, Michael, and Paul Alan Cox. *Plants, People, and Culture: The Science of Ethnobotany.* New York: Scientific American Library, 1996.
Bancroft, Edward. *An Essay on the Natural History of Guiana in South America.* London, 1769.
Barker, Francis, Peter Hulme, and Margaret Iversen, eds. *Colonial Discourse, Postcolonial Theory.* Manchester: Manchester University Press, 1994.
Barrera, Antonio. "Local Herbs, Global Medicines: Commerce, Knowledge, and Commodities in Spanish America." In *Merchants and Marvels,* ed. Smith and Findlen, 163–181.
Barrère, Pierre. *Essai sur l'histoire naturelle de la France Equinoxiale.* Paris, 1741.
——— *Nouvelle relation de la France équinoxiale.* Paris, 1743.

Beckford, William. *Remarks upon the Situation of Negroes in Jamaica*. London, 1788.
Beckles, Hilary McD. *Centering Woman: Gender Discourses in Caribbean Slave Society*. Kingston, Jamaica: I. Randle, 1999.
Bernard, Claude. *An Introduction to the Study of Experimental Medicine*. Trans. Henry Greene, 1865; New York: Dover, 1957.
Blackburn, Robin. *The Making of New World Slavery: From the Baroque to the Modern, 1492–1800*. London: Verso, 1997.
Blair, Patrick. *Pharmaco-Botanologia: or, An Alphabetical and Classical Dissertation on the British Indigenous and Garden Plants of the New London Dispensary*. London, 1723–1728.
Blunt, Wilfrid. *The Compleat Naturalist: A Life of Linnaeus*. London: William Collins Sons & Co., 1971.
Bodard, Pierre-Henri-Hippolyte. *Cours de botanique médicale comparée*. 2 vols. Paris, 1810.
Boord, Andrew. *The Breviarie of Health: Wherin doth folow, Remedies*. London, 1598.
Bourgeois, Louise. "Instructions . . . to her Daughter," In *The Compleat Midwife's Practice*, by T. C., I. D., M. S., and T. B. London, 1656.
[Bourgeois, Nicolas-Louis]. *Voyages intéressans dans différentes colonies Françaises, Espagnoles, Anglaises, etc*. London, 1788.
Bourguet, Marie-Noëlle, and Christophe Bonneuil, ed. "De L'Inventaire du monde à la mise en valeur du globe. Botanique et colonization." Special issue of *Revue française d'histoire d'outre-mer* 86 (1999).
Boxer, C. R. *The Dutch Seaborne Empire: 1600–1800*. New York: Knopf, 1965.
Boylston, Zabdiel. *An Historical Account of the Small-Pox Inoculated in New England upon all sorts of persons, Whites, Blacks, and of all Ages and Constitutions*. London, 1726.
Brande, William Thomas. *Dictionary of Materia Medica and Practical Pharmacy*. London, 1839.
Breyne, Jakob. *Exoticarum aliarumque minus cognitarium plantarum centuria prima*. Danzig, 1678.
Brockliss, Laurence, and Colin Jones. *The Medical World of Early Modern France*. Oxford: Clarendon Press, 1997.
Brondegaard, V. J. "Der Sadebaum als Abortivum." *Sudhoffs Archiv für Geschichte der Medizin und der Naturwissenschaften* 48 (1964): 331–351.
Broughton, Arthur. "Hortus Eastensis: A Catalogue of Exotic Plants, in the Garden of Hinton East, Esq., in the Mountains of Liguanea, Island of Jamaica." In *The History, Civil and Commercial, of the British West Indies*, Bryan Edwards. 2 vols. London, 1794, appendix to vol. 1.
Burton, June. "Human Rights Issues Affecting Women in Napoleonic Legal Medicine Textbooks." *History of European Ideas* 8 (1987): 427–434.
Bush, Barbara. *Slave Women in Caribbean Society, 1650–1832*. Bloomington: Indiana University Press, 1990.

Bynum, William. "Reflections on the History of Human Experimentation." In *The Use of Human Beings in Research*, ed. Stuart Spicker, Ilai Alon, Andre de Vries, and H. Tristram Engelhardt, Jr., 29–46. Dordrecht: Kluwer, 1988.

Cain, A. J. "Logic and Memory in Linnaeus's System of Taxonomy." *Proceedings of the Linnean Society of London* 169 (1958): 144–163.

Campet, Pierre. *Traité pratique des maladies graves.* Paris, 1802.

Cannon, John. "Botanical Collections." In *Sir Hans Sloane*, ed. MacGregor, 135–149.

Cauna, Jacques. "L'État sanitaire des esclaves sur une grand sucrerie (Habitation Fleuriau de Bellevue, 1777–1788)." *Revue Société Haitienne d'histoire et de geographie* 42 (1984): 18–78.

Chadwick, Derek, and Joan Marsh, eds. *Ethnobotany and the Search for New Drugs.* Chichester: J. Wiley, 1994.

Chaia, Jean. "A Propos de Fusée-Aublet: Apothicaire-Botaniste à Cayenne en 1762–1764." *90ᵉ Congrès des sociétés savantes* 3 (1965): 59–62.

[Chandler, John]. *Frauds Detected: or, Considerations Offered to the Public shewing the Necessity of some more Effectual Provision against Deceits, Difference, and Incertainties in Drugs and Compositions of Medicines.* London, 1748.

Chevalier, Jean Damien. *Lettres à M de Jean.* Paris, 1752.

Cole, Charles Woolsey. *Colbert and a Century of French Mercantilism.* 2 vols. New York: Columbia University Press, 1939.

[Collins, Dr.] *Practical Rules for the Management and Medical Treatment of Negro Slaves in the Sugar Colonies.* London, 1803.

Cope, Zachary. *William Cheselden, 1688–1752.* Edinburgh: Livingstone, 1953.

Cotte, E.-N. *Considérations médico-légales sur les causes de l'avortement prétendu criminel.* Aix, 1833.

Cottesloe, Gloria, and Doris Hunt. *The Duchess of Beaufort's Flowers.* Exeter: Webb & Bower, 1983.

Cowen, David. "Colonial Laws Pertaining to Pharmacy." *American Pharmaceutical Association* 23 (1937): 1236–1243.

——— *Pharmacopoeias and Related Literature in Britain and America, 1618–1847.* Aldershot: Ashgate, 2001.

Cox, Paul Alan. "The Ethnobotanical Approach to Drug Discovery." In *Ethnobotany*, ed. Chadwick and Marsh, 25–41.

Craton, Michael. *Searching for the Invisible Man: Slaves and Plantation Life in Jamaica.* Cambridge, Mass.: Harvard University Press, 1978.

——— and James Walvin. *A Jamaican Plantation: The History of Worthy Park, 1670–1970.* Toronto: University of Toronto Press, 1970.

Crosby, Alfred W. *The Columbian Exchange: Biological and Cultural Consequences of 1492.* Westport, Conn.: Greenwood Pub. Co., 1972.

Cullen, William. *A Treatise of the Materia Medica.* 2 vols. Edinburgh, 1789.

Dancer, Thomas. *Catalogue of Plants, Exotic and Indigenous, in the Botanical Garden, Jamaica.* St. Jago de la Vega, Jamaica, 1792.

——— *The Medical Assistant; or Jamaica Practice of Physic: Designed Chiefly for the Use of Families and Plantations.* Kingston, 1801.

——— *Some Observations Respecting the Botanical Garden.* Jamaica, 1804.
Dandy, J. E., ed. *The Sloane Herbarium.* London: British Museum, 1958.
Daubenton, Louis-Jean-Marie. *Histoire naturelle des animaux.* Vol. 1 of the *Encyclopédie méthodique.* Paris, 1782.
Davis, Natalie Zemon. *Women on the Margins: Three Seventeenth-Century Lives.* Cambridge, Mass.: Harvard University Press, 1995.
Dazille, Jean-Barthélemy. *Observations sur les maladies des negres.* 2 vols. 1776; Paris, 1792.
De Beer, Gavin. *Sir Hans Sloane and the British Museum.* 1953; New York: Arno Press, 1975.
Debien, Gabriel. *Les Esclaves aux Antilles françaises, XVIIe-XVIIIe siècle.* Basse-Terre: Société d'histoire de la Guadeloupe, 1974.
Descourtilz, Michel-Étienne. *Flore pittoresque et médicale des Antilles, ou Histoire naturelle des plantes usuelles des colonies Françaises, Anglaises, Espagnoles et Portugaises.* 8 vols. Paris, 1833.
——— *Guide sanitaire des voyageurs aux colonies, ou Conseils hygiéniques.* Paris, 1816.
Diderot, Denis, and Jean Le Rond d'Alembert, eds. *Encyclopédie, ou Dictionnaire raisonné des sciences, des arts et des métiers.* Paris, 1751–1776.
Dimsdale, Thomas. *The Present Method of Inoculating for the Small Pox.* London, 1779.
Dionis, Pierre. *Traité general des accouchemens.* Paris, 1718.
Direction des Archives de France, ed. *Voyage aux iles d'Amérique.* Paris: Archives Nationales, 1992.
Drayton, Richard. *Nature's Government: Science, Imperial Britain, and the 'Improvement' of the World.* New Haven: Yale University Press, 2000.
Duden, Barbara. *Disembodying Women: Perspectives on Pregnancy and the Unborn.* Trans. Lee Hoinacki. Cambridge, Mass.: Harvard University Press, 1993.
Du Tertre, Jean Baptiste. *Histoire générale des Ant-isles.* 4 vols. Paris, 1667–1671.
Duval, Marguerite. *The King's Garden.* Trans. Annette Tomarken and Claudine Cowen. Charlottesville: University Press of Virginia, 1982.
Engstrand, Iris. *Spanish Scientists in the New World: The Eighteenth-Century Expeditions.* Seattle: University of Washington Press, 1981.
Ersch, J. S., and J. G. Gruber. *Allgemeine Encyclopädie der Wissenschaften und Künste.* Leipzig, 1818.
Estes, J. Worth. "The European Reception of the First Drugs from the New World." *Pharmacy in History* 37 (1995): 3–23.
Evenden, Doreen. *The Midwives of Seventeenth-Century London.* Cambridge: Cambridge University Press, 2000.
Fermin, Philippe. *Traité des maladies les plus fréquentes à Surinam et des remèdes les plus propres à les guérir.* Amsterdam, 1765.
——— *Description générale, historique, géographique et physique de la colonie de Surinam.* 2 vols. Amsterdam, 1769.
Florkin, Marcel, ed. *Materia Medica in the Sixteenth Century.* Oxford: Pergamon Press, 1966.

Frank, Johann Peter. *System einer vollständigen medicinischen Polizey.* 4 vols. Mannheim, 1780–1790.

Freind, John. *Emmenologia.* Trans. Thomas Dale. London, 1729.

Fuller, Stephen. *The New Act of Assembly of the Island of Jamaica.* London, 1789.

Garrigus, John D. "Redrawing the Color Line: Gender and the Social Construction of Race in Pre-Revolutionary Haiti." *Journal of Caribbean History* 30 1–2 (1996): 28–50.

Gascoigne, John. *Science in the Service of Empire: Joseph Banks, the British State and the Uses of Science in the Age of Revolution.* Cambridge: Cambridge University Press, 1998.

Gaspar, David, and Darlene Hine, eds. *More than Chattel: Black Women and Slavery in the Americas.* Bloomington: Indiana University Press, 1996.

Gautier, Arlette. *Les Soeurs de solitude: La Condition féminine dans l'esclavage aux Antilles du XVIIe au XIXe siècle.* Paris: Éditions Caribéennes, 1985.

Geggus, David. "Slave and Free Colored Women in Saint Domingue." In *More than Chattel,* ed. Gaspar and Hine, 259–260.

Gelbart, Nina. *The King's Midwife: A History and Mystery of Madame du Coudray.* Berkeley: University of California Press, 1998.

Gerard, John. *The Herbal or Generall Historie of Plantes.* London, 1597.

Goslinga, Cornelis Christiaan. *The Dutch in the Caribbean and in the Guianas, 1680–1791.* Assen: Van Gorcum, 1985.

Grainger, James. *An Essay on the More Common West-India Diseases.* Edinburgh, 1802.

Grove, Richard. *Green Imperialism: Colonial Expansion, Tropical Island Edens and the Origins of Environmentalism, 1600–1860.* Cambridge: Cambridge University Press, 1995.

Guerra, Francisco. "Drugs from the Indies and the Political Economy of the Sixteenth Century." *Analecta Medico-Historica* 1 (1966): 29–54.

Guillot, Renée-Paule. "La vraie 'Bougainvillée': La première femme qui fit le tour du monde." *Historama* 1 (1984): 36–40.

Guy, Richard. *Practical Observations on Cancers and Disorders of the Breast.* London, 1762.

Haggis, A. W. "Fundamental Errors in the Early History of Cinchona." *Bulletin of the History of Medicine* 10 (1941): 417–459.

Hartley, David. *A Supplement to a Pamphlet Intitled, "A view of the Present Evidence for and against Mrs. Stephen's Medicines for Dissolving the Stone."* London, 1740.

Hawksworth, D. L., ed. *Improving the Stability of Names: Needs and Options.* Königstein: Koeltz, 1991.

Hein, Wolfgang-Hagen, ed. *Botanical Drugs of the Americas in the Old and New Worlds.* Stuttgart: Wissenschaftliche Verlagsgesellschaft, 1984.

Hélie, Théodore. "De l'Action vénéneuse de la rue." *Annales d'hygiène publique et de médecine légale* 20 (1836): 180–219.

Heniger, J. *Hendrik Adriaan van Reede tot Drakenstein and Hortus Malabaricus.* Rotterdam: A. A. Balkema, 1986.

Hérissant, M. "Experiments Made on a Great Number of Living Animals, with the Poison of Lamas, and of Ticunas." *Philosophical Transactions of the Royal Society of London* 47 (1751–1752): 75–92.
Hermann, Paul. *Horti academici Lugduno-Batavi catalogus.* Leiden, 1687.
——— *Materia medica.* London, 1727.
Hildebrandt, Georg Friedrich. *Versuch einer philosophischen Pharmakologie.* Braunschweig, 1786.
Hilliard d'Auberteuil, Michel-René. *Considérations sur l'état présent de la colonie française de Saint-Domingue.* Paris, 1776.
Hulme, Peter. *Colonial Encounters: Europe and the Native Caribbean, 1492–1797.* London: Methuen, 1986.
Humboldt, Alexander von. *Vues des Cordillères, et monumens des peuples indigènes de l'Amérique.* Paris, 1810.
——— (and Aimé Bonpland). *Personal Narrative of Travels to the Equinoctial Regions of the New Continent, during the Years 1799–1804.* Trans. Helen Williams. 7 vols. London, 1814–1829.
Hunter, Lynette, and Sarah Hutton, eds. *Women, Science and Medicine, 1500–1700.* Stroud, Gloucestershire: Sutton, 1997.
Jackson, B. D. "The New Index of Plant Names." *Journal of Botany* 25 (1887): 66–71, 150–151.
James, Robert. *A Medicinal Dictionary.* 3 vols. London, 1743–1745.
———. *The Modern Practice of Physic.* London, 1746.
Jaramillo-Arango, Jaime. *The Conquest of Malaria.* London: Heinemann, 1950.
Jarcho, Saul. *Quinine's Predecessor: Francesco Tori and the Early History of Cinchona.* Baltimore: Johns Hopkins University Press, 1993.
Jussieu, Antoine de. *Traité des vertus des plantes.* Nancy, 1771.
Jussieu, Joseph de. *Description de l'arbe à quinquina.* 1737; Paris: Société du Traitment des Quinquinas, 1936.
Keegan, William. "The Caribbean, Inclusion Northern South American and Lowland Central America: Early History." In *The Cambridge World History of Food.* 2 vols. Ed. Kenneth Kiple and Kriemhild Coneè Ornelas, vol. 2, 1260–1278. Cambridge: Cambridge University Press, 2000.
Klepp, Susan. "Lost, Hidden, Obstructed, and Repressed: Contraceptive and Abortive Technology in the Early Delaware Valley." In *Early American Technology,* ed. Judith A. McGaw, 68–113. Chapel Hill: University of North Carolina Press, 1994.
Koerner, Lisbet. *Linnaeus: Nature and Nation.* Cambridge, Mass.: Harvard University Press, 1999.
——— "Women and Utility in Enlightenment Science." *Configurations* 3 (1995): 233–255.
Labat, Jean-Baptiste. *Nouveau voyage aux isles de l'Amérique.* 6 vols. Paris, 1722.
La Condamine, Charles-Marie de. *The History of Inoculation.* New Haven, 1773.
——— *Relation abrégée d'un voyage fait dans l'interieur de L'Amérique Méridionale.* Paris, 1745.

——— *A Discourse on Inoculation, Read before the Royal Academy of Science at Paris, the 24th of April 1754*. Trans. Matthew Maty. London, 1755.
——— "Sur l'arbe du quinquina" (28 Mai 1737). *Histoire mémoires de l'Académie Royale des Sciences* (Amsterdam, 1706–1755): 319–346.
Lacroix, Alfred. *Figures de savants*. 4 vols. Paris: Gauthier-Villars, 1932–1938.
Lafuente, Antonio. "Enlightenment in an Imperial Context: Local Science in the Late-Eighteenth-Century Hispanic World." In *Nature and Empire*, ed. MacLeod, 155–173.
——— and Nuria Valverde. "Linnaean Botany and Spanish Imperial Biopolitics." In *Colonial Botany*, ed. Schiebinger and Swan.
Laissus, Yves, ed. *Les Naturalistes français en Amérique de Sud*. Paris: Édition du CTHS, 1995.
——— "Les Voyageurs naturalistes du Jardin du Roi et du Muséum d'Histoire Naturelle." *Revue d'histoire des sciences* 34 (1981): 259–317.
Lambert, Sheila, ed. *House of Commons Sessional Papers of the Eighteenth Century*. 147 vols. Wilmington, Del.: Scholarly Resources, 1975.
Latour, Bruno. *Science in Action: How to Follow Scientists and Engineers through Society*. Cambridge, Mass.: Harvard University Press, 1987.
Lawrence, George, ed. *Adanson: The Bicentennial of Michel Adanson's "Familles des plantes"*. Pittsburgh: The Hunt Botanical Library, 1963.
Leblond, Jean-Baptiste. *Voyage aux Antilles: D'île en île, de la Martinique à Trinidad 1767–1773*. Paris: Editions Karthala, 2000.
Le Boursier du Coudray, Angélique Marguerite. *Abrégé de l'art des accouchemens*. Paris, 1777.
Leibrock-Plehn, Larissa. *Hexenkräuter oder Arznei: Die Abtreibungsmittel im 16. und 17. Jahrhundert*. Stuttgart: Wissenschaftliche Verlagsgesellschaft, 1992.
Lewin, Louis. *Die Fruchtabtreibung durch Gifte und andere Mittel: Ein Handbuch für Ärzte, Juristen, Politiker, Nationalökonomen*. Berlin: Georg Stilke, 1925.
Ligon, Richard. *A True and Exact History of the Island of Barbados*. London, 1657.
Linnaeus, Carl. *Critica botanica*. Leiden, 1737.
——— *Amoenitates academicae*. Leiden, 1749–1790.
——— *Species Plantarum*. 1753; London: Ray Society, 1957, with introduction by William Stearn.
——— *The "Critica Botanica" of Linnaeus*. Trans. Arthur Hort and M. L. Green. London: Ray Society, 1938.
Löseke, Johann Ludwig Leberecht. *Materia Medica, oder Abhandlung von den auserlesenen Arzneymitteln*. 4th ed. Berlin, 1773.
Long, Edward. *The History of Jamaica*. 3 vols. London: 1774.
Lowthrop, John. *The Philosophical Transactions and Collections, to the End of the Year 1700*. 3 vols. London, 1722.
MacGregor, Arthur, ed. *Sir Hans Sloane: Collector, Scientist, Antiquary, Founding Father of the British Museum*. London: British Museum, 1994.
——— "The Life, Character and Career of Sir Hans Sloane." In *Sir Hans Sloane*, ed. MacGregor, 11–44.

MacKay, David. *In the Wake of Cook: Exploration, Science, and Empire, 1780–1801.* London: Croom Helm, 1985.

——— "Agents of Empire: The Banksian Collectors and Evaluation of New Land." In *Visions of Empire*, ed. Miller and Reill, 38–57.

MacLeod, Roy, ed. *Nature and Empire: Science and the Colonial Enterprise.* Special issue of *Osiris* 15 (2000).

Maehle, Andreas-Holger. *Drugs on Trial: Experimental Pharmacology and Therapeutic Innovation in the Eighteenth Century.* Amsterdam: Rodopi, 1999.

——— "The Ethical Discourse on Animal Experimentation, 1650–1900." In *Doctors and Ethics: The Earlier Historical Setting of Professional Ethics*, ed. Andrew Wear, Johanna Geyer-Kordesch, and Roger French, 203–251. Amsterdam: Rodopi, 1993.

Maitland, Charles. *Mr. Maitland's Account of Inoculating the Small Pox.* London, 1722.

Maupertuis, Pierre-Louis Moreau de. *Lettre sur le progrès des sciences.* Dresden, 1752.

Mauriceau, François. *Traité des maladies des femmes grosses,* 4th ed. Paris: 1694.

McClellan, James III. *Colonialism and Science: Saint Domingue in the Old Regime.* Baltimore: Johns Hopkins University Press, 1992.

McLaren, Angus. *A History of Contraception: From Antiquity to the Present Day.* Oxford: B. Blackwell, 1990.

McNeill, J. "Latin, the Renaissance Lingua Franca, and English, the 20th Century Language of Science: Their Role in Biotaxonomy." *Taxon* 46 (1997): 751–757.

McVaugh, Rogers. *Botanical Results of the Sessé and Mociño Expedition: 1787–1830.* Pittsburgh: Hunt Institute for Botanical Documentation, Carnegie Mellon University, 2000.

Mead, Richard. *The Medical Works.* 3 vols. Edinburgh, 1763.

Merian, Maria Sibylla. *Metamorphosis insectorum Surinamensium,* ed. Helmut Deckert. 1705; Leipzig: Insel Verlag, 1975.

Merson, John. "Bio-prospecting or Bio-piracy: Intellectual Property Rights and Biodiversity in a Colonial and Postcolonial Context." In *Nature and Empire*, ed. MacLeod, 282–296.

Miller, David Philip, and Peter Reill, eds. *Visions of Empire: Voyages, Botany, and Representations of Nature.* Cambridge: Cambridge University Press, 1996.

Miller, Genevieve. *The Adoption of Inoculation from Smallpox in England and France.* Philadelphia: University of Pennsylvania Press, 1957.

Miller, Philip. *The Gardener's Dictionary.* London, 1768.

Moitt, Bernard. *Women and Slavery in the French Antilles, 1635–1848.* Bloomington: Indiana University Press, 2001.

Monardes, Nicolás. *Joyfull Newes out of the Newe Founde Worlde.* 2 vols. Trans. John Frampton. 1577; London: Constable, 1925.

Monnier, Jeannine, Jean-Claude Jolinon, Anne Lavondes, and Pierre Elouard. *Philibert Commerson: Le Découvreur de Bougainvillier.* Châtillon-sur-Chalaronne: Association Saint-Guignefort, 1993.

Monro, Donald. *A Treatise on Medical and Pharmaceutical Chemistry, and the Materia Medica*. 3 vols. London, 1788.

——— ed. *Letters and Essays . . . by Different Practitioners*. London, 1778.

Moreau de Saint-Méry, Médéric-Louis-Elie. *Loix et constitutions des colonies françoises de l'Amérique sous le vent*. 6 vols. Paris, 1784–1790.

——— *Description topographique, physique, civile, politique et historique de la partie française de l'Isle Saint-Domingue*. 2 vols. Philadelphia, 1797–1798.

Morton, Julia. *Atlas of Medicinal Plants of Middle America: Bahamas to Yucatan*. Springfield, Ill.: Charles Thomas, 1981.

Moseley, Benjamin. *A Treatise on Sugar: With Miscellaneous Medical Observations*. London, 1800.

Nassy, David de Isaac Cohen. *Essai historique sur la colonie de Surinam*. Paramaribo, 1788.

Nicolas, Jean-Paul. "Adanson, the Man." In *Adanson*, ed. Lawrence, 1–122.

——— "Adanson et le mouvement colonial." In *Adanson*, ed. Lawrence, 393–451.

Noonan, John Thomas Jr. *Contraception: A History of its Treatment by the Catholic Theologians and Canonists*. Cambridge, Mass.: Belknap Press of Harvard University Press, 1986.

Nussbaum, Felicity. *Torrid Zones: Maternity, Sexuality, and Empire in Eighteenth-Century English Narratives*. Baltimore: Johns Hopkins University Press, 1995.

Olarte, Mauricio Nieto. "Remedies for the Empire: The Eighteenth Century Spanish Botanical Expeditions to the New World." Ph.D. Diss., History of Science and Technology, Imperial College, London, 1993.

Perry, G. "Nomenclatural Stability and the Botanical Code: A Historical Review." In *Improving the Stability of Names*, ed. Hawksworth, 79–93.

Pimentel, Juan. "The Iberian Vision: Science and Empire in the Framework of a Universal Monarchy, 1500–1800." In *Nature and Empire*, ed. MacLeod, 17–30.

Pinkerton, John, trans. and ed. *A General Collection of the Best and Most Interesting Voyages and Travels*. 17 vols. London, 1808–1814.

Pluchon, Pierre. "Le Cercle des philadelphes du Cap-Français à Saint-Domingue: Seule académie colonial de l'ancien régime." *Mondes et Cultures* 45 (1985): 157–191.

Pouliquen, Monique, ed. *Les voyages de Jean-Baptiste Leblond: Médecin naturaliste du Roi aux Antilles, en Amérique Espagnole et en Guyane, de 1767 à 1802*. Paris: Editions du C.T.H.S., 2001.

Pouppé-Desportes, Jean-Baptiste-René. *Histoire des maladies de Saint Domingue*. 3 vols. Paris, 1770.

Pratt, Mary Louise. *Imperial Eyes: Travel Writing and Transculturation*. London: Routledge, 1992.

Pycior, Helena, Nancy Slack, and Pnina Abir-Am, eds. *Creative Couples in the Sciences*. New Brunswick: Rutgers University Press, 1996.

Ramsay, James. *An Essay on the Treatment and Conversion of African Slaves in the British Sugar Colonies*. London, 1784.

Razzell, Peter. *The Conquest of Smallpox: The Impact of Inoculation on Smallpox Mortality in Eighteenth-Century Britain*. Firle: Caliban Books, 1977.
Reede tot Drakestein, Hendrik Adriaan van. *Hortus Indicus Malabaricus*. 12 vols. Amsterdam, 1678–1693.
Renny, Robert. *An History of Jamaica*. London, 1807.
Riddle, John M. *Contraception and Abortion from the Ancient World to the Renaissance*. Cambridge, Mass.: Harvard University Press, 1992.
——— *Eve's Herbs: A History of Contraception and Abortion in the West*. Cambridge, Mass.: Harvard University Press, 1997.
Risse, Guenter. "Transcending Cultural Barriers: The European Reception of Medicinal Plants from the Americas." In *Botanical Drugs*, ed. Hein, 31–42.
——— *Hospital Life in Enlightenment Scotland: Care and Teaching at the Royal Infirmary of Edinburgh*. Cambridge: Cambridge University Press, 1986.
Rochefort, Charles de. *Histoire naturelle et morale des Iles Antilles de l'Amérique*. Rotterdam, 1665.
Roger, Jacques. *Buffon: A Life in Natural History*. Trans. Sarah Bonnefoi. Ithaca: Cornell University Press, 1997.
Rouse, Irving. *The Taino: Rise and Decline of the People Who Greeted Columbus*. New Haven: Yale University Press, 1992.
Rublack, Ulinka. "The Public Body: Policing Abortion in Early Modern Germany." In *Gender Relations in German History: Power, Agency, and Experience from the Sixteenth to the Twentieth Century*, eds. Lynn Abrams and Elizabeth Harvey, 57–78. London: University of London Press, 1996.
Rücker, Elisabeth. *Maria Sibylla Merian, 1647–1717*. Nürnberg: Germanisches Nationalmuseum, 1967.
Ryan, Michael. *A Manual of Medical Jurisprudence*. London, 1831.
[Schaw, Janet]. *Journal of a Lady of Quality; Being a Narrative of a Journey from Scotland to the West Indies, North Carolina, and Portugal, in the years 1774 to 1776*, ed. Evangeline Andrews. New Haven: Yale University Press, 1922.
Schiebinger, Londa. *The Mind Has No Sex? Women in the Origins of Modern Science*. Cambridge, Mass.: Harvard University Press, 1989.
——— *Nature's Body: Gender and the Making of Modern Science*. Boston: Beacon Press, 1993.
——— *Has Feminism Changed Science?* Cambridge, Mass.: Harvard University Press, 1999.
——— and Claudia Swan, eds. *Colonial Botany: Science, Commerce, and Politics in the Early Modern World*. Philadelphia: University of Pennsylvania Press, 2004.
Segal, Sam. "Maria Sibylla Merian as a Flower Painter." In *Maria Sibylla Merian*, ed. Wettengl, 68–87.
Sheridan, Richard B. *Doctors and Slaves: A Medical and Demographic History of Slavery in the British West Indies, 1680–1834*. Cambridge: Cambridge University Press, 1985.
Shorter, Edward. *Women's Bodies: A Social History of Women's Encounter with Health, Ill-Health, and Medicine*. New Brunswick: Transaction, 1991.

Shteir, Ann. *Cultivating Women, Cultivating Science: Flora's Daughters and Botany in England 1760–1860.* Baltimore: Johns Hopkins University Press, 1996.
Sloane, Hans. *Catalogus plantarum quae in Insula Jamaica.* London, 1696.
——— *A Voyage to the Islands Madera, Barbados, Nieves, St. Christophers, and Jamaica; with the Natural History, etc . . .* 2 vols. London, 1707–1725.
——— *An Account of a Most Efficacious Medicine for Soreness, Weakness, and Several Other Distempers of the Eyes.* London, 1745.
——— "An Account of Inoculation." *Philosophical Transactions of the Royal Society of London* 49 (1756): 516–520.
Smellie, William, ed. *Encyclopedia Britannica.* Edinburgh, 1771.
Smith, Adam. *An Inquiry into the Nature and Causes of the Wealth of Nations,* ed. Edwin Cannan. 1776; Chicago: University of Chicago Press, 1976.
Smith, Pamela H., and Paula Findlen, eds. *Merchants and Marvels: Commerce, Science, and Art in Early Modern Europe.* New York: Routledge, 2002.
Sörlin, Sverker. "Ordering the World for Europe: Science as Intelligence and Information as Seen from the Northern Periphery." In *Nature and Empire,* ed. MacLeod, 51–69.
Spary, Emma. *Utopia's Garden: French Natural History from Old Regime to Revolution.* Chicago: University of Chicago Press, 2000.
Stafleu, Frans. *Linnaeus and the Linnaeans: The Spreading of Their Ideas in Systematic Botany, 1735–1789.* Utrecht: A. Oosthoek's Uitgeversmaatschappij, 1971.
——— "Adanson and the 'Familles des plantes.'" In *Adanson,* ed. Lawrence, 123–259.
Stagl, Justin. *A History of Curiosity: The Theory of Travel 1550–1800.* Chur, Switzerland: Harwood Academic Publishers, 1995.
Stearn, William Thomas. "Botanical Exploration to the Time of Linnaeus." *Proceedings of the Linnean Society of London* 169 (1958): 173–196.
——— "The Background of Linnaeus's Contributions to the Nomenclature and Methods of Systematic Biology." *Systematic Zoology* 8 (1959): 4–22.
——— *Botanical Latin: History, Grammar, Syntax, Terminology, and Vocabulary.* Newton Abbot, Devon: David and Charles, 1992.
——— ed. *Humboldt, Bonpland, Kunth and Tropical American Botany.* Lehre: J. Cramer Verlag, 1968.
——— "Carl Linnaeus's Acquaintance with Tropical Plants." *Taxon* 37 (1988): 776–781.
Stedman, John Gabriel. *Narrative of a Five years' Expedition against the Revolted Negroes of Surinam.* London, 1796.
——— *Stedman's Surinam: Life in an Eighteenth-Century Slave Society,* eds. Richard Price and Sally Price. Baltimore: Johns Hopkins Press, 1992.
Stofft, Henri. "Un Avortement criminel en 1660." *Histoire des sciences medicales* 20 (1986): 67–85.
Störck, Anton von. *An Essay on the Medicinal Nature of Hemlock.* London, 1760.
——— *Traité de l'inoculation de la petite vérole.* Vienna, 1771.
Stroup, Alice. *A Company of Scientists: Botany, Patronage, and Community at the*

Seventeenth-Century Parisian Royal Academy of Sciences. Berkeley: University of California Press, 1990.
Stukenbrock, Karin. *Abtreibung im ländlichen Raum Schleswig-Holsteins im 18. Jahrhundert.* Neumünster: K. Wachholtz Verlag, 1993.
Swartz, Olof. *Observationes botanicae quibus plantae Indiae Occidentalis.* Erlangen, 1791.
Taillemite, Étienne. *Bougainville et ses compagnons autour du monde, 1766–1769.* 2 vols. Paris: Imprimerie nationale, 1977.
Talbot, Alathea. *Natura Exenterata: or Nature Unbowelled by the most Exquisite Anatomizers of Her.* London, 1655.
Tardieu, Ambroise. *Étude médico-légale sur l'avortement.* 3rd ed. Paris, 1868.
Thiery de Menonville, Nicolas-Joseph. *Traité de la culture du nopal et de l'éducation de la cochenille dans les colonies Françaises de l'Amérique.* Cap-Français, 1787.
Thomson, James. *A Treatise on the Diseases of Negroes, as they occur in the Island of Jamaica; with Observations on the Country Remedies.* Jamaica, 1820.
Thunberg, Carl. *Travels in Europe, Africa and Asia, performed between the years 1770 and 1779.* 4 vols. London, 1795.
Tournefort, Joseph Pitton de. *Élémens de botanique, ou, Méthode pour connaître les plantes.* 6 vols. Paris, 1694.
Trapham, Thomas. *A Discourse of the State of Health in the Island of Jamaica.* London, 1679.
Wagstaffe, William. *A Letter to Dr. Freind; Shewing the Danger and Uncertainty of Inoculating the Small Pox.* London, 1722.
Walters, S. M. "The Shaping of Angiosperm Taxonomy." *New Phytologist* 60 (1961): 74–84.
Walvin, James. *Fruits of Empire: Exotic Produce and British Taste, 1660–1800.* New York: New York University Press, 1997.
Ward, J. R. *British West Indian Slavery, 1750–1834: A Process of Amelioration.* New York: Oxford University Press, 1988.
Wettengl, Kurt, ed. *Maria Sibylla Merian (1647–1717): Artist and Naturalist.* Ostfildern: G. Hatje, 1998.
——— "Maria Sibylla Merian: Artist and Naturalist between Frankfurt and Surinam." In *Merian,* ed. Wettengl, 13–36.
Wijnands, D. O. *The Botany of the Commelins.* Rotterdam: Balkema, 1983.
Williamson, John. *Medical and Miscellaneous Observations, Relative to the West India Islands.* 2 vols. Edinburgh, 1817.
Withering, William. *An Account of the Foxglove, and Some of its Medical Uses.* Birmingham: 1785.
Woodville, William. *Medical Botany.* 3 vols. London, 1790.
Zedler, Johann Heinrich. *Grosses vollständiges Universal Lexicon aller Wissenschaften und Künste.* Halle and Leipzig, 1732–1754.

Credits

Figure I.1. Maria Sibylla Merian, *Metamorphosis* (1705), plate 45. By permission of the Wellcome Library, London.

Figure I.2. Trade card. By permission of Cadbury Limited.

Figure I.3. Frontispiece to Charles de Rochefort's *Histoire naturelle et morale des Iles Antilles de l'Amérique* (Rotterdam, 1658). By permission of the Houghton Library, Harvard University.

Figure I.4. Sir Hans Sloane's *Voyage to the Islands Madera, Barbados, Nieves, St. Christophers, and Jamaica; with the Natural History* . . . (London, 1707–1725, vol. 1, pullout following cliv.) By permission of the Wellcome Library, London.

Figure 1.1. The King's Garden. Guy de La Brosse, *Advis defensif du Jardin royal, des plantes médecinales à Paris* (Paris, 1636). Copyright © Bibliothèque Centrale, Muséum National d'Histoire Naturelle (MNHN), Paris, 2003. Reprinted with permission.

Figure 1.2. Cochineal in Oaxaca, Mexico. From Sir Hans Sloane, *Voyage* (1725), vol. 2, plate ix. By permission of the Wellcome Library, London.

Figure 1.3. Eighteenth-century European boxes for transporting plants. John Ellis, *Directions for Bringing over Seeds and Plants* (London, 1770), frontispiece. By permission of the Wellcome Library, London.

Figure 1.4. Frontispiece to Jean-Baptiste-Christophe Fusée-Aublet's *Histoire des plantes de la Guiane Françoise* (1775), vol. 1. Copyright © Bibliothèque Centrale, MNHN, Paris 2003. Reprinted with permission.

Figure 1.5. A European gentleman carried "by man's back" over the Andes. Alexander von Humboldt, *Vues des Cordillères* (1810), plate 5. By permission of the Pennsylvania State University Libraries.

Figure 2.1. "The natural inhabitants of the Antilles of America, called savages" under a papaya tree. Jean-Baptiste du Tertre, *Histoire* (1667–1671), vol. 2, 356. By permission of the Wellcome Library, London.

Figure 2.2. A Carib woman points to the *roucou* tree. Charles de Rochefort, *Histoire naturelle* (1665), 93. Copyright © Bibliothèque Centrale, MNHN, Paris 2003. Reprinted with permission.

Figure 2.3. Typical European fortification in the Caribbean. Charles de Rochefort, *Histoire naturelle* (1665), 53. Copyright © Bibliothèque Centrale, MNHN, Paris 2003. Reprinted with permission.

Figure 2.4. Lady Mary Wortley Montagu in Turkish dress. By permission of the Boston Athenaeum.

Figure 3.1. A sugar mill circa 1660. Jean-Baptiste du Tertre, *Histoire* (1667–1671), vol. 2, 122. Reprinted by permission of the Bibliothèque Nationale de France.

Figure 3.2. Dissolute Surinam planter. John Stedman, *Narrative* (1796), vol. 2, 56. By permission of the Wellcome Library, London.

Figure 3.3. A female slave being punished. John Stedman, *Narrative* (1796), vol. 1, 15. By permission of the Wellcome Library, London.

Figure 4.1. Linnaeus' "Pharmacopoea." Carl Linnaeus, *Materia medica* (Amsterdam, 1749), frontispiece. By permission of the British Library, London.

Figure 5.1. Hendrik Adriaan van Reede's *Tsjétti Mandáru*. By permission of the Library of the Gray Herbarium Archives, Harvard University.

Figure 5.2. Quassi shown in European dress, by William Blake. John Stedman, *Narrative* (1796), vol. 2, 348. By permission of the Wellcome Library, London.

Figure C.1. A merchant selling simples. Musée de la Banane, Martinique.

Index

Abortifacients, 1–2, 4–5, 12, 18–19, 23, 30, 81, 104–105, 107–149, 150–153, 159, 166, 177–193, 206, 226–234, 238–241; hybrid, 231

Abortion, 1, 18, 109, 111, 113–149; conflict between doctors and women, 109–110; criminal, 113; in Europe, 113–128; European physicians' attitudes toward, 109–110, 113–128, 140–141; legal aspects, 115–119, 130; and the Napoleonic Code, 19, 40, 277n99; physician-induced, 131–141; as political resistance, 107, 130–131; and the slave trade, 142; and slaves, 107, 128, 130–131, 139–142; and smallpox, 110, 175, 184, 189; therapeutic, 19, 113, 121; in the West Indies, 128–149. *See also* Miscarriage

Académie Royale des Sciences, 12, 36, 43, 71, 180, 220

Acclimatization of plants, 4, 7, 11, 36, 45–46, 56, 60, 83

Adanson, Michel, 6, 10, 57, 68, 71, 92; and botanical nomenclature, 196–197, 208, 219–225

Agnotology, 3, 18, 226–241

Aiton, William, 60

Alembert, Jean Le Rond d', 6, 8, 114, 127, 182, 187

Alibert, Jean-Louis-Marie, 183, 187

Alston, Charles, 153

Amazons, 16, 38, 62–72; Amazon stone, 63; and Humboldt, 62–67, 70; and La Condamine, 38, 62–72

Amerindians, 4, 12, 14, 15, 25, 29, 40, 63–65, 74–76, 79–86, 88–89, 105, 107, 129–130, 132, 138–139, 171, 197, 199, 201, 206, 213, 217, 226, 228, 230–232, 238–239, 254n8; dictionaries, 79; pharmacopoeia, 79; secrets, 45, 90–93; and smallpox, 29. *See also* Arawaks; Caribs; Tainos; Zapotecs

Aphrodisiacs, 81

Arawaks, 2, 14, 19–20, 74, 76–78, 80, 87, 128, 172, 195, 197, 207–208, 211, 230–231. *See also* Amerindians

Arthaud, Charles, 23, 53, 161; on midwives, 112, 236. *See also* Cercle des Philadelphes

Assafoetida, 183

Astruc, Jean, 170, 188

Aublet, Jean-Baptiste-Christophe Fusée, 21, 23, 53–57, 139, 223–224; coins term *botaniste voyageur,* 14, 53; in Guiana, 56; in Isle de France, 53–56, 139, 223; and slavery, 56, 143; use of indigenous names of plants, 55

Balsam of Peru, 8, 215; of Tolu, 8

Bancroft, Edward, 84, 128, 139–140, 177, 231

Banks, Sir Joseph, 7, 11, 46, 57, 93, 236

Barbados Pride, 4, 87, 107, 241. *See also* Peacock flower

Baret (*also* Barret), Jeanne, 1, 24, 30, 46–51, 218

Baretia bonnafidia, 51, 218

Barrère, Pierre, 81, 197, 199, 247n3, 254n9

Basseporte, Madeleine, 33

Bauhin, Caspar, 194, 198, 202

Beame, Henry, 176

Beaufort, Mary Capel Somerset, Duchess of, 57–60, 218; gardens, 28, 58–60

Belon, Pierre, 45–46, 220

Bezoars, 90

Biocontact zones, 15, 82–84, 89, 104; "noise" in, 84, 87
Biopiracy, 35–44
Bioprospecting, 7, 15, 17, 73–104; in Europe, 93–100; in West Indies, 75–82
Blair, Patrick, 15, 99–100
Block, Agnes, 33, 60
Blumenbach, Johann, 32, 103
Bodard, Pierre-Henri-Hippolyte, 73
Boerhaave, Herman, 57, 110, 151, 185, 269n4
Bonpland, Aimé, 33, 58, 67, 227
Bontius, Jacobus, 27, 55
Boord, Andrew, 18
Botanical gardens, 5, 35, 43, 52, 57–60, 87, 126–127, 150, 204, 218, 229, 234, 244n12; Amsterdam, 33, 151, 207, 210; Edinburgh, 153; Jamaica, 231; Kew, 10, 57, 59–60, 83, 211; Leiden, 10, 57, 150–151; Madrid, 7, 151; Martinique, 234; Montpellier, 36; Paris, 10–12, 25–26, 31, 33, 36–37, 44, 46, 48, 55–58, 65, 83, 150, 181, 219, 202, 222–223; Saint Vincent, 53. *See also* Chelsea Physic Garden; Gardens, Slave; Gardens, Taino; Jardin du Roi
Botanical nomenclature, 5–7, 11, 19–21, 59, 194–225; Adansonian, 196–197, 208, 219–225; binomial, 205; Linnaean, 7, 197–198
Botanist, 6, 10–12, 16, 20–21, 23–72, 82, 89, 138, 156, 194–195, 198–205, 207, 210–211, 221, 223, 227, 229, 235–236; African, 14; Amerindian, 14, 74–75; assistants, 46–57; creole, 51–57, 211; *de cabinet* (armchair) 14, 23, 24, 57, 59–62; field practices, 11, 14, 24, 27, 75; missionary, 25, 37, 130, 223; *voyageur,* 14, 23–30, 35. *See also* Biopiracy; Bioprospecting
Botany, 5–8, 10–12, 20, 25, 78, 96, 194–195, 198, 200–201, 203–204, 206, 213, 217–226, 239; applied, 5–6; and colonial expansion, 7, 11, 19, 24, 74, 104, 210, 233; economic, 3, 5–12, 194, 210, 221, 234; medical, 6, 53, 57; theoretical, 61
Bougainville, Louis-Antoine de, 30, 47–51
Bouguer, Pierre, 38, 71
Bourgeois, Louise, 122, 159
Bourgeois, Nicolas-Louis, 14, 79, 91; on drugs in Saint Domingue, 31, 53, 74; on slave medicine, 80, 90
Boutavèry, 49
Bowdich, Sarah, 31, 51
Boylston, Zabdiel, 103, 157
Breyne, Jakob, 58, 207
Bromelia pinguin, 139
Brown, Robert, 218
Buchan, William, 183
Buffon, Georges-Louis Leclerc, comte de, 11, 36, 56–57, 208, 219–222

Cacao, 6, 7, 23, 30, 38–39, 53, 73, 178, 180, 253n3. *See also* Chocolate
Caesalpinia pulcherrima, 107, 209–210, 230. *See also* Poinciana pulcherrima
Campet, Pierre, 85–87, 145
Cancer, 163; breast, 154, 166, 168–169, 171, 273n47, 275n77; hemlock cure for, 169–170
Candolle, Augustin-Pyrame de, 195, 224, 230
Canning, Lady Charlotte, 1
Caoutchouc, 38
Caribs, 9, 19, 62, 74, 76–78, 80, 85, 87, 128, 141, 172, 197, 199, 211, 232–233, 240, 254n8. *See also* Amerindians
Cassava, 105, 155, 178, 233; bread, 35
Cayenne, 38, 46, 51, 63, 69, 71, 81, 155, 197, 220, 223, 235, 247n3
Cercle des Philadelphes, 43, 53, 161, 171, 250n62, 274n53
Chelsea Physic Garden, 4, 10, 59, 151
Cheselden, William, 97, 171
Chocolate, 4, 7–8, 32, 73, 150, 244n16. *See also* Cacao
Cinchona officinalis, 3, 5, 26, 37–39, 46, 53, 81, 90, 183, 190, 211, 214, 216, 234. *See also* Quinine
Clitoridis, 219
Clusius, Carolus, 96
Cochineal, 8, 11, 21, 39–45, 138; Spanish monopoly on, 39–44
Code noir, 12, 109, 137, 266n64
Coffee, 4, 6–7, 10–11, 23, 27, 39, 53, 59, 172, 178, 183, 234, 253n3
Colbert, Jean-Baptiste, 12, 35–37, 235
Columbus, Christopher, 3, 4, 19–20, 62, 64, 76, 132, 253n8
Commerson, Philibert, 1, 24, 30, 33, 47–48, 50–51, 218

Compagnie des Indes Occidentales, 10, 35, 53, 92, 139
Constantinus Africanus, 120, 126
Contraceptives, 105, 11–12, 116, 119, 121, 123, 125–126, 128–129, 150, 238, 241. *See also* Abortion
Cook, James, 46, 227, 236
Coudray, Angélique Marguerite Le Boursier du, 122, 235
Cougari, 141, 232
Creole, 12, 15, 24–25, 44, 75, 51–57, 86–87, 110, 133, 142–143, 223; corn, 15; language, 15, 197; naturalists, 51–57, 81, 211; pig, 15; science, 51–52, 165
Cross-dressing, 47–51
Cullen, William, 185, 187, 213
Culpeper, Nicholas, 122
Curare, 155; experiments in chickens, 38; experiments in humans, 159
Cuvier, Georges, 194

Dancer, Thomas, 111, 229, 230, 244n12, 273n53
Daubenton, Louis-Jean-Marie, 5, 6, 14
Dazille, Jean-Barthélemy, 146, 226, 235, 236
Death carrot, 120
Denis, Jean, 160, 163
Descourtilz, Michel-Étienne, 74, 81, 107, 108, 112–113, 141, 142, 146, 177, 179, 183, 219, 228, 232, 241
Diderot, Denis, 6, 8, 114, 127, 182, 187
Digitalis purpurea, 98, 162
Dimsdale, Thomas, 162, 173–174
Dionis, Pierre, 182
Dioscorides, 75, 120, 126, 141, 153, 201
Dodart, Denis, 163
Dominica, 19, 76–78, 130, 240–241
Drug testing, 161; and animals, 153–156; in colonies, 155, 171–172; of emmenagogues and abortifacients, 177–193; and humans, 156–166; and sex difference, 166–171; and slaves, 171–177
Dugée, Charlotte, 51
Du Tertre, Jean-Baptiste, 25, 77, 132–133, 151, 269n2, 280n38

El Dorado, 53
Emmenagogue, 19, 107, 109, 118, 120, 122; 124, 141–142, 159, 179, 181–187, 189, 228, 232, 259n7, 275n77

Eryngium foetidum, 142, 177, 232, 241, 262n31
Ethnobotany, 16, 93

Fermin, Philippe, 90–91, 112, 155, 213
Fernandez de Ribera, Francisca, 214
Fesche, Charles-Félix-Pierre, 48
Feuillée, Louis, 25
Flos pavonis. *See* Peacock flower
Flour fence, 107, 109, 113, 139. *See also* Peacock flower
Fowler, Thomas, 168
Frank, Johann Peter, 188
Freind, John, 19; experiments with emmenagogues, 118, 181, 183–185
Freycinet, Rose Marie Pinon de, 51

Gardens, 10–11, 23–24, 27–28, 33–38, 44, 46, 57–60, 74, 76, 78, 96, 105, 108, 126–127, 150–151, 153, 188, 222, 224, 229, 234; kitchen, 86, 126, 150, 159; public, 191; revictualing stations, 229; slave, 229; Taino, 76. *See also* Botanical gardens
Gender, 18, 62, 65, 67, 226, 234–35, 239, 243; and heroic narratives, 65–72
Gerard, John, 120, 164, 169, 199
Global commons, 45
Gmelin, Johann Friedrich, 156
Godin, Louis, 38, 71
Godin des Odonais, Isabelle de Grandmaison, 68–69
Godin des Odonais, Jean, 68–69
Grainger, James, 90
Grew, Nehemiah, 235
Guadeloupe, 11, 19, 53, 132–134, 137, 240
Guaiacum, 8, 38, 105, 178, 190, 218, 253n3
Gully-root, 130, 177, 231
Guy, Richard, 170–171

Haiti, 11, 13, 21, 44, 53, 74, 81, 139, 141, 144, 149, 195. *See also* Saint Domingue
Haller, Albrecht von, 157, 189
Hammock, 25, 49, 85, 105
Harlow, James, 58
Healers, 16, 94, 96, 99, 154, 191, 231; unlicensed, 94, 96, 99, 191
Health, 19, 80, 108, 111, 118–119, 121, 129, 157, 164, 166, 168, 234–235; of

indigenous populations, 77, 80–81, 239; of slave populations, 88, 131, 140, 143–144, 149, 234, 236; of women in the tropics, 31–32, 69
Hélie, Théodore, 190
Herbe à serpent, 82
Herb women, 1, 99, 100, 159; and physicians, 99–100; root cutters, 96
Hermann, Paul, 27, 32, 57–58, 110, 124, 202, 208, 210
Hernández, Francisco, 151, 165, 182, 202
Hibiscus esculentus, 177, 231
Hilliard d'Auberteuil, Michel-René, 142, 145, 236, 238
Hindostan, 44. *See also* India
Hispaniola, 4, 11, 13, 19, 76, 129, 195, 229
Home, Francis, 185–186
Hooker, Sir Joseph, 204, 211
Hope, John, 180, 183
Hospitals, 52, 55, 57, 73, 85, 111, 123, 127, 144, 154, 161, 167–168, 170–171, 173, 179, 181, 191, 235; Danzig, 167; Hôpital de la Charité, 55, 78; Hôpital des Dames Blanches, 237; Hôtel-Dieu, 165; Hôpital Général, 164–165; Royal Native, 165; Saint Bartholomew's, 103, 165; Saint George's, 173; Saint Thomas's, 94; San Andrés, 165; Westminster Lying-In Hospital for Women, 171
Hughes, Reverend Griffith, 179, 228
Humboldt, Alexander von, 15, 32–33, 52, 58, 62, 64–67, 70, 84–85, 89, 129–130, 197, 217, 227, 230, 238–239

Im Thurn, Sir Everard Ferdinand, 65
Incra, Manuel, 3
India, 44, 56, 69, 75, 183, 196, 230
Infanticide, 47, 116, 127, 129, 142, 146, 148, 234, 236–237
Intellectual property rights, 17
Ipecacuanha, 6, 8, 28, 38, 73, 150, 152, 165, 178, 190, 233–234
Isle de France (Mauritius), 46, 50, 53, 55–56, 139, 223, 225, 229
Ives, Edward, 91–92

Jalap, 6, 8, 41–43, 73, 150, 152, 178, 183–184, 221, 233–234, 253n3, 284n15
Jamaica (*also* Xaymaca), 12–15, 20, 25–31, 34, 52–53, 58, 60, 70, 73, 76, 80–82, 87–88, 95, 107, 109–113, 130–131, 133–135, 139–140, 142–145, 148–150, 153, 161, 165, 171–173, 175–180, 183, 205, 227–232, 236, 238–239; College of Physicians and Surgeons, 171; *Jamaica Physical Journal,* 171
James, Robert, 82, 114
Jameson, Mark, 126
Jardin du Roi (*also* Jardin Royal des Plantes Médicinales; Jardin Royal des Plantes), 10–12, 26, 31, 33, 36–37, 44, 48, 55–57, 59, 83, 150, 181, 202, 219, 222–223
Jaucourt, Louis de, 127–128
Jefferson, Thomas, 129
Jesuits, 56, 84–85, 90, 215, 238, 247n3
Johnson, Thomas, 99–100
Jones, Hugh, 27
Jones, Sir William, 222–223
Jumping seeds, 14
Jungfernpalme, 126. *See also* Savin
Juniperus sabina, 125. *See also* Savin
Jussieu, Antoine de, 25, 36, 53, 57, 87, 181, 224
Jussieu, Antione-Laurent de, 224
Jussieu, Bernard de, 57
Jussieu, Christophe de, 53
Jussieu, Joseph de, 38, 58, 71, 183, 215

Kaempfer, Engelbert, 27
Kalm, Peter, 6
Kindermord, 126. *See also* Savin

La Brosse, Guy de, 36–37
La Caille, Nicolas Louis de, 71
La Condamine, Charles-Marie de, 38–39, 58, 62–64, 68, 71, 81–82, 84, 89–90, 101–103, 155–156, 159, 214–216
Langley, Elizabeth, 30
Language, 14–15, 20–21, 63, 70, 79, 84–86, 87, 122, 196–197, 199–201, 206, 208, 214–215, 217–219, 221–222, 232, 238, 241, 282n72; creole, 15, 197; "Negro-English," 85; and power relations, 15, 200–201, 224
Leake, John, 171
Learned societies, 171; Barbados Society for the Encouragement of Natural History and Useful Arts, 53; College of Physicians and Surgeons (Jamaica), 171; London Society for the Encouragement of Natural History and Useful Arts, 53;

Learned societies *(continued)*
Physico-Medical Society of Grenada, 53; Royal College of Physicians (London), 7, 26, 99, 180, 183, 185; Surinaamse Lettervrienden, 53; Temple Coffee House Botany Club (London), 27. *See also* Académie Royale des Sciences; Cercle des Philadelphes; Royal Society of London

Leblond, Jean-Baptiste, 46, 58, 85, 139

Letterwood, 105

Lewis, William, 213

Ligon, Richard, 14, 229, 233

Linnaeus, Carl, 6–7, 11, 23, 26, 57–59, 62, 67, 87, 93, 117, 152, 210, 213; on abortifacients, 124, 126, 182; and binomial nomenclature, 7, 20, 59, 67, 87, 194–206, 211, 214–225

Long, Edward, 11, 73, 82–83, 132, 140, 145, 150, 179

Martinique, 10, 52–53, 59, 76, 82, 113, 135, 137, 179, 234, 237, 239–240

Masculinity, 67, 233; and heroic narratives, 65

Maupertuis, Pierre-Louis Moreau de, 73, 75, 82, 163, 164

Mauriceau, François, 110, 112, 114, 117

Medicine, and botany, 6, 8, 10–11, 16, 18–19, 25–29, 33, 35–37, 45, 55–57, 64, 70, 73–76, 78–83, 85, 87–90, 93–99, 101, 104–105, 108, 110–111, 113–115, 120, 122, 125, 140, 144–145, 150, 153, 158–159, 161–162, 165–169, 173–174, 178, 181, 183, 185–189, 191–194, 197, 211, 213–214, 216, 221, 226, 229, 231–234; Amerindian, 14, 29, 75, 79–81, 86, 90–93, 213, 228, 231; education, 23–24, 26, 50, 78, 271n23; slave, 29, 75, 80–81, 90–93, 228

Menstrual regulators. *See* Emmenagogue

Mercantilism, 5, 8, 17, 73, 234–236, 239

Merian, Maria Sibylla, 1, 23, 25, 30–35, 46, 48, 52, 56, 58, 68, 70, 83, 104, 130–131, 151, 179, 200–202, 204, 207–210, 218, 228, 241, 279n13; and abortifacients, 1, 2, 4, 19, 30, 104, 107–113, 153, 178–179, 193, 206–210, 230, 239, 241; daughters, 1, 30–31, 35, 61; field practices, 1, 76, 206–207, 210; and Labadists, 32; and silk, 34; and slaves, 1,
35, 107, 111, 131, 145–146, 228, 230, 239

Midwives and midwifery, 110, 117, 120–122, 144, 171, 178, 182, 187–189, 192, 232–233, 235–237; European, 96, 112, 114, 122–124, 127, 159, 188, 237; Quaker, 112; slave, 111, 137, 145–146, 148–149, 175, 236–237; women of color, 112

Miller, Philip, 4, 5, 151, 229

Miscarriage, 113–115, 117, 124, 130, 143–144, 148, 173–175, 182, 187, 189–190, 192, 237, 259n7. *See also* Abortion

Monardes, Nicolás, 90, 182

Moninckx, Jan, 33

Moninckx, Maria, 33

Monopoly, 6, 34, 38–39, 56, 90–93; in drug trade, 3, 17, 46, 95, 99, 104, 122

Monro, Donald, 161, 165, 173, 175–176

Monson, Lady Anne, 30, 217–218

Montagu, Lady Mary Wortley, 100–102, 164, 214, 216

Moreau de Saint Méry, Médéric-Louis-Elie, 52, 149

Morris, William, 176

Moseley, Benjamin, 27, 88

Naming practices, 20, 195–225. *See also* Botanical nomenclature

Nassy, David de Isaac Cohen, 81, 214

Nauchea pudica, 219

Newgate prison, 161, 164, 273n41

New Spain, 21, 39, 41, 43–44, 51–52, 138, 165, 197, 199, 223

Nicolson, Father, 145

Nicotiana tabacum, 162

Oaxaca, 39–42

Obeah, 87–88

Orphans, 30, 34, 49, 133, 160–161, 191, 237; and smallpox, 160, 164–166

Ortega, Gomez, 20, 210

Oviedo y Valdés, Gonzalo Fernández de, 64

Patin, Guy, 117, 123, 127

Patris, Jean-Baptiste, 51

Peacock flower (*also* Flour fence; *Poinciana pulcherrima*), 1, 4, 18–19, 30, 107–108, 111, 113, 131, 150–193, 196, 207–208, 230, 241

Pennyroyal, 120, 123–124, 150, 180, 184, 187, 263n38
Peruvian bark, 3, 6, 8, 11, 19, 26, 28, 34, 38, 73–74, 154, 164–165, 172, 178, 183–184, 213–215, 233. *See also* Quinine
Petiver, James, 58, 61, 108, 177, 231
Pharmacopoeia, 16, 73, 79, 178, 193; Amerindian, 79; European, 19, 26, 28, 74, 79, 98, 124, 128, 141, 150, 152, 180, 183, 191, 213, 231–232, 238
Pirates: Anne Bonny, 50; Mary Read, 50
Plumier, Charles, 25, 37, 58, 199, 210–202, 204
Poinciana pulcherrima, 4, 30, 86, 107, 141, 150–151, 153, 177, 179–180, 193, 196, 198, 209–210, 228–231, 241. *See also* Peacock flower
Poincy, Philippe de Lonvilliers, chevalier de, 86, 108, 151, 196, 210, 228–229, 233
Poivre, Pierre, 46, 55–56
Pouppé-Desportes, Jean-Baptiste-René, 32, 53, 55, 79, 82, 84, 142, 199, 208

Quassi, Graman, 211–214, 216–217
Quassia, 172, 190, 211–219
Quier, John, 53, 148, 161, 172–177
Quinine (*also* Peruvian bark and *Quinquina*), 3–4, 28, 84, 150, 179, 214–215
Quinquina, 38, 84, 178–179, 214–216. *See also* Quinine

Race, 15, 21, 81, 132, 137–139; and human experimentation, 137, 171–177; racism, 75, 82, 237
Ray, John, 25–26, 72, 220, 227
Reede tot Drakenstein, Hendrik Adriaan van, 27, 75, 199–200, 202, 204, 208–210, 218, 221, 225, 233
Richmond, Thomas, 46
Riddell, Maria Woodley, 30
Rochefort, Charles de, 9, 78, 87, 199
Royal Society of London, 7, 26, 61, 94, 99, 108, 167, 213, 247n5
Rue, 120–121, 124, 130, 150, 180, 184–185, 187, 189–190, 263nn36,38
Rugeley, Luke, 95

Saint Domingue, 11–13, 21, 23, 31, 39–42, 44, 46, 51, 53, 74, 78–82, 85, 87, 89–91, 105, 108, 112, 133, 134–135, 140, 143–144, 146, 149–150, 161, 171, 177, 183, 195, 199, 208, 228, 230, 232, 236, 241. *See also* Haiti
Saint Vincent, 10, 53, 58, 76, 85, 89, 236, 256n41
Samson, Elisabeth, 138
Savin, 19, 105, 109, 120–121, 123–128, 180, 184–192, 238
Schaw, Janet, 21, 68, 140
Secrets, 7, 17, 38, 41–42, 45, 56, 61, 72, 74, 79, 87, 112, 126, 137, 237–238; guild, 99, 226; medical, 45, 79–80, 86, 90–95, 97–99, 122–123, 146, 213, 215, 226, 241
Sherard, William, 58, 60, 67
Silkworms, 7, 34
Slaves and slavery, 1–2, 4, 12–14, 18, 22–25, 29, 43–44, 55, 63, 65, 79–81, 85–91, 98, 103–107, 109–112, 162, 171–176, 179, 185–197, 207–208, 213–214, 217–218, 226–233, 236–237; and abortion, 107, 128, 130–131, 139–142; African, 1, 24–25, 35, 75, 79, 85, 89, 143, 171, 197, 211, 228; birthing practices, 111–112, 130, 132, 137, 145–146, 148–149, 175, 236–237; human experimentation in, 171–177; medicine and medical practices, 29, 75, 80–81, 90–93, 228; names of, 195–196; and suicide, 129, 131, 146
Sloane, Sir Hans, 7–8, 13, 19, 21, 23, 25–30, 34–35, 52, 55, 58–61, 70, 72–73, 76, 80–81, 90–91, 94–95, 99, 101, 107–113, 130, 139–140, 160–161, 166, 180, 182–183, 202–203, 205–206, 208, 227–229
Smallpox, 29, 69, 82, 189, 239; and abortion, 110, 175, 184–185; and beauty, 102, 161; inoculation, 19, 100–103, 110, 160–161, 164–166, 170, 172–173, 176, 179, 235; Newgate Prison experiments, 161, 164
Smellie, William, 114, 187
Smith, Adam, 92, 127
Spanish flies, 124
Spice trade, 4, 56, 92, 150, 154, 194
Stedman, John, 79–80, 109, 131, 134–136, 138, 142–143, 213–214
Stephens, Joanna, 96–98
Störck, Anton von, 154
Styptic, 93–94

Sugar, 7–8, 10–13, 23, 27, 30, 39, 53, 79, 82–83, 105–106, 131, 133, 135, 143–144, 150, 155–156, 178–179; sugar cane, 4, 10–11, 83, 106, 234
Surinam marriage, 135, 265n61
Swartz, Olof, 202, 210
Swieten, Gerard van, 155, 164, 169–170
Sydenham, Thomas, 93, 96

Tahiti and Tahitians, 4, 47, 49
Tainos, 9, 14, 19, 74, 76, 80, 87, 128–129, 199, 208, 227, 230–231, 233. *See also* Amerindians
Talbor, Sir Robert, 92
Talbot, Lady Alathea, 121, 159
Tardieu, Ambroise, 188, 190
Tetanus, 86, 146, 148
Thiery de Menonville, Nicolas-Joseph, 58, 71, 91, 138; as biopirate, 39–46; and cochineal, 21, 39–46
Thistlewood, Thomas, 139
Thomas, Robert, 144
Thomson, James, 52–53, 88, 144–145, 149–150, 153, 172, 177, 179
Thornton, Robert, 219
Thouin, André, 83, 202
Thunberg, Carl, 10, 27, 31, 69, 83, 92, 135, 217
Timoni, Emanuele, 101
Tlacoxiloxochitl, 208. *See also* Peacock flower
Tournefort, Joseph Pitton de, 26, 36–37, 58, 67, 201–203, 209–210, 221
Trading companies, 23–24, 27, 32, 45, 72, 83, 154, 233, 238; Compagnie des Indes, 10, 35, 53, 55, 92, 139; Compagnie des Isles de l'Amérique, 35; Dutch East India, 10, 33, 57, 92, 208; Dutch West India, 10; English East India, 17, 34, 91
Trapham, Thomas, 14, 31
Tropics, 5–7, 13–14, 16, 23–24, 28–33, 46, 67–69, 72, 74–76, 79–80, 104–105, 190, 207, 214, 223, 225, 227, 230–231, 233; and female reproductive health, 31–32, 69
Tsjétti mandáru, 196, 208–209. *See also* Peacock flower
Tussac, Richard de, 21

Universities, 23, 24, 57, 96, 99, 164, 170, 209; Autonomous of Santo Domingo, 52; Cambridge, 225; Codrington College, 52; Harderwijk, 26; Harvard, 209; Edinburgh, 50, 150, 153, 158–159, 172, 180, 185; Glasgow, 126; Leipzig, 163; Michigan, 18; Nacional Mayor de San Marcos de Lima, 52, 197; Orange, 26; Oxford, 221; of Padua, 101; Royal Pontifical University, 52; Uppsala, 222; Vienna, 169; William and Mary, 52; Yale, 52

Vicat, Philippe, 188
Vicq-d'Azyr, Félix, 220
Voison, Cathérine, 127
Voodoo, 87

Wagstaffe, William, 103, 176, 177
Withering, William, 98, 162
Wolff, Christian Sigismund, 163
Woodville, William, 27, 183
Wright, William, 179–80

Zapotecs, 39, 84. *See also* Amerindians